Other Titles in This Series

174 A. A. Bolibruch, A. S. Merkur′ev, and N. Yu. Netsvetaev, Editors, Mathematics in St. Petersburg
173 V. Kharlamov, A. Korchagin, G. Polotovskiĭ, and O. Viro, Editors, Topology of Real Algebraic Varieties and Related Topics
172 K. Nomizu, Editor, Selected Papers on Number Theory and Algebraic Geometry
171 L. A. Bunimovich, B. M. Gurevich, and Ya. B. Pesin, Editors, Sinai's Moscow Seminar on Dynamical Systems
170 S. P. Novikov, Editor, Topics in Topology and Mathematical Physics
169 S. G. Gindikin and E. B. Vinberg, Editors, Lie Groups and Lie Algebras: E. B. Dynkin's Seminar
168 V. V. Kozlov, Editor, Dynamical Systems in Classical Mechanics
167 V. V. Lychagin, Editor, The Interplay between Differential Geometry and Differential Equations
166 O. A. Ladyzhenskaya, Editor, Proceedings of the St. Petersburg Mathematical Society, Volume III
165 Yu. Ilyashenko and S. Yakovenko, Editors, Concerning the Hilbert 16th Problem
164 N. N. Uraltseva, Editor, Nonlinear Evolution Equations
163 L. A. Bokut′, M. Hazewinkel, and Yu. G. Reshetnyak, Editors, Third Siberian School "Algebra and Analysis"
162 S. G. Gindikin, Editor, Applied Problems of Radon Transform
161 Katsumi Nomizu, Editor, Selected Papers on Analysis, Probability, and Statistics
160 K. Nomizu, Editor, Selected Papers on Number Theory, Algebraic Geometry, and Differential Geometry
159 O. A. Ladyzhenskaya, Editor, Proceedings of the St. Petersburg Mathematical Society, Volume II
158 A. K. Kelmans, Editor, Selected Topics in Discrete Mathematics: Proceedings of the Moscow Discrete Mathematics Seminar, 1972–1990
157 M. Sh. Birman, Editor, Wave Propagation. Scattering Theory
156 V. N. Gerasimov, N. G. Nesterenko, and A. I. Valitskas, Three Papers on Algebras and Their Representations
155 O. A. Ladyzhenskaya and A. M. Vershik, Editors, Proceedings of the St. Petersburg Mathematical Society, Volume I
154 V. A. Artamonov et al., Selected Papers in K-Theory
153 S. G. Gindikin, Editor, Singularity Theory and Some Problems of Functional Analysis
152 H. Draškovičová et al., Ordered Sets and Lattices II
151 I. A. Aleksandrov, L. A. Bokut′, and Yu. G. Reshetnyak, Editors, Second Siberian Winter School "Algebra and Analysis"
150 S. G. Gindikin, Editor, Spectral Theory of Operators
149 V. S. Afraĭmovich et al., Thirteen Papers in Algebra, Functional Analysis, Topology, and Probability, Translated from the Russian
148 A. D. Aleksandrov, O. V. Belegradek, L. A. Bokut′, and Yu. L. Ershov, Editors, First Siberian Winter School "Algebra and Analysis"
147 I. G. Bashmakova et al., Nine Papers from the International Congress of Mathematicians, 1986
146 L. A. Aĭzenberg et al., Fifteen Papers in Complex Analysis
145 S. G. Dalalyan et al., Eight Papers Translated from the Russian
144 S. D. Berman et al., Thirteen Papers Translated from the Russian
143 V. A. Belonogov et al., Eight Papers Translated from the Russian
142 M. B. Abalovich et al., Ten Papers Translated from the Russian
141 H. Draškovičová et al., Ordered Sets and Lattices
140 V. I. Bernik et al., Eleven Papers Translated from the Russian
139 A. Ya. Aĭzenshtat et al., Nineteen Papers on Algebraic Semigroups
138 I. V. Kovalishina and V. P. Potapov, Seven Papers Translated from the Russian
137 V. I. Arnol′d et al., Fourteen Papers Translated from the Russian

(See the AMS catalog for earlier titles)

Mathematics in St. Petersburg

American Mathematical Society

TRANSLATIONS

Series 2 • Volume 174

Advances in the Mathematical Sciences – 30

(Formerly Advances in Soviet Mathematics)

Mathematics in St. Petersburg

A. A. Bolibruch
A. S. Merkur'ev
N. Yu. Netsvetaev

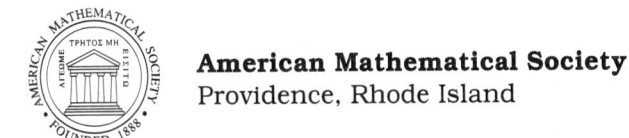

American Mathematical Society
Providence, Rhode Island

ADVANCES IN THE MATHEMATICAL SCIENCES
EDITORIAL COMMITTEE

V. I. ARNOLD

S. G. GINDIKIN

V. P. MASLOV

Translation edited by A. B. Sossinsky

1991 *Mathematics Subject Classification.* Primary 00B15; Secondary 12G05, 14D20, 14L05, 14P25, 19-XX, 34A30, 52A37, 57Mxx, 58F07, 68Q99, 81T70.

ABSTRACT. The papers in this collection are written by well-known mathematicians from St. Petersburg, and contain new results on a variety of topics in mathematics: algebra, K-theory, topology, differential equations, among others. One of the reasons these papers were brought together is that their authors graduated from the same high school: the Specialized Physico-Mathematical School #45 at Leningrad University.

The book is useful to graduate students and researchers in mathematics.

Library of Congress Card Number 91-640741
ISBN 0-8218-0559-2
ISSN 0065-9290

Copying and reprinting. Material in this book may be reproduced by any means for educational and scientific purposes without fee or permission with the exception of reproduction by services that collect fees for delivery of documents and provided that the customary acknowledgment of the source is given. This consent does not extend to other kinds of copying for general distribution, for advertising or promotional purposes, or for resale. Requests for permission for commercial use of material should be addressed to the Assistant to the Publisher, American Mathematical Society, P. O. Box 6248, Providence, Rhode Island 02940-6248. Requests can also be made by e-mail to reprint-permission@ams.org.

Excluded from these provisions is material in articles for which the author holds copyright. In such cases, requests for permission to use or reprint should be addressed directly to the author(s). (Copyright ownership is indicated in the notice in the lower right-hand corner of the first page of each article.)

© Copyright 1996 by the American Mathematical Society. All rights reserved.
The American Mathematical Society retains all rights
except those granted to the United States Government.
Printed in the United States of America.

∞ The paper used in this book is acid-free and falls within the guidelines
established to ensure permanence and durability.
♻ Printed on recycled paper.

10 9 8 7 6 5 4 3 2 1   01 00 99 98 97 96

Contents

Editors' Preface	ix
Foreword	xi
Some Memories of Boarding School #45 A. A. Bolibruch	1

Part I. Algebra

The Geometrical Genus of the Moduli Space of Abelian Varieties V. A. Gritsenko	9
On the Cohomology Groups of the Field of Rational Functions O. T. Izhboldin	21
On Topological Filtration for Severi–Brauer Varieties II N. A. Karpenko	45
On the Norm Residue Homomorphism for Fields A. S. Merkur′ev	49
Concerning Hall's Theorem A. G. Moshonkin	73
Simplicial Determinant Maps and the Second Term of Weight Filtrations A. Yu. Nenashev	79
On Common Zeros of Two Quadratic Forms A. S. Sivatsky	95
Isogenies of Height One and Filtrations of Formal Groups A. L. Smirnov	103
Homology Stability for H-Unital Q-Algebras A. A. Suslin	117

Part II. Geometry and Analysis

Willmore Surfaces, 4-Particle Toda Lattice and Double Coverings of Hyperelliptic Surfaces M. V. Babich	143
On the Birkhoff Standard Form of Linear Systems of ODE A. A. Bolibruch	169

NC Solving of a System of Linear Ordinary Differential Equations in
Several Unknowns
    D. Yu. Grigoriev   181

Hyperreflexivity of Contractions Close to Isometries
    V. V. Kapustin   193

On the Definition of the 2-Category of 2-Knots
    V. M. Kharlamov and V. G. Turaev   205

Application of Topology to Some Problems in Combinatorial Geometry
    V. V. Makeev   223

Homology and Cohomology of Hypersurfaces with Quadratic Singular
Points in Generic Position
    N. Yu. Netsvetaev   229

Asymptotic Solutions to the Quantized Knizhnik–Zamolodchikov
Equation and Bethe Vectors
    V. O. Tarasov and A. N. Varchenko   235

Editors' Preface

The present volume is dedicated to the 30th anniversary of the founding of the Academic Gymnasium of Saint Petersburg University. This is a specialized secondary school whose goals are a more intensive (advanced) study of mathematics and physics in the upper grades and preparation for further studies at the University. From the day when it was founded until recent times, the school was officially called the "Specialized Physico-Mathematical Boarding School #45 of Leningrad University" but was generally known as the "45th Boarding School".

The readers, who wish to have a better understanding of what the 45th Boarding School was, are referred to the Foreword, whose first version was written by Yu. I. Ionin, to whom the school is largely indebted for the system of mathematical education that was formed there and flourished, especially in the seventies.

Almost all the authors of this volume graduated at different times from the 45th Boarding School and then studied at Leningrad University. The exceptions are A. Suslin and A. Varchenko. Let us note, however, that A. Suslin taught at the school for a certain period of time and was the scientific adviser of several of the authors when they were undergraduate and graduate students at Leningrad University. A. Varchenko is an alumnus of a similar boarding school in Moscow.

On the other hand, this jubilee volume does not include work by all the alumni of the school presently active in mathematical research, far from it. Among the best known graduates of the 45th Boarding School not represented in this volume, let us mention A. B. Aleksandrov, S. V. Kislyakov, and V. V. Peller (mathematical analysis), I. B. Fesenko and I. A. Panin (algebraic K-theory), D. Yu. Burago (geometry "in the large"), G. Yu. Panina (integral geometry).

The initial impetus for the appearance of this volume was a three-day conference, also dedicated to the school's 30th anniversary, held at Saint Petersburg University in October 1993, with the financial support of the American Mathematical Society organized by professor Daniel Stroock, to whom we express our gratitude. The program of the conference differed somewhat from the contents of this volume. Here is this program:

Hour reports

V. P. Khavin. *From analytic functions to harmonic vector fields.*
A. L. Smirnov. *Fermat's Last Theorem is proved* (?)
A. A. Suslin. *Algebraic K-theory and motives.*
N. Yu. Netsvetaev. *Classification of differentiable manifolds.*

Half-hour reports

O. T. Izhboldin. *Decomposition of multiple sums.*
I. B. Fesenko. *Arithmetical description of Abelian p-extensions.*
Yu. D. Burago. *Spaces of Riemannian manifolds.*
M. N. Gusarov. *New knot invariants.*
I. A. Panin. *Vector bundles on homogeneous manifolds.*
A. G. Moshonkin. *Computationally difficult problems in number theory and their applications to cryptography.*
Andras Szücs. *Once more about turning the sphere inside out.*
G. Yu. Panina. *Mixed volumes of nonconvex bodies.*

The idea of conducting this conference is primarily due to Sasha Nenashev, who did much to implement it. The papers constituting this volume were mostly collected during the winter and spring of 1994. Its publication would not have been possible without Oleg Izhboldin, who prepared the electronic version of this volume. We are deeply grateful to him.

The Editors

Foreword

The papers constituting this collection have not been brought together because they belong to the same branch of mathematics, but because their authors (with two exceptions) graduated from the same high school: the Specialized Physico-Mathematical School #45 of Leningrad University. The collection itself is dedicated to the school's 30th anniversary. Clearly, this must be an unusual school. Here we try to explain why and in what sense it is so unusual.

Why were the specialized boarding schools organized

In the sixties the prestige of education, and mathematical education in particular, was extremely high in Russia. This is related to the remarkable achievements of science and technology (e.g., the sputnik) and the interest of the Soviet state in the development of science (a striking example is the founding in the late fifties and the early sixties of the Siberian Division of the USSR Academy of Sciences). There was a strong incentive for high school graduates to continue their education at universities. Throughout the country, even in the most remote places, there were many talented young people with a keen interest in mathematics. There were not enough qualified teachers to satisfy this interest.

At that time several leading scientists strived to move the country ahead primarily by developing science and culture. In the sixties, Soviet mathematicians created several new structures whose goal was to attract talented young people to the study of mathematics. Among these structures were mathematics-oriented and science-oriented boarding schools. This was an example of a successful implementation of a natural idea: to organize a nation-wide search for talented students in order to bring them together in special institutions where they would be able to develop their aptitudes. The success achieved here is apparently due to the fact that mathematical talent, just as talent in sports or music, usually manifests itself at an early stage.

How the boarding schools began

In the sixties, at the incentive of such outstanding scientists as M. A. Lavrent'ev and A. N. Kolmogorov, several unique secondary schools, sponsored by the leading universities, were founded – the so-called *specialized physico-mathematical boarding schools*, where talented young people from high schools located far from university centers could live and study.

The first such institution opened in Novosibirsk in 1961 under the auspices of Novosibirsk State University. It was followed by schools in Moscow, Leningrad, and Kiev.

The Leningrad boarding school opened in the fall of 1963. The main roles in its organization were played by D. K. Faddeev, A. D. Aleksandrov, and Yu. V. Linnik. For many years the educational and organizational policy of the 45th Boarding School was monitored and directed by its Board of Trustees. One of the most active younger mathematicians involved in the functioning of the school was M. I. Bashmakov.

Who studied there

The selection of future students was done by means of entrance examinations that took place nation-wide and were carried out by university teachers (who would travel to the corresponding regions). Usually these examinations took place immediately after mathematics olympiads, in which most high school students interested in math and science participated.

A few words about olympiads

The system of mathematical competitions (olympiads), now known throughout the world, originated in Leningrad in 1935 (and in Moscow the next year) at the incentive of B. N. Delone and spread throughout the country after World War II. Mathematical olympiads (as well as chess tournaments) gave intellectually gifted youngsters the chance to demonstrate their talents. At the same time, they provided an excellent opportunity for selecting the most talented ones for the specialized schools.

Achievements in mathematical competitions remained an important stimulus during the period of study (which was usually 2 or 3 years). The Leningrad boarding school conducted its own intramural olympiads, whose level was comparable to the All-Union competitions. Original problems were always proposed there, requiring nontrivial ideas for their solutions and rigorous proofs.

What was studied and why

Unlike the majority of Soviet schools (which worked along the strict guidelines of a unified syllabus), the boarding schools were allowed from the outset to form their own programs of study in mathematics, physics, chemistry, and biology. Each of these schools was supervised by an outstanding university, the syllabus would be worked out by university professors, who often also taught there.

What was the mathematical content of such a syllabus? One should keep in mind that our system of university education was drastically different from that, say, of the USA. In particular, all freshman university mathematics students took the same required courses with the same contents. Therefore there was no reason, even in a very advanced school, to study the standard Calculus I, II, and III courses. Since the syllabus at the boarding school was determined by the university, the teachers at the school had the opportunity of raising the level of mathematical education not by an early study of standard university courses, but rather by a deeper investigation of beautiful (often half forgotten) branches of mathematics and the creation of nonstandard, experimental approaches to the traditional topics of elementary and higher mathematics.

Another feature of the teaching process was related to the fact that in the USSR students must choose their area of study before entering the university. In order to enter the mathematical department (as well as different scientific and technological departments), one had to pass very difficult entrance examinations. At the oral exam, the applicant had to demonstrate the ability to prove various theorems in elementary algebra and geometry. The written exam consisted in solving rather superficial problems involving equations and inequalities, as well as planar and spacial geometry, often overloaded with technical and logical details.

The necessity of successfully passing these exams led the teachers at the boarding schools to pay particular attention to the acquisition of strong technical proficiency in algebra, plane and space geometry, trigonometry, and introductory calculus.

Who taught mathematics at the boarding schools

The boarding schools had a lot of autonomy not only in choosing their programs, but also in their choice of teachers.

Many lessons were conducted by graduate students and university faculty members. In the Leningrad boarding school, at different times, cycles of lectures were given by V. A. Rokhlin, D. K. Faddeev, A. M. Vershik, M. Z. Solomyak, O. A. Ladyzhenskaya, A. A. Suslin, and O. Ya. Viro.

Only a small number of professional high school teachers taught there, most of the instructors began teaching math at the boarding school when they were still university students, and their lack of professionalism in teaching methods was more than compensated by their enthusiasm and excellent knowledge of the subject matter. They loved mathematics and were quite capable of transmitting this feeling to their pupils.

Beginning with the early seventies, many graduates of Boarding School #45 returned to their alma mater to teach mathematics or science.

Probably the main reason for the success of many graduates of the Leningrad boarding school was the fact that talented youngsters and extremely enthusiastic young teachers were brought together under one roof in an atmosphere where the study of science was an end in itself, and mathematics was indeed regarded as the Queen of the sciences.

Conclusion

Let us summarize. In the thirty years or so of their existence, the specialized boarding schools have demonstrated their worth. Among the alumni of the Leningrad boarding school, there are many working high-level research mathematicians. An idea of their research activity may be obtained from the articles published in this volume.

Why did this specialized boarding school (and others) produce so many high-level research mathematicians? Without trying to give an exhaustive explanation of this phenomenon, we have only noted some characteristic properties of the mathematics teaching process in these schools.

The main idea that led to the founding of these schools, to help talented young people from provincial areas acquire a first class education, is more important today than ever before. Although science in our country is going through a very difficult period, the educational process is still alive and not doing so badly, but it is still in dire need of assistance.

October 1994   B. M. Bekker, Yu. I. Ionin, N. Yu. Netsvetaev

Some Memories of Boarding School #45

A. A. Bolibruch

The two years that I spent at Boarding School #45 were among the most interesting ones of my life. They determined, to a great extent, not only my professional destiny as a mathematician, but my character as a person, my attitude to life, and all of the things usually included in the notions of life's values or ethical position. The school syllabus contained algebra, calculus, history, literature, and all these subjects were superbly taught by our favorite teachers—Yuriĭ Iosifovich Ionin, Efim Emmanuilovich Naimark, Irina Georgievna Polubyarinova, and many others, whom I remember and hold in high esteem. But besides the required school syllabus, there were many extracurricular activities, which were of no less importance to us. It is several episodes of this nonacademic school life that I would like to recall in these notes.

In those days the Boarding School was located in Leningrad on Savushkina Street, beyond Chernaya Rechka. This was only about 30–45 minutes from the center, which meant that Leningrad's museums, theaters, and the University buildings (including the library) were easily accessible. In the winter of 1966 we had weekly excursions to the Hermitage Fine Arts Museum. With our excellent guide (unfortunately, I don't remember her name), we visited almost the entire modern art collection of the museum exposition, from the impressionists to Picasso. We were told about impressionism and pointillism, with Claude Monet and Paul Signac as examples, about cubism, Russian modernists, and learned how to appreciate modern art. I valued these lessons no less than those in algebra and calculus: they familiarized me with an entirely new world – the world of art.

* * *

Generally, I believe that a 14–15 year old only needs to learn about the existence of some world unknown to him or her (by observation, reading, or hearsay), be it the world of modern art, poetry or, say, Chinese literature. And, what is important, learn from a source worthy of trust, for example from a favorite teacher. Thus in our literature class I learned from Irina Georgievna about the existence of Leningrad's Bolshoĭ Drama Theater and its director G. A. Tovstonogov. I succeeded in getting tickets (a nearly impossible task then) to his staging of Dostoevskiĭ's *Idiot* with the admirable Innokentiĭ Smoktunovskiĭ in the title role of Prince Myshkin, and I still think that this was the best theatrical production that I have ever attended. Since then I have become an enthusiastic theatergoer. Later, in Moscow, I saw all the shows of the Sovremennik theater, the Taganka theater, and Obraztsov's puppet theater. I did not miss any of the plays directed by A. Efros in the Malaya Bronnaya theater.

How important a teacher's opinion can be, when one random phrase uttered by a teacher can exert a crucial influence on the outlook of his pupils' lives, is something I had the occasion of appreciating time and time again; in particular when I, then a fifth year university student, taught a mathematics class in Moscow's School #7 (a famous specialized mathematical school). I will give only one example here. During a lesson, when I was explaining to my students something about the cardinality of the set of continuous functions on the line, I noticed that one of them was not paying attention, but engrossed in a book that had obviously nothing to do the lesson. I was infuriated not so much by the fact that he was reading material unrelated to the lesson, but by *what* he was reading. He was not reading works by the modernist poet Khlebnikov, or Bulgakov, or, say, A. Belyĭ, but (from my point of view) a third-rate popular writer. I took the book away from him, and gave him the only book that I had in my briefcase, a volume of poetry by one of my favorite poets, Tao Yuang-Min. I would have forgotten about this episode, but several years later, during a class reunion to which my former students had invited me, I met that same student. He explained that after I had given him that book, he had become so absorbed in Chinese poetry that later, as a student of biology at Moscow University, he had organized a kind of *Lovers of classical Chinese literature* society and had become quite an expert in the field! This episode really impressed me, and since then during lectures and exercise classes I always try to find time to say a few words to the students about an interesting exposition, a book, or a theatrical production.

* * *

Our boarding school not only had classes in which mathematics was the major subject, but biology and chemistry classes as well. Kostya Raĭkin, the son of Arkadiĭ Raĭkin, the celebrated anti-establishment comedian, was in one of the latter. I doubt that Kostya seriously intended to study biology after graduating from our school, although he was certainly a talented biology student. But his artistic nature prevailed, and he continued his education in a theatrical school. (Today he is a very well-known theatrical and movie comedian and currently directs the *Moscow Theater of Miniatures*.) At the time, however, as a student of the 9th and 10th grades, he often amused us by his extraordinarily expressive body movements and his ability to transform even the most serious events into an amusing game. Thus at the oral final examination in mathematics, he was required to report on the decimal presentation of real numbers. The examination commission included members of the Board of Trustees of the school, in particular the famous algebraist D. K. Faddeev. During Kostya's answer, Faddeev began to talk with the other members of the Commission, who ceased paying attention to the answer. Kostya was quick to take advantage of this situation. His answer went something like this: "Suppose the number α lies in the closed interval $[0, 1]$. Divide it into ten equal subintervals and consider the one containing α. Now divide this subinterval into ten equal parts and again consider the part that contains $\alpha \ldots$". Usually at this point one says "and so on" and concludes the proof. But Kostya, in a monotonous but pleasantly modulated voice, continued subdividing the intervals into smaller and smaller ones to the general amusement (restrained with difficulty) of his fellow students. Finally, when Kostya had reached intervals whose size must have been comparable to that of an atomic nuclei, the commission noticed the suspicious silence in the auditorium and began looking at the blackboard again. A pause followed, and D. K. Faddeev said that he felt that a sufficient number of digits after the decimal point had been obtained and the subdivision process could be discontinued. Kostya was not asked any more questions and his examination was successfully completed.

* * *

While I was still studying at the boarding school, literary evenings were often organized there, with the participation of well-known artists, unofficial poets, and singers. These evenings were not limited to a formal presentation. The concert was usually followed by informal discussions, with very varied topics, from current Soviet poetry to the state of affairs in Algeria. Undoubtedly, the most unforgettable evening was the one conducted by Arkadiĭ Raĭkin. The school's main auditorium was overflowing, and for an hour and a half Raĭkin presented his best satirical mini-shows, including those that he was not allowed to give at official concerts. It was there that I first heard the expressions "genetics, the fallen woman of imperialism" or "paralysis in the Soviet Union is the most progressive in the world," and many other phrases that were to become anti-establishment classics of the Soviet stage.

* * *

There is nothing unusual about the fact that many students of physico-mathematical schools were not only interested in poetry but even wrote poetry themselves. At the time a tremendous amount of prestige was attached to the profession of physicist, the number of participants at the entrance examinations in the leading technical universities was huge, and many young people with interest in learning (including the humanities) would go out for mathematics or physics. But their literary interests also had to be exteriorized, and for this reason perhaps every other student of our school wrote poetry.

The peak of this poetic activity was reached in the winter of 1966, and it was then that two curious episodes related to it occurred. The first was preceded by the appearance of a school newspaper, displayed on one of the walls and containing verses by our boarding school poets, and the second, by the circulation of self-prepared literary journals in the school's dorms. Such a journal appeared in the room where I lived, commonly known as "Uncle Toms' Cabin", because the name of the room's president was Victor Toms. The title of the journal (whose Editor-in-Chief and only Editorial Board member was I) was "Quiet Pond", and its first issue contained the poetic production of my roommates. I should note that for most of us, these were our first literary endeavors, and that in our lexicon dismal words prevailed, interspersed with expletives that should have been deleted, so that our journal certainly contrasted with the optimistic verses that had been displayed in the school journal. There was nothing insincere in either publication, they simply reflected different aspects of our school life, which had its happy as well as its sad moments. Having finished editing our journal, we hung it on the door of the clothes closet (on the inside, just in case), and went to classes.

It would happen that on this very day, the Minister of Education of the Russian Federation himself came to visit our boarding school (this was the first of the two curious episodes mentioned above). The Minister was shown the school, physics lab (he sat in on one of the lessons), the school newspaper on the wall with our verses, and then was taken to visit the dormitory. I do not know how it happened, but the minister immediately went to our room, opened the door of the clothes closet, removed the tacks holding our newspaper, glanced through it and said: "Now I see what is really going on here", and left for Moscow. At the time we were still in the classrooms, but "good news spreads quickly", and we returned to the dorm with our heads down, in the worst mood imaginable.

It was not so much that we felt guilty and were afraid of punishment, but, more importantly, we had let our teachers as well as the entire school down. It was known that the minister was opposed to all kinds of specialized mathematical schools, because from his point of view they were in contradiction with the principles of social justice. So one of the consequences of what happened could be that no more new students would be accepted at our school, and eventually the whole school would be closed down. Then things developed in a way that I have not been able to forget since. Our teachers, who had undoubtedly been seriously reprimanded by the local scholastic authorities for what occurred, came to visit us in our room. Instead of the expected scolding or punishment, there was not a single word of blame addressed to us. The teachers explained that they had come to apologize for the behavior of some grown-ups incapable of understanding modern poetry. They asked us not to keep a grudge against such people, who did not know any better, adding that in the future we will often come in contact with people unable to understand certain things, but this should not justify our being angry at them. Also they told us not to stop writing poetry and invited us to a poetry reading of school poets that was soon to take place in the boarding school. On that day we received a lesson in a subject that does not appear in the school lesson timetable, but which we shall remember for all our lives.

* * *

The graduation examinations in school are a very special time. The last and most difficult exam was in physics. Our class had brought together some very strong mathematics students: there were several winners of the All-Russian olympiad, winners of the Leningrad city olympiad, and one future International Mathematics Olympiad winner (Victor Turchaninov, now a brilliant systems analyst). But in physics things did not go as well: there was only one winner of the All-Russian physics olympiad (Boris Rovner). Our physics teachers had decided to examine us quite severely. Before the examination, they told us that they would examine us very strictly and give us two grades: an official one for the school diploma and an unofficial one, which should in principle be equal to the one given at the Leningrad University entrance examinations in the case of a most severe session of questions. This was motivated by the necessity of a good training session before the University entrance exams.

This preamble to the exam did not worry me in the least, I was then a straight A student, with a good chance to be awarded the gold medal for excellence in studies, and I have never experienced problems in previous examinations. But to this day I remember the nightmare that followed. I began my answer well enough, having clearly explained the theoretical question and correctly solved the required problem. But the first additional problem spoiled my good mood. I was asked to describe the trajectory of a stone thrown from the surface of the earth in the absence of the atmosphere, and I immediately answered "along a parabola". But it was politely explained to me that my answer was erroneous, the correct answer being an ellipse (since in the statement of the problem it was not specified that the Earth's surface could be regarded as flat). And so on. Finally all ended well, but since then I have developed a slight mistrust of physics.

* * *

Today our boarding school is housed in a different part of the city, closer to the new university campus. But each time I come to Saint Petersburg (which does not happen as often as I would like), I get on the #80 bus and ride all the way past Chernaya Rechka, to 61 Savushkina Street.

Translated by A. SOSSINSKY

PART I

Algebra

The Geometrical Genus of the Moduli Space of Abelian Varieties

V. A. Gritsenko

Dedicated to Boarding School #45 on the occasion of its 30th anniversary

ABSTRACT. We prove that the geometrical genus of the moduli space of polarized abelian varieties over $\mathbb{C}$ with polarization of type $\mathrm{diag}(1,\dots,1,t)$, t a positive integer, is greater than or equal to the geometrical genus of the moduli space of principally polarized abelian varieties.

Let X be an abelian variety over $\mathbb{C}$. The underlying complex manifold of X is a complex torus W/Λ, with W a complex vector space of dimension g and Λ a $\mathbb{Z}$-lattice of rank $2g$. By definition, a *polarization* on X is the first Chern class $c_1(L) \in H^2(X, \mathbb{Z}) = \bigwedge^2 \mathrm{Hom}(\Lambda, \mathbb{Z})$ of a positive definite line bundle L on X. Let H be the Hermitian form on W corresponding to $c_1(L)$. In accordance with the elementary divisor theorem, there is a canonical (symplectic) basis of the $\mathbb{Z}$-lattice Λ, with respect to which the integral-valued alternating form im H is given by the integral matrix

$$S_T = \begin{pmatrix} 0 & T \\ -T & 0 \end{pmatrix},$$

where $T = \mathrm{diag}(t_1, \dots, t_g)$ for some positive integers t_i with $t_i | t_{i+1}$ for any i. The matrix T is uniquely determined by L and is called the *type* of the polarization.

The period matrix Π of the abelian variety X written in the basis

$$\{e_1, \dots, e_g, t_1^{-1} e_{g+1}, \dots, t_g^{-1} e_{2g}\},$$

where e_i are the elements of the canonical basis of Λ, has the form $\pi = (Z, T)$, where $Z \in M_g(\mathbb{C})$. According to the Riemann relations, the matrix Z belongs to the Siegel upper half-plane of degree g,

$$\mathbb{H}_g = \{Z \in M_g(\mathbb{C}) : {}^t Z = Z, \, \mathrm{im}\, Z > 0\}$$

and $(\mathrm{im}\, Z)^{-1}$ is the matrix of the Hermitian form H.

1991 *Mathematics Subject Classification.* Primary 14K10.
Key words and phrases. Abelian varieties, geometrical genus, moduli space.

©1996, American Mathematical Society

Conversely, for any $Z \in \mathbb{H}_g$, the pair
$$(X_Z, (\operatorname{im} Z)^{-1}), \qquad X_Z := \mathbb{C}^g/(Z,T)\mathbb{Z}^{2g}$$
is a polarized abelian variety of type T. Two such varieties $(X_{Z_1}, (\operatorname{im} Z_1)^{-1})$ and $(X_{Z_2}, (\operatorname{im} Z_2)^{-1})$ are isomorphic iff there is an element in the subgroup $\Gamma(T)$ of the symplectic group $Sp_{2g}(\mathbb{Q})$, which maps Z_1 to Z_2. Recall that the real symplectic group $Sp_{2g}(\mathbb{R})$ acts on the Siegel upper half-plane $\mathbb{H}_g$ as the group of fractional linear transformations
$$g\langle Z\rangle := (AZ+B)(CZ+D)^{-1} \quad \text{for } g = \begin{pmatrix} A & B \\ C & D \end{pmatrix} \in Sp_{2g}(\mathbb{R}).$$

The group $\Gamma(T)$ is conjugate to the so-called integral paramodular group
$$Sp(T,\mathbb{Z}) = \{g \in M_{2g}(\mathbb{Z}) : gS_T{}^tg = S_T\}.$$

By definition
$$\Gamma(T) = I_T^{-1} Sp(S_T, \mathbb{Z}) I_T, \quad \text{where } I_T = \begin{pmatrix} E_g & 0 \\ 0 & T \end{pmatrix}.$$

The quotient
$$\mathcal{A}(T) = \Gamma(T) \setminus \mathbb{H}_g$$
is a normal complex analytic space of dimension $g(g+1)/2$. $\mathcal{A}(T)$ has the structure of a quasi-projective algebraic variety and is the coarse moduli space of polarized abelian varieties of type T (see [**M, I, LB**]).

In this paper, we restrict ourselves to the case of a polarization of type
$$T = T_t = \operatorname{diag}(1,\ldots,1,t),$$
i.e., $t_1 = \cdots = t_{g-1} = 1$ and $t_g = t$. For such T, we put
$$\Gamma_g^{(t)} = \Gamma(T_t), \qquad \mathcal{A}_g^{(t)} = \mathcal{A}(T_t).$$

For $t = 1$ we have $\Gamma_g = Sp_{2g}(\mathbb{Z})$ and the quotient $\mathcal{A}_g = Sp_{2g}(\mathbb{Z}) \setminus \mathbb{H}_g$ is the moduli space of principally polarized abelian varieties.

Let V be any complex variety of dimension n. We denote by $\Omega_n(V)$ the space of global differential n-forms on V,
$$\Omega_n(V) = H^0(V, O(K_V)),$$
where K_V is the canonical line bundle on V. An element from $\Omega_n(V)$ is called a *canonical differential form*. A *pluri-canonical form* is defined as a global section of mK_V,
$$\Omega_n^{\otimes m}(V) = H^0(V, O(mK_V)).$$

The dimension
$$\dim_{\mathbb{C}} \Omega_n^{\otimes m}(V) = \dim_{\mathbb{C}} H^0(V, O(mK_V))$$
is called the *m-genus* of V.

For $m = 1$,
$$\dim_{\mathbb{C}} \Omega_n(V) = h^{n,0}(V) = \dim_{\mathbb{C}} H^{n,0}(V)$$
is the *geometrical genus* of V, where $h^{n,0}(V)$ the $(n,0)$-Hodge number of the complex variety V.

Let us consider a smooth model $\overline{\mathcal{A}}_g^{(t)}$ of the moduli space $\mathcal{A}_g^{(t)}$. For example, one can construct a toroidal compactification of $\Gamma_g^{(t)} \setminus \mathbb{H}_g$ ([SC]), which has only finite quotient singularities. By resolving those singularities, one gets a smooth compact variety.

It is known that the m-genus does not depend on the choice of a smooth model. We shall use the following notation for the geometrical genus

$$p_g(\mathcal{A}_g^{(t)}) = h^{l,0}(\overline{\mathcal{A}}_g^{(t)}),$$

where $l = \frac{1}{2}g(g+1)$.

The structure of the moduli space of polarized abelian varieties looks "more complicated" in the case of nonprincipal polarization ($t > 1$) than in the case of principal polarization. For example, the variety $\mathcal{A}_g^{(t)}$ has more boundary components. The rigorous mathematical meaning of this "feeling" is illustrated by the following

THEOREM 1. *For any natural number t there exists an embedding of the canonical differential forms on $\mathcal{A}_g$ in the space of canonical differential forms on $\mathcal{A}_g^{(t)}$,*

$$\Omega_l(\overline{\mathcal{A}}_g) \hookrightarrow \Omega_l(\overline{\mathcal{A}}_g^{(t)}),$$

where $l = \frac{1}{2}g(g+1)$. In particular, the geometrical genus of the moduli space $\mathcal{A}_t^{(g)}$ of the complex abelian varieties of dimension g with a polarization of type $T = \mathrm{diag}(1, \ldots, 1, t)$ is greater than or equal to the geometrical genus of the moduli space of the complex abelian varieties with principal polarization

$$h^{l,0}(\mathcal{A}_g^{(t)}) \geqslant h^{l,0}(\mathcal{A}_g).$$

To prove this theorem, we construct a correspondence between the sections of the canonical line bundles of both varieties using the theory of modular forms on paramodular groups. We consider the variety

$$\widetilde{\mathcal{A}}_g^{(t)} = (\Gamma_g \cap \Gamma_g^{(t)}) \setminus \mathbb{H}_g$$

which covers the varieties $\mathcal{A}_g$ and $\mathcal{A}_g^{(t)}$:

(1)
$$\begin{array}{ccc} & \widetilde{\mathcal{A}}_g^{(t)} & \\ \swarrow & & \searrow \\ \mathcal{A}_g & & \mathcal{A}_g^{(t)}. \end{array}$$

We would like to define the canonical differential forms on the varieties $\mathcal{A}_g$ and $\mathcal{A}_g^{(t)}$ in terms of modular forms on Γ_g and $\Gamma_g^{(t)}$.

By a modular form of weight k on the group $G \subset Sp_{2g}(\mathbb{Q})$ ($g > 1$), we mean a holomorphic function F on the Siegel upper half-plane $\mathbb{H}_g$ that satisfies the functional equation

(2)
$$F|_k M(Z) := J(M,Z)^{-k} F(M\langle Z \rangle) = F(Z),$$

where $J(M,Z) = \det(CZ + D)$ for $M = \begin{pmatrix} A & B \\ C & D \end{pmatrix} \in G.$

The modular form F is called a *cusp form* if the restriction of F to the boundary of the quotient space $G \setminus \mathbb{H}_g$ is identically equal to zero. We shall denote the space of all cusp forms of weight k on G by $\mathfrak{M}_k^0(G)$.

Let $\mathbb{H}_g^\circ$ be the subset of all points of the Siegel upper-half plane, where the map
$$\mathbb{H}_g \to G \setminus \mathbb{H}_g$$
is unbranched. A holomorphic differential form
$$\omega_F = F(Z) \bigwedge_{i \leqslant j} dZ_{ij}, \quad \text{where } Z = (Z_{ij}) \in \mathbb{H}_g,$$
is G-invariant if and only if the function F satisfies (2) with $k = g + 1$. Moreover, as it was proved by Freitag (see [**F**, Chapter 3]) that the differential form ω_F can be extended to any smooth model of the quotient space $G \setminus \mathbb{H}_g^\circ$ iff Ω_F is square-integrable on $G \setminus \mathbb{H}_g^\circ$, or equivalently iff F is a cusp form of weight $g + 1$ with respect to the group G. Thus we have the following isomorphisms of linear spaces
$$(3) \qquad \Omega_l(\overline{\Gamma_g^{(t)} \setminus \mathbb{H}_g}) \approx \Omega_l^2(\Gamma_g^{(t)} \setminus \mathbb{H}_g^\circ) \approx \mathfrak{M}_{g+1}^{(\text{cusp})}(G).$$

Let us consider the spaces of modular forms $\mathfrak{M}_k^0(\Gamma_g)$ and $\mathfrak{M}_k^0(\Gamma_g^{(t)})$. They are subspaces of $\mathfrak{M}_k^0(\Gamma_g \cap \Gamma_g^{(t)})$. For the coverings (1), we may define correspondences

$$\mathfrak{M}_k^{(\text{cusp})}(\Gamma_g) \xrightarrow{\text{Sym}} \mathfrak{M}_k^{(\text{cusp})}(\Gamma_g^{(t)})$$
$$\searrow \qquad \nearrow$$
$$\mathfrak{M}_k^{(\text{cusp})}(\Gamma_g \cap \Gamma_g^{(t)})$$

where Sym is the restriction of the symmetrization operator Sym_t, which we are going to define, to the space of Siegel modular forms on Γ_g.

DEFINITION. Let be $F \in \mathfrak{M}_k(\Gamma_g \cap \Gamma_g^{(t)})$. We define the operator
$$\text{Sym}_t : \mathfrak{M}_k(\Gamma_g \cap \Gamma_g^{(t)}) \to \mathfrak{M}_k(\Gamma_g^{(t)})$$
by the following formula
$$\text{Sym}_t(F) = \sum_{g \in \Gamma_g \cap \Gamma_g^{(t)} \setminus \Gamma_g^{(t)}} F|_k g.$$

THEOREM 2. *Let us assume that $g \geqslant 3$ and t is square-free. The restriction of Sym_t to the space $\mathfrak{M}_k^0(\Gamma_g)$ is injective.*

Let us prove that Theorem 1 follows from Theorem 2 and the next lemma, which we shall prove below.

LEMMA 1. *If $s = td^2$, then the paramodular group $\Gamma_g^{(s)}$ can be regarded as a subgroup of $\Gamma_g^{(t)}$ of a finite index.*

PROOF OF THEOREM 1. The embedding of the paramodular group $\Gamma_g^{(s)}$ gives us a finite covering of the moduli spaces $\mathcal{A}_g^{(s)} \to \mathcal{A}_g^{(t)}$. Thus every canonical differential form on $\mathcal{A}_g^{(td^2)}$ may be regarded as a canonical differential form on $\mathcal{A}_g^{(t)}$ and we have to show the existence of the embedding of the space of canonical differential forms only for square-free t. The isomorphism (3) and the injectivity of the symmetrization gives us the required embedding
$$\Omega_l(\overline{\mathcal{A}_g}) \to \Omega_l(\overline{\mathcal{A}_g^{(t)}}). \qquad \square$$

In some cases, we have information about the geometrical genus: for $g = 24n$ the genus of $\mathcal{A}_g$ is positive (see [**F**]).

COROLLARY 1. *For $g = 24n$, the geometrical genus of the variety $\mathcal{A}_g^{(t)}$ is positive for any t.*

We have applied Theorem 2 to produce the canonical differential forms on the variety $\overline{\mathcal{A}}_t^{(g)}$. For large g, we can construct pluri-canonical differential forms

$$F(Z)\left(\bigwedge_{i \leq j} dZ_{ij}\right)^{\otimes m} \in H^{lm,0}(\overline{\mathcal{A}}_g^{(t)}, \mathbb{C}).$$

This gives us the following particular result about the geometrical type (the Kodaira dimension) of $\mathcal{A}_g^{(t)}$.

COROLLARY 2. *For any natural t the moduli space of abelian varieties of dimension g with a polarization of type $\mathrm{diag}(1,\ldots,1,t)$ is of general type if g is large enough. For instance, $\mathcal{A}_g^{(2)}$ is of general type for $g \geq 13$.*

PROOF. Denote by $\mathfrak{M}_{m(g+1)}^{(mt)}(\Gamma_g)$ the subspace of modular forms of weight $m(g+1)$ with respect to Γ_g that vanish at infinity with order tm. This means that the Fourier–Jacobi extension of F (see below) starts from the index mt. According to Theorem 2.8 of [**T**],

$$\dim_{\mathbb{C}} \mathfrak{M}_{m(g+1)}^{(mt)}(\Gamma_g) \sim Cm^l,$$

where $l = \frac{1}{2}g(g+1)$ and C is a positive constant, if the dimension g satisfies the inequality

$$(4) \qquad \frac{g!(g+1)^g}{(2g)!} B_g > t^g,$$

where B_g is the Bernoulli number $B_g = 2(2g!)(2\pi)^{-2g}\zeta(2g)$. If $F \in \mathfrak{M}_{m(g+1)}^{(mt)}(\Gamma_g)$, then the modular form $\mathrm{Sym}_t(F)$ with respect to the paramodular group $\Gamma_g^{(t)}$ has order of vanishing at least m at infinity. Therefore the corresponding pluri-canonical differential form can be continued to a smooth compactification of $\mathcal{A}_g^{(t)}$ with accordance with Tai's criterion and the fact that all quotient singularities are canonical for $g \geq 5$. The left-hand side of (4) grows faster then any function t^g.

If $t = 2$, then the inequality (4) is valid for $g \geq 13$. □

LEMMA 2. *The paramodular group $\Gamma_g^{(t)}$ is equal to the following subgroup of the rational symplectic group*

$$\Gamma_g^{(t)} = \left\{ \begin{pmatrix} A & * & B & t* \\ t* & a & t* & tb \\ C & * & D & t* \\ * & t^{-1}c & * & d \end{pmatrix} \in Sp_g(\mathbb{Q}) \right\},$$

where $A,\ldots,D$ denote integral square matrices of size $g-1$, $*$ denotes an integral column or row of length $g-1$ or an integer. Moreover, the following inclusions are valid

$$\begin{pmatrix} A & B \\ C & D \end{pmatrix} \in Sp_2(\mathbb{Z}/t\mathbb{Z}), \qquad \begin{pmatrix} a & b \\ c & d \end{pmatrix} \in SL_2(\mathbb{Z}/t\mathbb{Z}).$$

PROOF. Let $M = \begin{pmatrix} A & B \\ C & D \end{pmatrix}$ belong to the group $Sp(S_T, \mathbb{Z})$, then

$$M^{-1} = \begin{pmatrix} T^{-1}\,{}^tDT & -T^{-1}\,{}^tBT \\ -T^{-1}\,{}^tCT & T^{-1}\,{}^tAT \end{pmatrix} \in M_{2g}(\mathbb{Z}).$$

Thus

$$I_T^{-1} M I_T = \begin{pmatrix} A & B \\ T^{-1}C & T^{-1}DT \end{pmatrix} \in \Gamma_g^{(t)}$$

has the form indicated in the lemma.

For certain reasons we prefer to work with the group conjugate to $\Gamma_g^{(t)}$:

$$\widetilde{\Gamma}_g^{(t)} = V_t \Gamma_g^{(t)} V_t^{-1} = \left\{ M = \begin{pmatrix} A & t* & Bt & * \\ * & a & * & t^{-1}b \\ *C & t* & D & * \\ t* & tc & t* & d \end{pmatrix} \in Sp_g(\mathbb{Q}) \right\},$$

with $V_t = \mathrm{diag}(E_{g-1}, t^{-1}, E_{g-1}, t)$, where the notation is like that in Lemma 2. It is easy to see that

$$I_t^{-1} \widetilde{\Gamma}_g^{(t)} I_t = \{ g \in M_{2g}(\mathbb{Z}) : {}^t g S_t g = S_t \}.$$

We keep the term "the paramodular group" for $\widetilde{\Gamma}_g^{(t)}$. Since the groups $\Gamma_g^{(t)}$ and $\widetilde{\Gamma}_g^{(t)}$ are conjugate, we may prove the analog of Theorem 2 for the symmetrization operator

$$\mathrm{Sym}'_t : F \to \sum_{g \in \Gamma_g \cap \widetilde{\Gamma}_g^{(t)} \backslash \widetilde{\Gamma}_g^{(t)}} F|_k g$$

from $\mathfrak{M}_k^{(\mathrm{cusp})}(\Gamma_g)$ to $\mathfrak{M}_k^{(\mathrm{cusp})}(\widetilde{\Gamma}_g^{(t)})$.

The main idea of proving Theorem 2 is to represent the symmetrization operator as a Hecke operator with respect to a maximal parabolic subgroup. Such operators are defined in the author's paper [G1] and are connected with a special Fourier–Jacobi expansion of modular forms. The Fourier–Jacobi expansions of modular forms are connected in turn with boundary components of the variety $\Gamma_g \backslash \mathbb{H}_g$ corresponding to maximal parabolic subgroups of Γ_g (see [PS]).

We fix the following decomposition of the points of the Siegel upper-half plane

$$Z' = \begin{pmatrix} Z & \mathfrak{z} \\ {}^t\mathfrak{z} & \omega \end{pmatrix} \in \mathbb{H}_g, \quad \text{where } Z \in \mathbb{H}_{g-1},\ \mathfrak{z} \in \mathbb{C}^{g-1},\ \omega \in \mathbb{H}_1.$$

Let us consider the Fourier expansion of a Siegel modular form $F \in \mathfrak{M}_k(\Gamma_g)$ with respect to ω,

$$F(Z, \mathfrak{z}, \omega) = f_0(Z) + \sum_{m \geqslant 1} f_m(Z, \mathfrak{z}) \exp(2\pi i m \omega).$$

This is the so-called Fourier–Jacobi expansion corresponding to the parabolic subgroup

$$\Gamma_\infty = \left\{ \begin{pmatrix} A & 0 & B & * \\ * & * & * & * \\ C & 0 & D & * \\ 0 & 0 & 0 & * \end{pmatrix} \in \Gamma_g, \begin{pmatrix} A & B \\ B & D \end{pmatrix} \in \Gamma_{g-1} \right\},$$

preserving an isotropic line. The summands $f_m(Z, \mathfrak{z}) \exp(2\pi i m\omega)$ are invariant with respect to the action of the elements $\gamma \in \Gamma_\infty$,

(5) $$f_m(Z, \mathfrak{z}) \exp(2\pi i m\omega)|_k \gamma = f_m(Z, \mathfrak{z}) \exp(2\pi i m\omega)$$

(see (2)). The parabolic subgroup modulo $\pm E_{2g}$ is isomorphic to the semidirect product
$$\Gamma_\infty / \{\pm E_{2g}\} = \Gamma_{g-1} \ltimes H_{2g-1}(\mathbb{Z}),$$
where $H_{2g-1}(\mathbb{Z})$ is the Heisenberg group of rank $2g - 1$, i.e., the central extension
$$0 \to \mathbb{Z} \to H_{2g-1}(\mathbb{Z}) \to \mathbb{Z}^{g-1} \times \mathbb{Z}^{g-1} \to 0.$$

The group $\Gamma_\infty / \{\pm E_{2g}\} = \Gamma_{g-1} \ltimes H_{2n+1}(\mathbb{Z})$ is called the *Jacobi group*. The equation (5) shows that the Fourier–Jacobi coefficients $f_m(Z, \mathfrak{z}) \exp(2\pi i m\omega)$ of a modular form F are examples of modular forms of weight k with respect to the parabolic subgroup Γ_∞. Such functions are called *Jacobi modular forms*. The functional equation (5) is equivalent to the following two equations

$$f_m(Z, \mathfrak{z}) = J(\gamma, Z)^{-k} \exp(-2\pi i m\,{}^t\mathfrak{z}(CZ+D)^{-1}C\mathfrak{z}) f_m(\gamma\langle Z\rangle, {}^t(CZ+D)^{-1}\mathfrak{z}),$$
$$f_m(Z, \mathfrak{z}) = \exp(2\pi i m (2\,{}^t q\mathfrak{z} + {}^t q Z q)) f_m(Z, \mathfrak{z} + l + Zq)$$

for any $\gamma = \begin{pmatrix} A & B \\ C & D \end{pmatrix} \in \Gamma_{g-1}$ and $l, q \in \mathbb{Z}^{g-1}$.

In the paper [G1], the Hecke operators with respect to the parabolic subgroup Γ_∞ are investigated. First let us define the *Hecke ring* of the parabolic subgroup
$$\mathcal{H}(\Gamma_\infty) = \mathcal{H}(\Gamma_\infty, \Gamma_\infty(\mathbb{Q}))$$
that consists of all formal finite sums $\sum_i a_i \Gamma_\infty N_i$, where $N_i \in \Gamma_\infty(\mathbb{Q})$ and $a_i \in \mathbb{Q}$, which are invariant with respect to the right multiplication by $\gamma \in \Gamma_\infty$. The multiplication in $\mathcal{H}(\Gamma_\infty)$ is defined by

$$\left(\sum_i a_i \Gamma_\infty N_i\right) \cdot \left(\sum_j b_j \Gamma_\infty M_j\right) = \sum_{i,j} a_i b_j \Gamma_\infty N_i M_j.$$

$\mathcal{H}(\Gamma_\infty)$ is an associative ring. According to the elementary divisor theorem, any $X \in \mathcal{H}(\Gamma_g) = \mathcal{H}(Sp_{2g}(\mathbb{Z}), GSp_{2g}(\mathbb{Q}))$ can be represented in the form $X = \sum_i a_i \Gamma_g M_i$, where $M_i \in G\Gamma_\infty(\mathbb{Q})$. One can easily prove that

$$\mathrm{Im}\colon X = \sum_i a_i \Gamma_g M_i \mapsto \sum_i a_i \Gamma_\infty M_i$$

is an embedding of $\mathcal{H}(\Gamma^{(g)})$ in $\mathcal{H}(\Gamma_\infty)$ (see [G1]). Thus the Hecke ring of the parabolic subgroup can be taken as a noncommutative extension of the usual Hecke ring.

We have the following representation of the ring $\mathcal{H}(\Gamma_\infty)$ on the space of functions, which are invariant with respect to the $|_k$-action of the parabolic subgroup Γ_∞,

$$F \to F|_k X = \sum_i a_i J(g_i, Z)^{-k} F(g_i \langle Z \rangle) \qquad \left(X = \sum_i a_i \Gamma_\infty g_i \in \mathcal{H}(\Gamma_\infty)\right).$$

For instance, $|_k X$ acts on the space of all Fourier–Jacobi coefficients of Siegel modular forms. We note that if $F \in \mathfrak{M}_k^{(\mathrm{cusp})}(\Gamma_g)$, then $F|_k X$ is not a modular form on Γ_g, but this function still has a Fourier–Jacobi extension of the same type.

In order to represent the symmetrization operator as a Hecke operator, we consider the following embedding the multiplicative semigroup of natural numbers in this ring.

LEMMA 3. *The map*
$$n \to \Lambda_n = \Gamma_\infty \operatorname{diag}(1_{g-1}, n, 1_{g-1}, n^{-1}) \Gamma_\infty$$
is an embedding of $\mathbb{N}^\times$ *in* $\mathcal{H}(\Gamma_\infty)$.

To prove this, it suffices to note that $\Lambda(n)$ has only one left coset:
$$\Lambda_n = \Gamma_\infty \operatorname{diag}(1_{g-1}, n, 1_{g-1}, n^{-1}).$$

REMARK. One can regard $\mathbb{N}^\times$ as a trivial Hecke ring $\mathcal{H}(\{1\}, \mathbb{N}^\times)$. For a theory of Hecke rings of parabolic subgroups, see [**G1, G3**].

For any
$$F(Z, \mathfrak{z}, \omega) = \sum_{m \geqslant 1} f_m(Z, \mathfrak{z}) \exp(2\pi i m \omega) \in \mathfrak{M}_k(\Gamma_g),$$
one gets
$$(6) \qquad (F|_k \Lambda(n))(Z') = n^k \sum_{m \geqslant 1} f_m(Z, n\mathfrak{z}) \exp(2\pi i n^2 m \omega).$$

LEMMA 4. *For any $F \in \mathfrak{M}_k(\Gamma_g)$ one can represent the symmetrization operator*
$$\operatorname{Sym}'_t : \mathfrak{M}_k(\Gamma_g) \to \mathfrak{M}_k(\widetilde{\Gamma}_g^{(t)})$$
as a Γ_∞-Hecke operator:
$$\operatorname{Sym}'_t(F) = \sum_{ad=t} F|_k \nabla_a \Lambda_d,$$
where the summation is over all positive divisors of t and
$$\nabla_a = \sum_{r \in \mathbb{Z}/a^{-1}\mathbb{Z}} \Gamma_\infty \begin{pmatrix} 1_{g-1} & 0 & 0 & 0 \\ 0 & 1 & 0 & r \\ 0 & 0 & 1_{g-1} & 0 \\ 0 & 0 & 0 & 1 \end{pmatrix} \in \mathcal{H}(\Gamma_\infty).$$

PROOF. Let
$$M = \begin{pmatrix} A & t* & B & * \\ * & a & * & t^{-1}b \\ *C & t* & D & * \\ t* & tc & t* & d \end{pmatrix} \in \widetilde{\Gamma}_g^{(t)}$$
(see Lemma 2). To find a system of the representatives $\Gamma_{00}(t) \backslash \widetilde{\Gamma}_g^{(t)}$, $\Gamma_{00}(t) = \Gamma_g \cap \widetilde{\Gamma}_g^{(t)}$, we must control only the SL_2-part of the elements M. We put
$$\phi_2(M) = \begin{pmatrix} a & b \\ c & d \end{pmatrix} \in SL_2(\mathbb{Z}/t\mathbb{Z}) \quad \text{and} \quad \psi_2 \begin{pmatrix} a & b \\ c & d \end{pmatrix} = \begin{pmatrix} E_{g-1} & 0 & 0 & 0 \\ 0 & a & 0 & b \\ 0 & 0 & E_{g-1} & 0 \\ 0 & c & 0 & d \end{pmatrix},$$
where ψ_2 is an embedding of the group $SL_2(\mathbb{Q})$ in $Sp_g(\mathbb{Q})$. For any two elements γ_1 and γ_2 from the system $\{\gamma_i\}_i = \Gamma_{00}(t) \backslash \widetilde{\Gamma}_g^{(t)}$, the corresponding left cosets with respect to Γ_g are still disjoint: $\Gamma_g \gamma_1 \neq \Gamma_g \gamma_2$. In the lemma, we consider the restriction of

the symmetrization operator to the space of modular forms with respect to the full modular group; thus it suffices to describe the set of left cosets $\{\Gamma_g \gamma_i\}_i$.

The natural projection $SL_2(\mathbb{Z}) \to SL_2(\mathbb{Z}/t\mathbb{Z})$ is surjective. Therefore for any $M \in \widetilde{\Gamma}_g^{(t)}$ we may find an element $h \in SL_2(\mathbb{Z})$ such that $M\tilde{h} \in \Gamma_{00}(t)$, where $\tilde{h} = I_t \psi_2(h) I_t^{-1}$ with $I_t = \operatorname{diag}(1, \ldots, 1, t)$. Thus

$$\Gamma_{00}(t) M \subset \Gamma_{00}(t) I_t \phi_2(SL_2(\mathbb{Z})) I_t^{-1}.$$

The ϕ_2-part of the element of the class $\Gamma_g M$ belongs to the set

$$SL_2(\mathbb{Z}) \begin{pmatrix} 1 & 0 \\ 0 & t \end{pmatrix} SL_2(\mathbb{Z}) \begin{pmatrix} 1 & 0 \\ 0 & t^{-1} \end{pmatrix},$$

which is, up to a factor, a double class of the Hecke ring of $SL_2(\mathbb{Z})$. Taking its system of representatives and bearing in mind that t is square-free, we see that if $\{\gamma_i\} = \Gamma_{00}(t) \setminus \widetilde{\Gamma}_g^{(t)}$, then

$$\bigcup_i \Gamma_g \gamma_i = \bigcup_{\substack{ad=t \\ b \bmod t}} \Gamma_g \psi_2 \left(\begin{pmatrix} 1 & bd^{-1} \\ 0 & 1 \end{pmatrix} \right) \psi_2 \left(\begin{pmatrix} a & 0 \\ 0 & a^{-1} \end{pmatrix} \right).$$

This system gives us the Γ_∞-Hecke operator of the lemma. $\square$

Using the previous lemma, we can calculate the Fourier–Jacobi expansion of $\operatorname{Sym}'_t(F)$.

LEMMA 5. *Let*

$$F(Z, \mathfrak{z}, \omega) = \sum_{m \geq 1} f_m(Z, \mathfrak{z}) \exp(2\pi i m \omega) \in \mathfrak{M}_k^{(\mathrm{cusp})}(\Gamma_g).$$

The Fourier–Jacobi expansion of the symmetrization of F has the following form

$$\operatorname{Sym}'_t(F)(Z, \mathfrak{z}, \omega) = \sum_{m \geq 1} f_{tm}^{(t)}(Z, \mathfrak{z}) \exp(2\pi i t m \omega),$$

$$f_{tm}^{(t)}(Z, \mathfrak{z}) = \sum_{ad=t} d^k \varepsilon_a(tm/d^2) f_{tm/d^2}(Z, d\mathfrak{z}) \quad \text{with } \varepsilon_a(tm/d^2) = \begin{cases} a, & \text{if } m/d \in \mathbb{N}, \\ 0, & \text{otherwise.} \end{cases}$$

PROOF. This follows from Lemma 4 and (6).

LEMMA 6. *Let $g \geq 3$, $F \in \mathfrak{M}_k^0(\Gamma_g)$. There exists an index $m = tqd^2$, where $(t, q) = 1$, such that the Fourier–Jacobi coefficient $f_m(Z, \mathfrak{z})$ of F is not identically zero.*

PROOF. Let us take the Fourier expansion of the cusp form F,

$$F(Z) = \sum_{N > 0} a(N) \exp(2\pi i \operatorname{tr}(NZ)),$$

where $Z \in \mathbb{H}_g$ and the sum is taken over all symmetric positive definite semi-integral matrices N. It is well known that

$$a(N) = a({}^t UNU), \quad U \in GL_g(\mathbb{Z}).$$

Let us take a matrix N such that $a(N) \neq 0$. For $g \geq 4$, according to the Hasse–Minkowski theorem, any positive number is represented by the quadratic form N over

$\mathbb{Q}$. For instance, there is an $x \in \mathbb{Q}^g$ such that ${}^t x N x = t$. After multiplication by an integer, we get a primitive integral representation $\tilde{x}$ of a number of type td^2. Let us add the integral column $\tilde{x}$ to the $SL_g(\mathbb{Z})$ matrix U_x. We have

$$a(M) = a({}^t U_x N U_x) = a(N) \neq 0 \quad \text{and} \quad (M)_{gg} = td^2.$$

Thus the Fourier–Jacobi coefficient $f_{td^2}(Z, \mathfrak{z})$ does not vanish identically. For $g = 3$, the same method works with a small modification. One has to change t in a finite number of p-places to get a number that can be represented by a ternary quadratic form N. This gives us a factor q in the formulation of the lemma. $\square$

END OF THE PROOF OF THEOREM 2. If $t = p_1 \cdots p_\nu$ is the decomposition of square-free t in the product of primes, then according to the last lemma the modular form F has a nontrivial Fourier–Jacobi coefficients with index of type

$$m = p_1^{2\delta_1+1} \cdots p_\nu^{2\delta_\nu+1} m_1 \qquad (m_1, t) = 1.$$

Let us take such index m with the minimal possible sum $\delta_1 + \cdots + \delta_\nu$. In Lemma 5 we have calculated the Fourier–Jacobi coefficients of the symmetrization $\mathrm{Sym}'_t(F)$. In particular

$$f_m^{(t)}(Z, \mathfrak{z}) = t f_m(Z, \mathfrak{z}) \neq 0.$$

Therefore the function $\mathrm{Sym}'_t(F)$ has at least one nontrivial Fourier–Jacobi coefficient. Theorem 2 is proved. $\square$

REMARK. It is possible to use the same method to prove the analog of Theorem 2 for a pair of the groups $\Gamma_g^{(t)}$ and $\Gamma_g^{(tp)}$, where p is a prime that does not divide t. This will give us an embedding

$$\Omega_l(\overline{\mathcal{A}_g^{(t)}}) \to \Omega_l(\overline{\mathcal{A}_g^{(tn)}}),$$

and the inequality

$$p_g(\mathcal{A}_g^{(tn)}) \geq p_g(\mathcal{A}_g^{(t)})$$

for the geometrical genus for any natural number n, which is coprime to t.

This result is valid for $g = 2$ as well. To prove the injectivity of the symmetrization in this case, we use another arithmetical method (see [G2]).

References

[SC] A. Ash, D. Mumford, M. Rapoport, and Y. Tai, *Smooth compactification of locally symmetric varieties*, MSP, Brookline, 1975.
[F] E. Freitag, *Siegelsche Modulfunktionen*, Grundlehren Math. Wissensch., vol. 254, Springer-Verlag, Berlin–Heidelberg–New York, 1983.
[G1] V. A. Gritsenko, *The action of modular operators on the Fourier–Jacobi coefficients of modular forms*, Mat. Sb. **119** (1982), 248–277; English transl., Math. USSR-Sb. **47** (1984), 237–268.
[G2] _____, *Modulformen zur Paramodulgruppe und Modulräume der Abelschen Varietäten*, Vorlesungen, Heidelberg, 1994.
[I] J. Igusa, *Theta function*, Grundlehren Math. Wissensch., vol. 254, Springer-Verlag, Berlin–Heidelberg–New York, 1972.
[LB] H. Lange and Ch. Birkenhake, *Complex abelian varieties*, Grundlehren Math. Wissensch., vol. 302, Springer-Verlag, Berlin–Heidelberg–New York, 1992.
[M] D. Mumford, *Geometric invariant theory*, Ergeb. Math., vol. 34, Springer-Verlag, Berlin, 1965.

[PS] I. I. Pyatetskiĭ-Shapiro, *Automorphic functions and the geometry of classical domains*, Gordon and Breach, New York, 1969.
[T] Y. Tai, *On the Kodaira dimension of the moduli spaces of abelian varieties*, Invent. Math. **68** (1982), 425–439.

Translated by THE AUTHOR

POMI, FONTANKA 27, ST. PETERSBURG 191011, RUSSIA

On the Cohomology Groups of the Field of Rational Functions

O. T. Izhboldin

Dedicated to the memory of D. K. Faddeev, the founder of the 45th Boarding School

Introduction

In this paper we consider the groups

$$H_p^0(F) = \mu_p(F), \quad H_p^1(F) = {}_p X(F), \quad H_p^2(F) = {}_p \operatorname{Br}(F),$$

and a generalization of these groups in the special case $\operatorname{char}(F) = p$. The case under consideration is essentially different from the "classical" case $\operatorname{char}(F) \neq p$.

It is well known (see for example [Se, Fa]) that for any field F of characteristic different from p we have isomorphisms

$$(1) \qquad {}_p \operatorname{Br}(F((t))) \cong {}_p \operatorname{Br}(F) \oplus {}_p X(F),$$

$$(2) \qquad {}_p \operatorname{Br}(F(t)) \cong {}_p \operatorname{Br}(F) \oplus \coprod_{v \neq \infty} {}_p X(F_v).$$

On the other hand, if $\operatorname{char}(F) = p > 0$ and F is not a perfect field, the isomorphisms (1) and (2) do not exist.

The main results of the present paper are as follows:

In §§2 and 3, we compute the group $H_p^n(L)$, where L is a complete discrete valuation field of characteristic p. We prove that

$$H_{p,\mathrm{ur}}^n(L) \cong H_p^n(F) \oplus H_p^{n-1}(F),$$

where $H_{p,\mathrm{ur}}^n(L) = \ker(H_p^n(L) \to H_p^n(L^{\mathrm{ur}}))$, and describe the factor group $\widetilde{H}_p^n(L) = H_p^n(L)/H_{p,\mathrm{ur}}^n(L)$ in terms of invariants of F.

In §§4 and 5, we compute the group $H_p^n(F(t))$, where $\operatorname{char}(F) = p > 0$. We prove that

$$H_{p,\mathrm{sep}}^n(F(t)) \cong H_p^n(F) \oplus \coprod_{v \neq \infty} H_p^{n-1}(F_v),$$

1991 *Mathematics Subject Classification.* Primary 13A20; Secondary 12G99.

Key words and phrases. Cohomology groups, field of rational functions, complete discrete valuation field.

where $H^n_{p,\text{sep}}(F(t)) = \ker(H^n_p(F(t)) \to H^n_p(F^{\text{sep}}(t)))$, and compute the factor group $\widetilde{H}^n_p(F(t)) = H^n_p(F(t))/H^n_{p,\text{sep}}(F(t))$. The main result of §6 describes the groups $H^n_m(F(t))$ for arbitrary m.

Here are the most interesting corollaries of the main theorem: for any separably closed field F, the homomorphism $H^n_m(F(t)) \to \coprod_v H^n_m(F(t)_v)$ is an isomorphism; for any field F, the homomorphism $H^n_m(F(t)) \to \coprod_v H^n_m(F(t)_v)$ is a monomorphism.

§1. Main definitions

1.1. Basic notation. We use the following notation throughout the paper.
- F is a field.
- p is a prime integer.
- $\mu(F)$ is the group of roots of unity in F.
- $G_F = \text{Gal}(F^{\text{sep}}/F)$ is the absolute Galois group of F.
- $X(F) = \text{Hom}_c(\text{Gal}(F), \mathbb{Q}/\mathbb{Z})$ is the group of all continuous characters of the compact group $\text{Gal}(F)$.
- $\text{Br}(F)$ is the Brauer group of F.
- $\mu_p(F)$ (resp. $_pX(F)$, $_p\text{Br}(F)$) denotes the subgroup of all elements of exponent p in $\mu(F)$ (resp. $X(F)$, $\text{Br}(F)$).

1.2. Galois cohomology. If M is a G_F-module such that the stabilizer of each element is open in G, then the cohomology groups $H^n(F, M)$ are defined as the right derived functor in M of the functor $M \to M^{G_F}$.

If F is a field of characteristic different from p, then we denote the group $H^n(F, \mu_p^{\otimes(n-1)})$ by $H^n_p(F)$, where $\mu_p^{\otimes(n-1)}$ is the tensor power of the G_F-module $\mu_p = \mu_p(F^{\text{sep}})$ on which $G_F = \text{Gal}(F^{\text{sep}}/F)$ acts in natural way.

It is well known that $H^0_p(F) \cong \mu_p(F)$, $H^1_p(F) \cong {_pX(F)}$, and $H^2_p(F) \cong {_p\text{Br}(F)}$. On the other hand, if the characteristic of F is p, then μ_p is trivial and $\text{cd}_p(F) \leq 1$. Hence we need a new definition of $H^n_p(F)$ in the case $\text{char}(F) = p$.

1.3. For any field F of characteristic $p > 0$, let Ω^n_F be the nth exterior power over F of the absolute differential module $\Omega^1_F = \Omega_{F/\mathbb{Z}}$. Let $d\Omega^{n-1}_F$ be the image of the exterior derivation $d_{n-1}\colon \Omega^{n-1}_F \to \Omega^n_F$. Denote the group $\ker(d_n) = \ker(\Omega^n_F \to \Omega^{n+1}_F)$ by $\Omega^n_{F,d=0}$. Now let $H^{n+1}_p(F)$ be the cokernel of Cartier's homomorphism

$$\Omega^n_F \xrightarrow{\wp = \Phi - 1} \Omega^n_F/d\Omega^{n-1}_F,$$
$$a\frac{db_1}{b_1} \wedge \cdots \wedge \frac{db_n}{b_n} \longrightarrow (a^p - a)\frac{db_1}{b_1} \wedge \cdots \wedge \frac{db_n}{b_n}.$$

1.4. Definition. Let F be a field and p a prime integer. We set

$$H^{n+1}_p(F) = \begin{cases} H^{n+1}(F, \mu_p^{\otimes n}) & \text{if } \text{char}(F) \neq p, \\ \text{coker}(\Omega^n_F \xrightarrow{\wp} \Omega^n_F/d\Omega^{n-1}_F) & \text{if } \text{char}(F) = p. \end{cases}$$

1.5. The main properties of the groups Ω_F^n ($n \geq 0$).

1.5.1. LEMMA. *The homology groups $H^n(\Omega_F^*) = \ker(d_n)/\ker(d_{n-1})$ of the complex $\cdots \to \Omega_F^{n-1} \xrightarrow{d_{n-1}} \Omega_F^n \xrightarrow{d_n} \Omega_F^{n=1} \to \cdots$ are isomorphic to Ω_F^n:*

$$\Omega_F^n \xrightarrow[\sim]{\Phi} H^n\Omega_F^*,$$
$$a\frac{db_1}{b_1} \wedge \cdots \wedge \frac{db_n}{b_n} \longrightarrow a^p \frac{db_1}{b_1} \wedge \cdots \wedge \frac{db_n}{b_n}.$$

PROOF. See [IL, K2].

1.5.2. COROLLARY. *The Frobenius homomorphism $\Omega_F^n \xrightarrow{\Phi} \Omega_F^n/d\Omega_F^{n-1}$ is injective.*

1.5.3. COROLLARY. *The cokernel of the homomorphism $\Omega_F^n \xrightarrow{\Phi} \Omega_F^n/d\Omega_F^{n-1}$ is isomorphic to $\Omega_F^n/\Omega_{F,d=0}^n$.*

1.5.4. LEMMA. *Let L/F be a separable extension of fields. Then $\Omega_L^n \cong L \otimes_F \Omega_F^n$.*

1.5.5. COROLLARY. *The homomorphisms*

$$\Omega_F^n \to \Omega_{F^{\mathrm{sep}}}^n, \quad \Omega_{F,d=0}^n \to \Omega_{F^{\mathrm{sep}},d=0}^n, \quad \Omega_F^n/d\Omega_F^{n-1} \to \Omega_{F^{\mathrm{sep}}}^n/d\Omega_{F^{\mathrm{sep}}}^{n-1}$$

induced by the inclusion of fields are injective.

§2. Computation of $H_p^n(L)$ for a complete discrete valuation field L

Let L be a complete discrete valuation field with residue field F. Let v be the normalized additive discrete valuation of L, and let

$$\mathcal{O}_L = \{x \in L : v(x) \geq 0\}, \qquad U_L = \{x \in L : v(x) = 0\}.$$

Fix a prime element π of L. Denote by L^{ur} the maximal unramified extension of L. It is clear that F^{sep} is the residue field of L^{ur}.

Denote by $H_{p,\mathrm{ur}}^n(L)$ the kernel of the homomorphism $H_p^n(L) \to H_p^n(L^{\mathrm{ur}})$ and by $\widetilde{H}_p^n(L)$ the factor group $H_p^n(L)/H_{p,\mathrm{ur}}^n(L)$.

2.1. If $\mathrm{char}(F) \neq p$, then the p-primary part of the Galois cohomology groups of L can often be represented in terms of similar Galois cohomology groups of F. There exists an isomorphism

$$(3) \qquad H_p^n(L, \mu_p^{\otimes r}) \cong H^n(F, \mu_p^{\otimes r}) \oplus H^{n-1}(F, \mu_p^{\otimes(r-1)}).$$

In the case $r = n - 1$, the isomorphism (3) yields an isomorphism

$$(4) \qquad H_p^n(L) \cong H_p^n(F) \oplus H_p^{n-1}(F) \quad \text{if } \mathrm{char}(F) \neq p.$$

Note that in this case $H_{p,\mathrm{ur}}^n(L) = H_p^n(L)$ and $\widetilde{H}_p^n(L) = 0$ since

$$H_p^n(L^{\mathrm{ur}}) \cong H_p^n(F^{\mathrm{sep}}) \oplus H_p^{n-1}(F^{\mathrm{sep}}) = 0.$$

2.2. Now let $\operatorname{char}(F) = p$. In this case, the isomorphism (4) does not exist. On the other hand, the group $H^n_{p,\mathrm{ur}}(L)$ has the following similar description [**Se, K2**]:

$$H^n_{p,\mathrm{ur}}(L) \cong H^n_p(F) \oplus H^{n-1}_p(F).$$

The most interesting and complicated case, where $\operatorname{char}(L) = 0$, $\operatorname{char}(F) = p$, was studied by Kato [**K2**].

In this section we consider the case $\operatorname{char}(L) = p$ and compute the groups $H^n_p(L)$, $H^n_{p,\mathrm{ur}}(L)$, $\widetilde{H}^n_p(L)$.

2.3. From now on we assume that L is a complete discrete valuation field of characteristic $p > 0$ with residue field F.

For an integer i, let $U_i = U_i H^{n+1}_p(L)$ be the subgroup of $H^{n+1}_p(L)$ generated by elements of the form

$$f \frac{dg_1}{g_1} \wedge \cdots \wedge \frac{dg_n}{g_n} \qquad (f \in L, \ g_1, \ldots, g_n \in L^*, \ v(f) \geq -i).$$

It is obvious that $H^{n+1}_p(L) = \bigcup_i U_i$, and it follows from (3.3) below that $U_{-1} = 0$. Hence

$$0 = U_{-1} \subset U_0 \subset U_1 \subset \cdots \subset U_i \subset \cdots, \qquad \bigcup_{i=0}^{\infty} U_i = H^{n+1}_p(L).$$

2.4. As easily seen, the following homomorphisms ρ_i $(i \geq 0)$ are well defined.
(i) If $i = 0$, then

$$\rho_i = \rho_0 \colon H^{n+1}_p(F) \oplus H^n_p(F) \to U_0/U_{-1} = U_0,$$

$$\left(\overline{a}\frac{d\overline{b}_1}{\overline{b}_1} \wedge \cdots \wedge \frac{d\overline{b}_n}{\overline{b}_n}, 0\right) \mapsto a \frac{db_1}{b_1} \wedge \cdots \wedge \frac{db_n}{b_n},$$

$$\left(0, \overline{a}\frac{d\overline{b}_1}{\overline{b}_1} \wedge \cdots \wedge \frac{d\overline{b}_{n-1}}{\overline{b}_{n-1}}\right) \mapsto a \frac{d\pi}{\pi} \frac{db_1}{b_1} \wedge \cdots \wedge \frac{db_{n-1}}{b_{n-1}},$$

where $a \in \mathcal{O}_L$, $b_1, \ldots, b_n \in U_L$, and $\overline{a} \in F$ denotes the residue class of $a \in \mathcal{O}_L$.

(ii) If $i > 0$ and i is prime to p, let ρ_i be the homomorphism

$$\rho_i \colon \Omega^n_F \to U_i/U_{i-1},$$

$$\overline{a}\frac{d\overline{b}_1}{\overline{b}_1} \wedge \cdots \wedge \frac{d\overline{b}_n}{\overline{b}_n} \mapsto \frac{a}{\pi^i} \frac{db_1}{b_1} \wedge \cdots \wedge \frac{db_n}{b_n} \quad (\operatorname{mod} U_{i-1}),$$

where $a \in \mathcal{O}_L$, $b_1, \ldots, b_n \in U_L$.

(iii) If $i > 0$ and $i \vdots p$, let ρ_i be the homomorphism

$$\rho_i \colon \Omega^n_F/\Omega^n_{F,d=0} \oplus \Omega^{n-1}_F/\Omega^{n-1}_{F,d=0} \to U_i/U_{i-1},$$

$$\left(\overline{a}\frac{d\overline{b}_1}{\overline{b}_1} \wedge \cdots \wedge \frac{d\overline{b}_n}{\overline{b}_n}, 0\right) \mapsto \frac{a}{\pi^i} \frac{db_1}{b_1} \wedge \cdots \wedge \frac{db_n}{b_n} \quad (\operatorname{mod} U_{i-1}),$$

$$\left(0, \overline{a}\frac{d\overline{b}_1}{\overline{b}_1} \wedge \cdots \wedge \frac{d\overline{b}_{n-1}}{\overline{b}_{n-1}}\right) \mapsto a \frac{d\pi}{\pi} \frac{db_1}{b_1} \wedge \cdots \wedge \frac{db_{n-1}}{b_{n-1}} \quad (\operatorname{mod} U_{i-1}),$$

where $a \in \mathcal{O}_l$, $b_1, \ldots, b_n \in U_L$.

2.5. THEOREM. *Let L be a complete discrete valuation field of characteristic p with residue field F, and let $0 = U_{-1} \subset U_0 \subset U_1 \subset \cdots \subset U_i \subset \cdots \subset H_p^{n+1}(L)$ be the subgroups of $H_p^{n+1}(L)$ defined in (2.3).*

Then all the homomorphisms ρ_i ($i \geq 0$) are bijective and hence

$$U_i/U_{i-1} \cong \begin{cases} H_p^{n+1}(F) \oplus H_p^n(F) & \text{if } i = 0, \\ \Omega_F^n & \text{if } i > 0,\ i \not\vdots p, \\ \Omega_F^n/\Omega_{F,d=0}^n \oplus \Omega_F^{n-1}/\Omega_{F,d=0}^{n-1} & \text{if } i > 0,\ i \vdots p. \end{cases}$$

We prove this theorem below in §3.

2.6. COROLLARY. *The group $H_{p,\mathrm{ur}}^{n+1}(L)$ coincides with U_0, and hence*

$$H_{p,\mathrm{ur}}^{n+1}(L) \cong H_p^{n+1}(F) \oplus H_p^n(F).$$

PROOF. In view of (1.5.5), the homomorphisms $\Omega_F^n \to \Omega_{F^{\mathrm{sep}}}^n$ and

$$\Omega_F^n/\Omega_{F,d=0}^n \oplus \Omega_F^{n-1}/\Omega_{F,d=0}^{n-1} \to \Omega_{F^{\mathrm{sep}}}^n/\Omega_{F^{\mathrm{sep}},d=0}^n \oplus \Omega_{F^{\mathrm{sep}}}^{n-1}/\Omega_{F^{\mathrm{sep}},d=0}^{n-1}$$

are injective, and hence the homomorphism

$$U_i H_p^{n+1}(L)/U_{i-1} H_p^{n+1}(L) \to U_i H_p^{n+1}(L^{\mathrm{ur}})/U_{i-1} H_p^{n+1}(L^{\mathrm{ur}})$$

is injective for any $i > 0$.

Thus, we conclude that

$$H_{p,\mathrm{ur}}^{n+1}(L) = \ker(H_p^{n+1}(L) \to H_p^{n+1}(L^{\mathrm{ur}})) \subset U_0 H_p^{n+1}(L) = U_0.$$

The inclusion $U_0 \subset H_{p,\mathrm{ur}}^{n+1}(L)$ is obvious since

$$U_0 H_p^{n+1}(L^{\mathrm{ur}}) = H_p^{n+1}(F^{\mathrm{sep}}) \oplus H_p^n(F^{\mathrm{sep}}) = 0.$$

Finally, we obtain $H_{p,\mathrm{ur}}^{n+1}(L) = U_0 \cong H_p^{n+1}(F) \oplus H_p^n(F)$. □

2.7. COROLLARY. *The isomorphism $H_{p,\mathrm{ur}}^{n+1}(L) \cong H_p^{n+1}(F) \oplus H_p^n(F)$ gives rise to the split exact sequence*

$$0 \to H_p^{n+1}(F) \xrightarrow{i} H_{p,\mathrm{ur}}^{n+1}(L) \xrightarrow{\partial} H_p^n(F) \to 0$$

where the homomorphisms i and ∂ do not depend on the choice of π.

PROOF. The homomorphism i coincides with the composition

$$H_p^{n+1}(F) \xrightarrow{(\mathrm{id},0)} H_p^{n+1}(F) \oplus H_p^n(F) \xrightarrow{\rho_0} U_0 = H_{p,\mathrm{ur}}(L),$$

and so it does not depend on π (cf. Definition (2.4)(i)).

Now it is sufficient to prove that the homomorphism $H_{p,\mathrm{ur}}^{n+1}(L)/\mathrm{im}(i) \xrightarrow[\sim]{\partial} H_p^n(F)$ does not depend on π. Consider the inverse map $H_p^n(F) \xrightarrow[\sim]{\partial^{-1}} H_{p,\mathrm{ur}}^{n+1}(L)/\mathrm{im}(i) = U_0/\mathrm{im}(i)$. In view of (2.4)(i), this map is given by

$$\overline{a} \frac{d\overline{b}_1}{\overline{b}_1} \wedge \cdots \wedge \frac{d\overline{b}_{n-1}}{\overline{b}_{n-1}} \mapsto a \frac{d\pi}{\pi} \wedge \frac{db_1}{b_1} \wedge \cdots \wedge \frac{db_{n-1}}{b_{n-1}} \in U_0/\mathrm{im}(i).$$

If π' is a new prime element of L, then the element

$$a\frac{d\pi}{\pi}\wedge\frac{db_1}{b_1}\wedge\cdots\wedge\frac{db_{n-1}}{b_{n-1}} - a\frac{d\pi'}{\pi'}\wedge\frac{db_1}{b_1}\wedge\cdots\wedge\frac{db_{n-1}}{b_{n-1}} = a\frac{d(\pi/\pi')}{\pi/\pi'}\wedge\frac{db_1}{b_1}\wedge\cdots\wedge\frac{db_{n-1}}{b_{n-1}}$$

belongs to $\text{im}(i)$ since $\pi'/\pi, b_1, \ldots, b_n \in U_L$.

Therefore the map $H_p^n(F) \xrightarrow[\sim]{\partial^{-1}} U_0/\text{im}(i)$ does not depend on π and, finally, the map $H_{p,\text{ur}}^{n+1}(L) \xrightarrow{\partial} H_p^n(F)$ does not depend on π too. $\square$

§3. Proof of Theorem 2.5

It is well known that there exists an $\mathcal{O}_L$-homomorphism $F \to \mathcal{O}_L$, and hence we may consider F as a subfield in L such that $L = F((\pi))$. This enables us to identify L with the formal power series field $F((t))$.

3.1. Filtration on Ω_L^n. Let M be one of the groups Ω_L^n, $\Omega_L^n/d\Omega_L^{n-1}$, $H_p^{n+1}(L)$. For an integer i, let $U_i M$ denote the subgroup of M generated by all the elements of the form

$$l\frac{dg_1}{g_1}\wedge\cdots\wedge\frac{dg_n}{g_n}, \quad \text{where } b \in L, \ v(l) \geqslant -i, \ g_1,\ldots,g_n \in L^*,$$

and let $U_{(i/i-1)}M$ denote the factor group $U_i M/U_{i-1} M$. It is clear that $U_{-1}M$ is generated by the elements of the form

$$l\frac{dg_1}{g_1}\wedge\cdots\wedge\frac{dg_n}{g_n},$$

where $v(l) > 0$. Denote by $\widetilde{U}M$ the factor group $M/U_{-1}M$, and let $\widetilde{U}_i M$ be the group $U_i M/U_{-1}M$.

Since

$$d\left(l\frac{dg_1}{g_1}\wedge\cdots\wedge\frac{dg_{n-1}}{g_{n-1}}\right) = l\frac{dl}{l}\wedge\frac{dg_1}{g_1}\wedge\cdots\wedge\frac{dg_{n-1}}{g_{n-1}},$$

we see that $d(U_i\Omega_L^{n-1}) \subset U_i\Omega_L^n$, and hence we get well-defined functions

$$d: U_{(i/i-1)}\Omega_L^{n-1} \to U_{(i/i-1)}\Omega_L^n, \quad d: \widetilde{U}\Omega_L^{n-1} \to \widetilde{U}\Omega_L^n, \quad d: \widetilde{U}_i\Omega_L^{n-1} \to \widetilde{U}_i\Omega_L^n.$$

3.2. LEMMA. *Let $\wp = \Phi - 1: \Omega_L^n \to \Omega_L^n/d\Omega_L^{n-1}$ be the Cartier map. Then $\wp(U_{-1}\Omega_L^n) = U_{-1}(\Omega_L^n/d\Omega_L^{n-1})$.*

PROOF. The inclusion $\wp(U_{-1}\Omega_L^n) \subset U_{-1}(\Omega_L^n/d\Omega_L^{n-1})$ is obvious. Now let

$$l\frac{dg_1}{g_1}\wedge\cdots\wedge\frac{dg_n}{g_n}$$

be one of the generators of $U_{-1}(\Omega_L^n/d\Omega_L^{n-1})$, i.e., $v(l) > 0$. Consider the element

$$\tilde{l} = \sum_{k\geqslant 0} l^{p^k} \in L.$$

Since $\tilde{l}^p - \tilde{l} = l$, we see that

$$l\frac{dg_1}{g_1}\wedge\cdots\wedge\frac{dg_n}{g_n} = \wp\left(\tilde{l}\frac{dg_1}{g_1}\wedge\cdots\wedge\frac{dg_n}{g_n}\right) \in \wp(U_{-1}\Omega_L^n)$$

and hence $\wp(U_{-1}\Omega_L^n) = U_{-1}(\Omega_L^n/d\Omega_L^{n-1})$. $\square$

3.3. COROLLARY. *The group $U_{-1} = U_{-1}H_p^{n+1}(L)$ is trivial.* □

3.4. COROLLARY. $H_p^{n+1}(L) \cong \operatorname{coker}(\widetilde{U}\Omega_L^n \xrightarrow{\wp} \widetilde{U}(\Omega_L^n/d\Omega_L^{n-1}))$.

PROOF. This easily follows from the definition of
$$H_p^{n+1}(L) = \operatorname{coker}(\Omega_L^n \xrightarrow{\wp} U(\Omega_L^n/d\Omega_L^{n-1}))$$
and the fact that $U_{-1}(\Omega_L^n/d\Omega_L^{n-1}) \subset \operatorname{im}(\wp)$. □

3.5. PROPOSITION. *Let $L = F((t))$. Then*
$$\widetilde{U}\Omega_L^n = \coprod_{i \geqslant 0} \left(t^{-i}\Omega_F^n \oplus t^{-i}\frac{dt}{t} \wedge \Omega_F^{n-1} \right), \quad \widetilde{U}_k\Omega_L^n = \coprod_{i=0}^k \left(t^{-i}\Omega_F^n \oplus t^{-i}\frac{dt}{t} \wedge \Omega_F^{n-1} \right).$$

PROOF. It suffices to show that the homomorphism
$$\coprod_{i \geqslant 0} \left(t^{-i}\Omega_F^n \oplus t^{-i}\frac{dt}{t} \wedge \Omega_F^{n-1} \right) \to \widetilde{U}\Omega_L^n$$
is bijective. To do this we explicitly construct the inverse map.

Let $u = f_0 df_1 \wedge \cdots \wedge df_n$ be one of the generators of $\widetilde{U}\Omega_L^n$. We may write $f_k \in L = F((t))$ in the form $f_k = \sum_i a_{k,i} t^i$, where $a_{k,i} \in F$ and $a_{k,i} = 0$ for $i \ll 0$. Note that if $\sum_{i=0}^n v(h_i) = v(h_0 h_1 \cdots h_n) > 0$, then $h_0 dh_1 \wedge \cdots \wedge dh_n = 0 \in \widetilde{U}\Omega_L^n$ since
$$h_0 dh_1 \wedge \cdots \wedge dh_n = h_0 h_1 \cdots h_n \frac{dh_1}{h_1} \wedge \cdots \wedge \frac{dh_n}{h_n} \in U_{-1}\Omega_L^n.$$

Hence in the group $\widetilde{U}\Omega_L^n$ we have the equation
$$u = f_0 df_1 \wedge \cdots \wedge df_n$$
$$= \left(\sum_{i_0} a_{0,i_0} t^{i_0}\right) d\left(\sum_{i_1} a_{1,i_1} t^{i_1}\right) \wedge \cdots \wedge d\left(\sum_{i_n} a_{n,i_n} t^{i_n}\right)$$
$$= \sum_{i_0+\cdots+i_n \leqslant 0} a_{0,i_0} t^{i_0} d(a_{1,i_1} t^{i_1}) \wedge \cdots \wedge d(a_{n,i_n} t^{i_n})$$
$$= \sum_{i_0+\cdots+i_n \leqslant 0} t^{i_0+\cdots+i_n} a_{0,i_0} da_{1,i_1} \wedge \cdots \wedge da_{n,i_n}$$
$$+ \sum_{k=1}^n \sum_{i_0+\cdots+i_n \leqslant 0} t^{i_0+\cdots+i_n} \frac{dt}{t} i_k (-1)^{k+1} a_{0,i_0}$$
$$\times da_{1,i_1} \wedge \cdots \wedge \widehat{da_{k,i_k}} \wedge \cdots \wedge da_{n,i_n}.$$

For any $i \geqslant 0$ we consider the maps $s_i \colon \widetilde{U}\Omega_L^n \xrightarrow{s_i} \Omega_F^n$, $r_i \colon \widetilde{U}\Omega_L^n \xrightarrow{r_i} \Omega_F^{n-1}$
$$s_i \colon u \mapsto \sum_{i_0+\cdots+i_n=-i} a_{0,i_0} da_{1,i_1} \wedge \cdots \wedge da_{n,i_n},$$
$$r_i \colon u \mapsto \sum_{k=1}^n \sum_{i_0+\cdots+i_n=-i} (-1)^{k+1} i_k a_{0,i_0} da_{1,i_1} \wedge \cdots \wedge \widehat{da_{k,i_k}} \wedge \cdots \wedge da_{n,i_n}.$$

Thus we get a well-defined function
$$\widetilde{U}\Omega_L^n \xrightarrow{\amalg(t^{-i}s_i \oplus t^{-i}\frac{dt}{t}\wedge r_i)} \coprod_{i\geq 0}\left(t^{-i}\Omega_F^n \oplus t^{-i}\frac{dt}{t}\wedge\Omega_F^{n-1}\right).$$

This implies the required result. □

REMARK. It is not true that
$$\Omega_L^n = \coprod_{i\in\mathbb{Z}}\left(t^{-i}\Omega_F^n \oplus t^{-i}\frac{dt}{t}\wedge\Omega_F^{n-1}\right)$$
and, moreover, it is not true that
$$\Omega_L^n = L\otimes_F \Omega_F^n \oplus L\otimes_F \Omega_F^{n-1}\wedge\frac{dt}{t}.$$

3.6. COROLLARY. $U_{(i/i-1)}\Omega_L^n \cong \Omega_F^n \oplus \Omega_F^{n-1}$. □

3.7. PROPOSITION. *The group* $\widetilde{U}(\Omega_L^n/d\Omega_L^{n-1})$ *is isomorphic to*
$$\coprod_{i\geq 0,\, p\nmid i} t^{-i}\Omega_F^n \oplus \coprod_{i\geq 0,\, p\mid i}\left(t^{-i}\Omega_F^n/d\Omega_F^{n-1} \oplus t^{-i}\frac{dt}{t}\wedge\Omega_F^{n-1}/d\Omega_F^{n-2}\right).$$

PROOF. In view of Proposition 3.5 it is sufficient to calculate the cokernel of the homomorphism
$$\coprod_{i\geq 0}\left(t^{-i}\Omega_F^{n-1} \oplus t^{-i}\frac{dt}{t}\wedge\Omega_F^{n-2}\right) \xrightarrow{d} \coprod_{i\geq 0}\left(t^{-i}\Omega_F^n \oplus t^{-i}\frac{dt}{t}\wedge\Omega_F^{n-1}\right).$$

This homomorphism takes the group $t^{-i}\Omega_F^{n-1}\oplus t^{-i}(dt/t)\wedge\Omega_F^{n-2}$ to the group $t^{-i}\Omega_F^n \oplus t^{-i}(dt/t)\wedge\Omega_F^{n-1}$. Indeed, let us choose $t^{-i}u + t^{-i}(dt/t)\wedge v$ so that $u\in\Omega_F^{n-1}$ and $v\in\Omega_F^{n-2}$. Then
$$d\left(t^{-i}u + t^{-i}\frac{dt}{t}\wedge v\right) = t^{-i}du + t^{-i}\frac{dt}{t}\wedge(-iu - dv) \in \left(t^{-i}\Omega_F^n \oplus t^{-i}\frac{dt}{t}\wedge\Omega_F^{n-1}\right).$$

Thus, $D_i = \begin{pmatrix} d & 0 \\ -i & -d \end{pmatrix}$ is the matrix of the homomorphism
$$\Omega_F^{n-1}\oplus\Omega_F^{n-2} \cong t^{-i}\Omega_F^{n-1}\oplus t^{-i}\frac{dt}{t}\wedge\Omega_F^{n-2} \xrightarrow{d} t^{-i}\Omega_F^n \oplus t^{-i}\frac{dt}{t}\wedge\Omega_F^{n-1} \cong \Omega_F^n\oplus\Omega_F^{n-1}.$$

If $i \vdots p$, then $D_i = \begin{pmatrix} d & 0 \\ 0 & -d \end{pmatrix}$ and hence
$$\operatorname{coker}(D_i) = \operatorname{coker}\begin{pmatrix} d & 0 \\ 0 & -d \end{pmatrix} = \Omega_F^n/d\Omega_F^{n-1} \oplus \Omega_F^{n-1}/d\Omega_F^{n-2}.$$

If $p\nmid i$, then
$$D_i = \begin{pmatrix} d & 0 \\ -i & -d \end{pmatrix} = \begin{pmatrix} 1 & -i^{-1}d \\ 0 & 1 \end{pmatrix}\begin{pmatrix} 0 & 0 \\ 1 & 0 \end{pmatrix}\begin{pmatrix} -i & -d \\ 0 & 1 \end{pmatrix}.$$

Since the matrices $\begin{pmatrix} 1 & -i^{-1} \\ 0 & 1 \end{pmatrix}$ and $\begin{pmatrix} -i & -d \\ 0 & 1 \end{pmatrix}$ are invertible, we obtain

$$\mathrm{coker}(D_i) \cong \mathrm{coker}(\Omega_F^{n-1} \oplus \Omega_F^{n-2} \xrightarrow{\begin{pmatrix} 0 & 0 \\ 1 & 0 \end{pmatrix}} \Omega_F^n \oplus \Omega_F^{n-1}) \cong \Omega_F^n. \qquad \square$$

3.8. COROLLARY. *Let $M = \Omega_L^n/d\Omega_L^{n-1}$. Then*

$$U_{(i/i-1)}M = \begin{cases} \Omega_F^n & \text{if } i \not\vdots p, \\ \Omega_F^n/d\Omega_F^{n-1} \oplus \Omega_F^{n-1}/d\Omega_F^{n-2} & \text{if } i \vdots p. \end{cases}$$

3.9. AUXILIARY LEMMA. *Let M, N be two abelian groups, and*

$$0 = M_{-1} \subset M_0 \subset M_1 \subset \cdots \subset M_i \subset \cdots ,$$
$$0 = N_{-1} \subset N_0 \subset N_1 \subset \cdots \subset N_i \subset \cdots$$

their filtrations. Let $\Phi, I : M \to N$ be homomorphisms such that
 (i) $\Phi(M_i) \subset N_{pi}$ *if* $i \geqslant 0$,
 (ii) $I(M_i) \subset N_i$ *if* $i \geqslant 0$,
 (iii) $\Phi : M_i/M_{i-1} \to N_{pi}/N_{pi-1}$ *is injective for all* $i \geqslant 0$.
Let H denote the cokernel of $M \xrightarrow{\Phi-I} N$, and let H_i be the image of N_i in H. Then we have

$$H_i/H_{i-1} \cong \begin{cases} \mathrm{coker}(M_0 \xrightarrow{\Phi-I} N_0) & \text{if } i = 0, \\ N_i/N_{i-1} & \text{if } i \geqslant 0, i \not\vdots p, \\ \mathrm{coker}(M_{i/p}/M_{i/p-1} \xrightarrow{\Phi} N_i/N_{i-1}) & \text{if } i > 0, i \vdots p. \end{cases}$$

PROOF. For arbitrary $u \in M$ (resp. $u \in N$) we denote by $v(u)$ the minimal integer $i \geqslant 0$ such that $u \in M_i$ (resp. $u \in N_i$). In view of the injectivity of $\Phi : M_i/M_{i-1} \to N_{pi}/N_{pi-1}$ we conclude that $v(\Phi(u)) = p \cdot v(u)$ for any $u \in M$.

It follows from $I(M_i) \subset N_i$ that $v(I(u)) \leqslant v(u) < p \cdot v(u) = v(\Phi(u))$ for any $u \in M - M_0$ and hence $v((\Phi - I)u) = p \cdot v(u)$.

Consequently,

$$\mathrm{im}(\Phi - I) \cap M_i + M_{i-1} = \begin{cases} M_{i-1} & \text{if } i \not\vdots p, \\ (\Phi - I)(M_{i/p}) + M_{i-1} & \text{if } i \vdots p, \end{cases}$$
$$= \begin{cases} M_{i-1} & \text{if } i \not\vdots p, \\ (\Phi - I)M_0 & \text{if } i = 0, \\ \Phi(M_{i/p}) + M_{i-1} & \text{if } i \vdots p, i \neq 0. \end{cases}$$

Now the assertion of the lemma immediately follows from the isomorphism

$$H_i/H_{i-1} \cong M_i/(\mathrm{im}(\Phi - I) \cap M_i + M_{i-1}). \qquad \square$$

3.10. Proof of Theorem 2.5. Let $M = \widetilde{U}\Omega_L^n$, $M_i = \widetilde{U}_i\Omega_L^n$, $N = \widetilde{U}(\Omega_L^n/d\Omega_L^{n-1})$, and $N_i = \widetilde{U}_i(\Omega_L^n/d\Omega_L^{n-1})$. Denote $\Phi\colon M \to N$ be the Frobenius map and $I\colon M \to N$ the natural projection. Conditions (i), (ii) of Lemma 3.9 are obviously satisfied. To prove condition (iii), consider the commutative diagram

$$\begin{array}{ccc} M_i/M_{i-1} & \xrightarrow{\Phi} & N_{pi}/N_{pi-1} \\ \downarrow\wr & & \downarrow\wr \\ \Omega_F^n \oplus \Omega_F^{n-1} & \xrightarrow{\Phi\oplus\Phi} & \Omega_F^n/d\Omega_F^{n-1} \oplus \Omega_F^{n-1}/d\Omega_F^{n-2} \end{array}$$

Here the lower row is injective by (1.5.2) and hence the upper row is injective, too.

Now using Lemma 3.9 we see that
- if $i = 0$, then

$$\begin{aligned} U_0 = U_0/U_{-1} &\cong \operatorname{coker}(M_0 \xrightarrow{\Phi-I} N_0) \\ &\cong \operatorname{coker}(\Omega_F^n \oplus \Omega_F^{n-1} \xrightarrow{\Phi-I} \Omega_F^n/d\Omega_F^{n-1} \oplus \Omega_F^{n-1}/d\Omega_F^{n-2}) \\ &\cong H_p^{n+1}(F) \oplus H_p^n(F), \end{aligned}$$

- if $i > 0$, $i \not\vdots p$, then

$$U_i/U_{i-1} \cong N_i/N_{i-1} = U_{i/i-1}(\Omega_L^n/d\Omega_L^{n-1}) \cong \Omega_F^n,$$

- if $i > 0$, $i \vdots p$, then, using (1.5.3), we see that

$$\begin{aligned} U_i/U_{i-1} &\cong \operatorname{coker}(M_{i/p}/M_{i/p-1} \xrightarrow{\Phi} N_i/N_{i-1}) \\ &= \operatorname{coker}(U_{(i/p,i/p-1)}\Omega_L^n \xrightarrow{\Phi} U_{(i/i-1)}(\Omega_L^n/d\Omega_L^{n-1})) \\ &\cong \operatorname{coker}(\Omega_F^n \oplus \Omega_F^{n-1} \xrightarrow{\Phi\oplus\Phi} \Omega_F^n/d\Omega_F^{n-1} \oplus \Omega_F^{n-1}/d\Omega_F^{n-2}) \\ &\cong \Omega_F^n/\Omega_{F,d=0}^n \oplus \Omega_F^{n-1}/\Omega_{F,d=0}^{n-1}. \end{aligned}$$

This completes the proof of Theorem 2.5.

3.11. REMARKS. 1. It is clear from Proposition 3.5 that there exists a natural map

$$\widetilde{U}\Omega_{F((t))}^n = \coprod_{i\geq 0}\left(t^{-i}\Omega_F^n \oplus t^{-i}\frac{dt}{t}\wedge\Omega_F^{n-1}\right) \xrightarrow{\mathrm{pr}} \left(t^{-0}\Omega_F^n \oplus t^{-0}\frac{dt}{t}\wedge\Omega_F^{n-1}\right) = \widetilde{U}_0\Omega_{F((t))}^n.$$

Moreover, this map defines a homomorphism

$$H_p^{n+1}(F((t))) \xrightarrow{\varphi} U_0 H_p^{n+1}(F((t))) = H_{p,\mathrm{ur}}^{n+1}(F((t))).$$

Hence φ is a split map for the exact sequence

$$0 \to H_{p,\mathrm{sep}}^n(F((t))) \to H_p^n(F((t))) \to \widetilde{H}_p^n(F((t))) \to 0,$$

and hence there is a natural homomorphism

$$H_p^n(F((t))) \cong H_{p,\mathrm{sep}}^n(F((t))) \oplus \widetilde{H}_p^n(F((t))).$$

2. We can define a specialization map $H_p^n(F((t))) \to H_p^n(F)$ as the composition

$$H_p^n(F((t))) \cong H_{p,\text{sep}}^n(F((t))) \oplus \widetilde{H}_p^n(F((t))) \xrightarrow{\text{pr}_1} H_{p,\text{sep}}^n(F((t)))$$
$$\xrightarrow{\sim} H_p^n(F) \oplus H_p^{n-1}(F) \xrightarrow{\text{pr}_1} H_p^n(F).$$

3. For a complete discrete valuation field L of characteristic p there is no *natural* isomorphism $H_p^n(L) \cong H_{p,\text{sep}}^n(L) \oplus \widetilde{H}_p^n(L)$.

§4. Cohomology groups of the field of rational functions in one variable

In this section we state some results about cohomology groups of the field $F(t)$ of rational functions in one variable t.

4.1. We use the following notation:
- V is the set of all nontrivial normalized additive valuations of $F(t)$ which are trivial on F;
- ∞ is the valuation $F(t) \xrightarrow{\infty} \mathbb{Z}$: $\dfrac{f(t)}{g(t)} \mapsto (\deg f(t) - \deg g(t))$;
- for $v \in V$ let $F(t)_v$ denote the v-completion of $F(t)$ and F_v the residue field of $F(t)_v$.

If $q(t) \in F[t]$ is a prime and monic polynomial, then we denote the corresponding discrete valuation of $F(t)$ by v_q. The field $F(t)_{v_q}$ we denote by $F(t)_q$ and the field F_{v_q} by F_q. It is clear that $F_q \cong F[t]/(q(t))$.

It is well known that $V = \{\infty\} \cup \{v_q \mid q \text{ is a prime monic polynomial}\}$.

Now let $H_{p,\text{sep}}^n(F(t))$ denote the kernel of the homomorphism $H_p^n(F(t)) \to H_p^n(F^{\text{sep}}(t))$, and let $\widetilde{H}_p^n(F(t))$ be the factor group $H_p^n(F(t))/H_{p,\text{sep}}^n(F(t))$.

4.2. The homomorphism i_v. Let $v \in V$ and let $i_v \colon H_p^n(F(t)) \to H_p^n(F(t)_v)$ be the inclusion homomorphism.

Since the residue field of $(F(t)_v)^{\text{ur}}$ contains F^{sep}, there exists an embedding $F^{\text{sep}}(t) \xrightarrow{\varphi} (F(t)_v)^{\text{ur}}$ such that the diagram

$$\begin{array}{ccc} F^{\text{sep}}(t) & \xrightarrow{\varphi} & (F(t)_v)^{\text{ur}} \\ \uparrow & & \uparrow \\ F(t) & \longrightarrow & F(t)_v \end{array}$$

is commutative.

Therefore, $i_v(H_{p,\text{sep}}^n(F(t))) \subset H_{p,\text{ur}}^n(F(t)_v)$, and hence we get a well-defined map $i_v \colon \widetilde{H}_p^n(F(t)) \to \widetilde{H}_p^n(F(t)_v)$, where $\widetilde{H}_p^n(F(t)) = H_p^n(F(t))/H_{p,\text{sep}}^n(F(t))$, $\widetilde{H}_p^n(F(t)_v) = H_p^n(F(t)_v)/H_{p,\text{ur}}^n(F(t)_v)$.

4.3. The connecting homomorphism ∂_v. Let $v \in V$. Since $F(t)_v$ is a complete discrete valuation field, the exact sequence of Corollary 2.7 yields the exact sequence

$$0 \to H_p^n(F_v) \to H_{p,\text{ur}}^n(F(t)_v) \xrightarrow{\partial} H_p^{n-1}(F_v) \to 0.$$

Denote by $\partial_v \colon H_{p,\text{sep}}^n(F(t)) \to H_p^{n-1}(F_v)$ the composition

$$\widetilde{H}_{p,\text{sep}}^n(F(t)) \xrightarrow{i_v} H_{p,\text{ur}}^n(F(t)_v) \xrightarrow{\partial} H_p^{n-1}(F_v).$$

4.4. Specialization. Let $f \in F$. We consider $q(t) = t - f \in F[t]$ as a prime monic polynomial. It is clear that $F(t)_q = F((q))$ and $F_q = F$.

In view of (2.6), one has the isomorphism $\rho_0 \colon H_p^n(F) \oplus H_p^{n-1}(F) \xrightarrow{\sim} H_{p,\mathrm{ur}}^n(F(t)_q)$. Hence we can consider the composition

$$H_{p,\mathrm{sep}}^n(F(t)) \xrightarrow{i_v} H_{p,\mathrm{ur}}^n(F(t)_q) \xrightarrow[\sim]{\rho} H_p^n(F) \oplus H_p^{n-1}(F) \xrightarrow{\mathrm{pr}_1} H_p^n(F),$$

which we denote by $s_f \colon H_{p,\mathrm{sep}}^n(F(t)) \to H_p^n(F)$.

REMARK. In view of (3.11), we can define a homomorphism $s \colon H_p^n(F(t)) \to H_p^n(F)$ as the composition

$$H_p^n(F(t)) \xrightarrow{i_q} H_p^n(F(t)_q) = H_p^n(F((q))) \xrightarrow{\varphi} H_{p,\mathrm{sep}}^n(F((q))) \xrightarrow{s} H_p^n(F).$$

4.5. THE MAIN THEOREM. *We have the following isomorphisms*
(i) $H_{p,\mathrm{sep}}^n(F(t)) \cong H_p^n(F) \oplus \coprod_{v \neq \infty} H_p^{n-1}(F)$,
(ii) $\widetilde{H}_p^n(F(t)) \cong \coprod_{v \in V} \widetilde{H}^n(F(t)_v)$.

This theorem is proved below in §5.

4.6. COROLLARY. *The sequence*

$$0 \to H_p^n(F) \xrightarrow{\mathrm{in}_*} H_{p,\mathrm{sep}}^n(F(t)) \xrightarrow{\coprod_{v \neq \infty} \partial_v} \coprod_{v \neq \infty} H_p^{n-1}(F_v) \to 0$$

is exact. □

4.7. COROLLARY. *If F is a perfect field, then*

$$H_p^n(F(t)) \cong H_p^n(F) \oplus \coprod_{v \neq \infty} H_p^{n-1}(F).$$

PROOF. In this case $\widetilde{H}_p^n(F(t)) \cong \coprod_{v \in V} \widetilde{H}^n(F(t)_v) = 0$, and hence $H_p^n(F(t)) = H_{p,\mathrm{sep}}^n(F(t))$. □

4.8. COROLLARY. *If F is a separably closed field, then*

$$H_p^n(F(t)) \cong \coprod_{v \in V} H_p^n(F(t)_v).$$

PROOF. In this case $H_{p,\mathrm{sep}}^n(F(t)) = 0$, $\coprod_{v \in V} H_{p,\mathrm{ur}}^n(F(t)_v) = 0$, and hence

$$H_p^n(F(t)) - \widetilde{H}_p^n(F(t)) \cong \coprod_{v \in V} \widetilde{H}_p^n(F(t)_v) = \coprod_{v \in V} H_p^n(F(t)_v). \quad \square$$

4.9. COROLLARY. *The map $\coprod i_v \colon H_p^n(F(t)) \to \coprod_{v \in V} H_p^n(F(t)_v)$ is injective.*

PROOF. Let $u \in \ker(H_p^n(F(t)) \to \coprod_v H_p^n(F(t)_v))$. Since the homomorphism

$$\coprod i_v \colon \widetilde{H}_p^n(F(t)) \to \coprod_v \widetilde{H}_p^n(F(t)_v)$$

is injective, it follows that $u \in \ker(H_p^n(F(t)) \to \widetilde{H}_p^n(F(t))) = H_{p,\text{sep}}^n(F(t))$.
Since $i_v(u) = 0$ for any $v \in V$, we have $\partial_v(u) = \partial(i_v(u)) = 0$ and, therefore,

$$u \in \ker\bigl(H_{p,\text{sep}}^n(F(t)) \to \coprod_{v \neq \infty} H_p^{n-1}(F_v)\bigr) = \operatorname{im}(H_p^n(F)) \xrightarrow{\operatorname{in}_*} H_{p,\text{sep}}^n(F(t)).$$

Now note that the composition

$$H_p^n(F) \xrightarrow{\operatorname{in}_*} H_{p,\text{sep}}^n(F(t)) \xrightarrow{i_\infty} H_p^n(F(t)_\infty) = H_p^n(F((t)))$$

is injective, and hence $u = 0$. $\square$

§5. Proof of Theorem 4.4

Let $q = q(t) \in F[t]$ be a prime monic polynomial. Denote by Γ_q the subgroup of $H_p^{n+1}(F(t))$ generated by elements of the form

$$\frac{a}{q^i} \frac{db_1}{b_1} \wedge \cdots \wedge \frac{db_n}{b_n}, \quad \text{where } a \in F[t],\ b_1, \ldots, b_n \in F(t)^*,\ i \in \mathbb{Z}.$$

Let Γ be the subgroup of $H_p^{n+1}(F(t))$ generated by elements of the form

$$a \frac{db_1}{b_1} \wedge \cdots \wedge \frac{db_n}{b_n}, \quad \text{where } a \in F[t],\ b_1, \ldots, b_n \in F(t)^*.$$

5.1. PROPOSITION. *The sequence* $0 \to \Gamma \xrightarrow{\operatorname{in}} \Gamma_q \xrightarrow{i_q} \widetilde{H}_p^{n+1}(F(t)_q) \to 0$ *is exact.*

PROOF. Denote by N_i ($i \geqslant 0$) the subgroup of Γ_q generated by elements of the form

$$\frac{a}{q^i} \frac{db_1}{b_1} \wedge \cdots \wedge \frac{db_n}{b_n}, \quad \text{where } a \in F[t],\ b_1, \ldots, b_n \in F(t)^*$$

(and $N_{-1} = 0$ by convention).
It is obvious that

$$0 \subset N_0 \subset N_1 \subset \cdots \subset N_i \subset \cdots, \quad \bigcup_{i=0}^\infty N_i = \Gamma_q, \quad N_0 = \Gamma.$$

In (2.3) we defined the subgroups $U_i = U_i H_p^{n+1}(F(t)_q)$ of $H_p^{n+1}(F(t)_q)$ such that

$$0 \subset U_0 \subset U_1 \subset \cdots \subset U_i \subset \cdots, \quad \bigcup_{i=0}^\infty U_i = H_p^{n+1}(F(t)_q), \quad U_0 = H_{p,\text{ur}}^{n+1}(F(t)_q).$$

It is obvious that $i_q(N_i) \subset U_i$ for any $i \geqslant 0$ and hence it is sufficient to prove that the homomorphism $i_q \colon N_i/N_{i-1} \to U_i/U_{i-1}$ is bijective for $i \geqslant 1$.
Consider two cases, $i \nmid p$ and $i \vdots p$.

Case 1. $i \geqslant 1$ and $i \not\vdots\, p$.

Using Theorem 2.5, we see that the homomorphism

$$\rho_i \colon \Omega^n_{F_q} \to U_i/U_{i-1},$$

$$\bar{a}\,\frac{d\bar{b}_1}{\bar{b}_1} \wedge \cdots \wedge \frac{d\bar{b}_n}{\bar{b}_n} \mapsto \frac{a}{q^i}\,\frac{db_1}{b_1} \wedge \cdots \wedge \frac{db_n}{b_n} \qquad (a \in \mathcal{O}_{F(t)_q},\ b_1,\ldots,b_n \in U_{F(t)_q})$$

is bijective.

Let $\rho'_i \colon \Omega^n_{F_q} \to N_i/N_{i-1}$ denote the homomorphism defined by the same rule

$$\bar{a}\,\frac{d\bar{b}_1}{\bar{b}_1} \wedge \cdots \wedge \frac{d\bar{b}_n}{\bar{b}_n} \mapsto \frac{a}{q^i}\,\frac{db_1}{b_1} \wedge \cdots \wedge \frac{db_n}{b_n},$$

but now $a \in F[t]$, $b_1,\ldots,b_n \in F[t]$ are such that $v_q(b_1) = \cdots = v_q(b_n) = 0$; here $\bar{a}, \bar{b}_1,\ldots,\bar{b}_n$ denote the residue classes of $a, b_1,\ldots,b_n$ in $F_q = F[t]/(q)$.

First we prove that ρ' is well defined. It is sufficient to show that if $a \equiv a'$, $b_1 \equiv b'_1, \ldots, b_n \equiv b'_n \pmod{q}$, then

$$\frac{a}{q^i}\,\frac{db_1}{b_1} \wedge \cdots \wedge \frac{db_n}{b_n} \equiv \frac{a'}{q^i}\,\frac{db'_1}{b'_1} \wedge \cdots \wedge \frac{db'_n}{b'_n} \pmod{N_{i-1}}.$$

We consider the case $n = 1$ only (the general case is absolutely similar). Then we must prove that

$$\frac{a+qr}{q^i}\,\frac{db}{b} \equiv \frac{a}{q^i}\,\frac{db}{b} \pmod{N_{i-1}}, \qquad \frac{a}{q^i}\,\frac{d(b+qr)}{b+qr} \equiv \frac{a}{q^i}\,\frac{db}{b} \pmod{N_{i-1}},$$

where $a, r, b \in F[t]$, $b \not\vdots\, q$.

The first congruence is obvious since

$$\frac{qr}{q^i} = \frac{r}{q^{i-1}}\,\frac{db}{b} \in N_{i-1}.$$

To prove the second, note that $(b+qr, q) = (b, q) = 1$ and hence there are $\alpha, \beta \in F[t]$ such that $a = \alpha(b+qr) + \beta q$. Therefore,

$$\frac{a}{q^i}\,\frac{d(b+qr)}{b+qr} = \frac{\alpha(b+qr) + \beta q}{q^i}\,\frac{d(b+qr)}{b+qr} \equiv \frac{\alpha(b+qr)}{q^i}\,\frac{d(b+qr)}{b+qr}$$

$$= \frac{\alpha}{q^i}\,d(b+qr) = \frac{\alpha}{q^i}\,db + \frac{\alpha}{q^i}\,d(qr) = \frac{\alpha b}{q^i}\,\frac{db}{b} + \frac{\alpha r}{q^{i-1}}\,\frac{(qr)}{qr}$$

$$\equiv \frac{\alpha b}{q^i}\,\frac{db}{b} = \frac{a - \alpha qr - \beta q}{q^i}\,\frac{db}{b} \equiv \frac{a}{q^i}\,\frac{db}{b} \pmod{N_{i-1}}.$$

Thus we have proved that ρ'_i is well defined.

Now we prove that ρ'_i is surjective. Let

$$u = \frac{a}{q^i}\,\frac{db_1}{b_1} \wedge \cdots \wedge \frac{db_n}{b_n} \pmod{N_{i-1}}, \quad \text{where } a \in F[t],\ b_1,\ldots,b_2 \in F(t)^*$$

be one of the generators of N_i/N_{i-1}. We must show that $u \in \operatorname{im}\rho'_i$.

It is sufficient to consider the case where $a \not\vdots\, q$ and each of $b_1,\ldots,b_n$ is a prime monic polynomial or belongs to F^* (since $F(t)^* \to \Omega^1_{F(t)} \colon b \mapsto db/b$ is a homomorphism).

If each of the b_i's is prime to q, then

$$u = \frac{a}{q^i} \frac{db_1}{b_1} \wedge \cdots \wedge \frac{db_n}{b_n} \pmod{N_{i-1}} = q'_i\left(\bar{a} \frac{d\bar{b}_1}{\bar{b}_1} \wedge \cdots \wedge \frac{d\bar{b}_n}{\bar{b}_n}\right) \in \mathrm{im}(q'_i).$$

If one of the elements $b_1, \ldots, b_n$ is equal to q (for example $b_1 = q$), then

$$\begin{aligned}
u &= \frac{a}{q^i} \frac{dq}{q} \wedge \frac{db_2}{b_2} \wedge \cdots \wedge \frac{db_n}{b_n} = -i^{-1} a dq^{-i} \wedge \frac{db_2}{b_2} \wedge \cdots \wedge \frac{db_n}{b_n} \\
&= i^{-1} q^{-i} da \wedge \frac{dd_2}{d_2} \wedge \cdots \wedge \frac{dd_b}{d_b} - d\left(i^{-1} a q^{-i} \frac{db_2}{b_2} \wedge \cdots \wedge \frac{db_n}{b_n}\right) \\
&= i^{-1} \frac{a}{q^i} \frac{da}{a} \wedge \frac{db_2}{b_2} \wedge \cdots \wedge \frac{db_n}{b_n} = \rho'_i\left(i^{-1} \bar{a} \frac{d\bar{a}}{\bar{a}} \wedge \frac{d\bar{b}_2}{\bar{b}_2} \wedge \cdots \wedge \frac{d\bar{b}_n}{\bar{b}_n}\right) \in \mathrm{im}(q'_i).
\end{aligned}$$

Note that in the last line we use the fact that $i \nmid p$ and that for any $v \in \Omega^{n-1}_{F(t)}$, the element dv is equal to 0 in $H_p^{n+1}(F(t))$.

Since ρ'_i is surjective, ρ_i is bijective, and the composition $\Omega^n_{F_q} \xrightarrow{\rho'_i} N_i/N_{i-1} \xrightarrow{i_q} U_i/U_{i-1}$ coincides with ρ_i, it follows that $i_q: N_i/N_{i-1} \to U_i/U_{i-1}$ is an isomorphism.

Case 2. $i \geqslant 1$, $i \,\vdots\, p$.

Using Theorem 2.5, we see that the homomorphism

$$\rho_i: \Omega^n_{F_q}/\Omega^n_{F_q, d=0} \oplus \Omega^{n+1}_{F_q}/\Omega^{n+1}_{F_q, d=0} \to U_i/U_{i-1}$$

is bijective. As in Case 1, we can construct a homomorphism

$$\rho'_i: \Omega^n_{F_q}/\Omega^n_{F_q, d=0} \oplus \Omega^{n+1}_{F_q}/\Omega^{n+1}_{F_q, d=0} \to N_i/N_{i-1},$$

which is easily seen to be surjective.

As in Case 1, we conclude that $N_i/N_{i-1} \to U_i/U_{i-1}$ is bijective for $i \geqslant 1$. This completes the proof of Proposition 5.1. $\square$

5.2. Proposition. *The sequence*

$$0 \to \Gamma \to H_p^{n+1}(F(t)) \xrightarrow{\coprod i_v} \coprod_{v \neq \infty} \widetilde{H}_p^{n+1}(F(t)_v) \to 0$$

is exact.

PROOF. Consider the homomorphisms

$$\alpha: \coprod_{v \neq \infty} \Gamma_v/\Gamma \xrightarrow{\sum \mathrm{in}} H_p^{n+1}(F(t))/\Gamma, \qquad \beta: H_p^{n+1}(F(t)) \xrightarrow{\coprod i_v} \coprod_{v \neq \infty} \widetilde{H}^{n+1}(F(t)_v).$$

The composition $\beta \circ \alpha$ coincides with

$$\coprod_{v \neq \infty} i_v: \coprod_{v \neq \infty} \Gamma_v/\Gamma \to \coprod_{v \neq \infty} \widetilde{H}^n(F(t)_v),$$

and hence $\beta \circ \alpha$ is bijective in view of (5.1).

The homomorphism α is surjective because an arbitrary element of $F(t)$ can be represented as a sum of elements of the form a/q^i, where $a \in F[t]$ and q is prime monic polynomial.

Since $\beta \circ \alpha$ is bijective and α is surjective, β is bijective. Hence the sequence

$$0 \to \Gamma \to H_p^{n+1}(F(t)) \xrightarrow{\coprod i_v} \coprod_{v \neq \infty} \widetilde{H}_p^{n+1}(F(t)_v) \to 0$$

is exact. □

5.3. Let $q(t) \in F[t]$ be a prime monic polynomial. Let D_q denote the subgroup of $H_p^{n+1}(F(t))$ generated by elements of the form

$$a \frac{dq}{q} \wedge db_2 \wedge \cdots \wedge db_n, \quad \text{where } a, b_2, \ldots, b_n \in F[t].$$

Let D be the subgroup of $H_p^{n+1}(F(t))$ generated by elements of the form

$$a\, db_1 \wedge \cdots \wedge db_n, \quad \text{where } a, b_1, \ldots, b_n \in F[t].$$

5.4. PROPOSITION. *The sequence*

$$0 \to D \to D_q \xrightarrow{\partial} H_p^n(F_q) \to 0$$

is exact.

PROOF. Let $j: H_p^n(F_q) \to D_q/D$ be the homomorphism such that

$$j(\overline{a}\, d\overline{b}_2 \wedge \cdots \wedge d\overline{b}_n) = a \frac{dq}{q} \wedge db_2 \wedge \cdots \wedge db_n \quad (\text{where } a, b_2, \ldots, b_n \in F[t]).$$

It is not hard to see that j is well defined and the composition

$$H_p^n(F_q) \xrightarrow{j} D_q/D \xrightarrow{\partial_q} H_p^n(F_q)$$

is the identity map. Since j is surjective, $D_q/D \xrightarrow{\partial_q} H_p^n(F_q)$ is bijective. □

5.5. PROPOSITION. *The sequence*

$$0 \to D \to \Gamma \xrightarrow{\coprod_{v \neq \infty} \partial_v} \coprod_{v \neq \infty} H_p^n(F_v) \to 0$$

is exact.

PROOF. Consider the homomorphisms

$$\alpha: \coprod_{v \neq \infty} D_v/D \xrightarrow{\sum \text{in}} \Gamma/D, \qquad \beta: \Gamma/D \xrightarrow{\coprod_{v \neq \infty} \partial_v} \coprod_{v \neq \infty} H_p^n(F_v).$$

The homomorphism $\beta \circ \alpha$ coincides with

$$\coprod_{v \neq \infty} \partial_v : \coprod_{v \neq \infty} D_v/D \to \coprod_{v \neq \infty} H_p^n(F_v)$$

and hence is bijective in view of (5.3). Now we prove that α is surjective. Let

$$u = a \frac{db_1}{b_1} \wedge \cdots \wedge \frac{db_n}{b_n}, \quad \text{where } a, b_1, \ldots, b_n \in F[t],$$

be one of the generators of Γ. As in the proof of Proposition 5.1, we can assume that each of the elements $b_1, \ldots, b_n$ is a prime monic polynomial or belongs to F^*. Moreover, we can assume that $b_1, \ldots, b_n$ are pairwise distinct (otherwise $u = 0$).

In this case there exist $\alpha_1, \ldots, \alpha_n \in F[t]$ such that

$$\frac{a}{b_1 \cdots b_n} = \frac{\alpha_1}{b_1} + \cdots + \frac{\alpha_n}{b_n},$$

and hence

$$u = \frac{a}{b_1 \cdots b_n} db_1 \wedge \cdots \wedge db_n = \left(\sum_{i=1}^n \frac{\alpha_i}{b_i}\right) db_1 \wedge \cdots \wedge db_n$$

$$= \sum_{i=1}^n \alpha_i (-1)^{i+1} \frac{db_i}{b_i} \wedge db_1 \wedge \cdots \wedge \widehat{db_i} \wedge \cdots \wedge db_n \in \mathrm{im}(\alpha).$$

Since $\beta \circ \alpha$ is bijective and α is surjective, β is bijective and therefore the sequence

$$0 \to D \to \Gamma \to \coprod_{v \neq \infty} H_p^n(F_v) \to 0$$

is exact. $\square$

5.6. PROPOSITION. *The sequence*

$$0 \to H_p^{n+1}(F) \xrightarrow{\mathrm{in}_*} D \xrightarrow{i_\infty} \widetilde{H}_p^{n+1}(F(t)_\infty) \to 0$$

is exact. Here $\mathrm{in}_*\colon H_p^{n+1}(F) \to D \subset H_p^{n+1}(F(t))$ *is induced by the inclusion of the fields.*

PROOF. The injectivity of in_* follows from that of the composition

$$H_p^{n+1}(F) \xrightarrow{\mathrm{in}_*} D \subset H_p^{n+1}(F(t)) \xrightarrow{i_t} H_p^{n+1}(F(t)_{(t)}) = H_p^{n+1}(F((t))).$$

Hence we can consider $H_p^{n+1}(F)$ as a subgroup in $D \subset H_p^{n+1}(F(t))$ and it is sufficient to prove that $i_\infty\colon D/H_p^{n+1}(F) \to \widetilde{H}_p^{n+1}(F(t)_\infty)$ is bijective.

Denote by N_i the subgroup of D generated by elements of the form $a\,db_1 \wedge \cdots \wedge db_n$ (where $a, b_1, \ldots, b_n \in F[t]$, $\deg(ab_1 \cdots b_n) \leq i$). It is obvious that

$$0 \subset N_0 \subset N_1 \subset \cdots \subset N_i \subset \cdots, \quad \bigcup_{i=0}^\infty N_i = D, \quad N_0 = H_p^{n+1}(F).$$

Since $F(t)_\infty$ is a complete discrete valuation field, we can consider the groups $U_i = U_i H_p^{n+1}(F(t)_\infty)$ that were defined in (2.3).

Since $i_\infty(N_i) \subset U_i$, $N_0 = H_p^{n+1}$, and $U_0 = H_{p,\mathrm{ur}}^{n+1}(F(t)_\infty)$, it is sufficient to prove that $i_\infty\colon N_i/N_{i-1} \to U_i/U_{i-1}$ is bijective for any $i \geq 1$.

In the case $i \not\equiv p$, we consider the homomorphism $\rho_i'\colon \Omega_F^n \to N_i/N_{i-1}$ defined by

$$a \frac{db_1}{b_1} \wedge \cdots \wedge \frac{db_n}{b_n} \mapsto at^i \frac{db_1}{b_1} \wedge \cdots \wedge \frac{db_n}{b_n} \quad (a, b_1, \ldots, b_n \in F).$$

If $i \stackrel{\cdot}{\cdot} p$, we consider the homomorphism $\rho_i': \Omega_F^n/\Omega_{F,d=0}^n \oplus \Omega_F^{n-1}/\Omega_{F,d=0}^{n-1} \to N_i/N_{i-1}$ given by

$$\left(a \frac{db_1}{b_1} \wedge \cdots \wedge \frac{db_n}{b_n}, 0\right) \mapsto at^i \frac{db_1}{b_1} \wedge \cdots \wedge \frac{db_n}{b_n},$$

$$\left(0, a \frac{db_2}{b_2} \wedge \cdots \wedge \frac{db_n}{b_n}\right) \mapsto at^i \frac{dt}{t} \frac{db_2}{b_2} \wedge \cdots \wedge \frac{db_n}{b_n}$$

$(a, b_1, \ldots, b_n \in F)$. The composition $i_\infty \circ \rho_i'$ coincides with the homomorphism ρ_i defined in (2.4). It is easy to verify that ρ_i' is surjective for any $i \geq 1$. By Theorem 2.5, all the ρ_i's ($i \geq 0$) are bijective. Therefore $i_\infty: N_i/N_{i-1} \to U_i/U_{i-1}$ is bijective for any $i \geq 1$. □

5.7. COROLLARY. *There is an isomorphism*

$$i_\infty \oplus \coprod_{v \neq \infty} \partial_v : \Gamma/H_p^{n+1}(F) \xrightarrow{\sim} \widetilde{H}_p^{n+1}(F(t)_\infty) \oplus \coprod_{v \neq \infty} H_p^n(F_v).$$

PROOF. Let $A = \widetilde{H}_p^{n+1}(F(t)_\infty)$, $B = \coprod_{v \neq \infty} H_p^n(F_v)$. The proof follows if we examine the commutative diagram with exact rows,

$$\begin{array}{ccccccccc}
0 & \longrightarrow & D/H_p^{n+1}(F) & \longrightarrow & \Gamma/H_p^{n+1}(F) & \longrightarrow & \Gamma/D & \longrightarrow & 0 \\
& & \downarrow i_\infty & & \downarrow i_\infty \oplus \coprod_{v \neq \infty} \partial_v & & \downarrow \coprod_{v \neq \infty} \partial_v & & \\
0 & \longrightarrow & A & \xrightarrow{\text{in}_1} & A \oplus B & \xrightarrow{\text{pr}_1} & B & \longrightarrow & 0
\end{array}$$

and take into account the bijectivity of the left and the right vertical arrow. □

5.8. Denote by $R(F)$ the kernel of the homomorphism $\Gamma \xrightarrow{i_\infty} \widetilde{H}_p^{n+1}(F(t)_\infty)$.

PROPOSITION. *The following sequences are exact*

(i) $$0 \to H_p^{n+1}(F) \to R(F) \xrightarrow{\coprod_{v \neq \infty} \partial_v} \coprod_{v \neq \infty} H_p^n(F_v) \to 0,$$

(ii) $$0 \to R(F) \to H_p^{n+1}(F(t)) \xrightarrow{\coprod i_v} \coprod_v \widetilde{H}_p^{n+1}(F(t)_v) \to 0.$$

PROOF. Using (5.7), we easily see that

$$R(F)/H_p^{n+1}(F) \xrightarrow{\coprod_{v \neq \infty} \partial_v} \coprod_{v \neq \infty} H^n(F_v)$$

is bijective. Hence the sequence (i) is exact. Now, (ii) is exact since

$$\Gamma/R(F) \xrightarrow{i_\infty} \widetilde{H}_p^{n+1}(F(t)_\infty) \quad \text{and} \quad H_p^{n+1}(F)/\Gamma \xrightarrow{\coprod_{v \neq \infty} i_v} \coprod_{v \neq \infty} \widetilde{H}_p^{n+1}(F(t)_v)$$

are bijective. □

5.9. COROLLARY. $R(F^{\text{sep}}) = 0$.

PROOF. This follows from (5.8)(i) since the cohomology groups are trivial for separably closed fields. □

5.10. LEMMA. $R(F) = H^{n+1}_{p,\text{sep}}(F(t))$.

PROOF. Using (5.8)(ii), we see that

$$H^{n+1}_{p,\text{sep}}(F(t)) = \ker(H^{n+1}_p(F(t)) \to \widetilde{H}^{n+1}_p(F(t)))$$
$$\subset \ker(H^{n+1}_p(F(t)) \xrightarrow{\amalg i_v} \widetilde{H}^{n+1}_p(F(t)_v)) = R(F).$$

On the other hand,

$$R(F) = \ker(R(F) \to R(F^{\text{sep}}))$$
$$\subset \ker(H^{n+1}_p(F(t)) \to H^{n+1}_p(F^{\text{sep}}(t))) = H^{n+1}_{p,\text{sep}}(F(t)).$$

Therefore $R(F) = H^{n+1}_{p,\text{sep}}(F(t))$. □

5.11. It follows from (5.8) and (5.10) that the sequences

$$0 \to H^{n+1}_p(F) \to H^{n+1}_{p,\text{sep}}(F(t)) \to \coprod_{v \neq \infty} H^n_p(F_v) \to 0,$$

$$0 \to H^{n+1}_{p,\text{sep}}(F(t)) \to H^{n+1}_p(F(t)) \to \coprod_v \widetilde{H}^{n+1}_p(F(t)_v) \to 0$$

are exact.

Since s_0 splits the first sequence, we have

$$H^{n+1}_{p,\text{sep}}(F(T)) \cong H^{n+1}_p(F) \oplus \coprod_{v \neq \infty} H^n_p(F_v).$$

Using the second sequence, we see that

$$\widetilde{H}^{n+1}_p(F(t)) = H^{n+1}_p(F(t))/H^{n+1}_{p,\text{sep}}(F(t)) \cong \coprod_v \widetilde{H}^{n+1}_p(F(t)_v).$$

This competes the proof of Theorem 4.5. □

§6. Computation of $H^n_m(F(t))$ for arbitrary m

6.1. Definition of $H^n_m(F(t))$. For a field F of characteristic p, we denote by $H^{n+1}_{p^k}(F)$ the cokernel of the map

$$\Phi - 1 \colon C^n_k(F) \to C^n_k(F)/dC^{n-1}_k(F),$$

where $C^n_k(F)$ is the Bloch group and Φ is the Frobenius homomorphism [**Bl, Mi**].

By [**K1**], $H^{n+1}_{p^k}(F)$ is isomorphic to $W_k(F) \otimes F^* \otimes \cdots \otimes F^*/J$, where $W_k(F)$ is the group of all the p-Witt vectors over F of length k, and J is the subgroup (of the tensor product) generated by all elements of the form
  (i) $w \otimes b_1 \otimes \cdots \otimes b_n$ with $b_i = b_j$ for some i and j, $i \neq j$,
  (ii) $(0,0,\ldots,0,a,0,\ldots,0) \otimes a \otimes b_2 \otimes \cdots \otimes b_n$,
  (iii) $(\Phi(w) - w) \otimes b_1 \otimes \ldots b_n$ where $\Phi(a_1,\ldots,a_k) = (a_1{}^p,\ldots,a_k{}^p)$.

6.2. LEMMA [K2]. *Let F be a field of characteristic p. Then we have an exact sequence*
$$0 \to H^{n+1}_{p^k}(F) \xrightarrow{r} H^{n+1}_{p^{k+l}}(F) \xrightarrow{\mu} H^{n+1}_{p^l}(F) \to 0,$$
where $r((a_1, \ldots, a_k) \otimes b_1 \otimes \cdots \otimes b_n) = (0, \ldots, 0, a_1, \ldots, a_k) \otimes b_1 \otimes \cdots \otimes b_n$ and $\mu((a_1, \ldots, a_l, a_{l+1}, \ldots, a_{k+l}) \otimes b_1 \otimes \cdots \otimes b_n) = (a_1, \ldots, a_l) \otimes b_1 \otimes \cdots \otimes b_n$.

6.3. DEFINITION [K2]. Let F be a field, $m \geqslant 1$, $n \geqslant 0$. Set
$$H^{n+1}_m(F) = \begin{cases} H^{n+1}(F, \mu_m^{\otimes n}) & \text{if char } F = 0, \\ H^{n+1}(F, \mu_{m'}^{\otimes n}) \oplus H^{n+1}_{p^k}(F) & \text{if char } F = p, \, m = m'p^k, \, m' \not| \, p. \end{cases}$$

6.4. DEFINITION. Let L be a complete discrete valuation field. Then $H^{n+1}_{m,\mathrm{ur}}(L)$ denotes the kernel of $H^{n+1}_m(L) \to H^{n+1}_m(L^{\mathrm{ur}})$ and $\widetilde{H}^{n+1}_m(L)$ is the factor group $H^{n+1}_m(L)/H^{n+1}_{m,\mathrm{ur}}(L)$.

6.5. DEFINITION. Denote by $H^{n+1}_{m,\mathrm{sep}}(F(t))$ the kernel of the homomorphim $H^{n+1}_m(F(t)) \to H^{n+1}_m(F^{\mathrm{sep}}(t))$, and let $\widetilde{H}^{n+1}_m(F(t))$ be the factor group $H^{n+1}_m(F(t))/H^{n+1}_{m,\mathrm{sep}}(F(t))$.

6.6. PROPOSITION. *Let $\mathrm{char}(F) = p$. Then the map $H^{n+1}_{p^k}(F) \oplus H^n_{p^k}(F) \to H^{n+1}_{p^k,\mathrm{ur}}(F((t)))$ given by*
$$(w \otimes b_1 \otimes \cdots \otimes b_n, 0) \mapsto w \otimes b_1 \otimes b_2 \otimes \cdots \otimes b_n,$$
$$(0, w \otimes b_2 \otimes \cdots \otimes b_n) \mapsto w \otimes t \otimes b_2 \otimes \cdots \otimes b_n$$
($w \in W_k(F)$, $b_1, \ldots, b_n \in F^*$) is bijective.

PROOF. Induction on k. The case $k = 1$ follows from (2.6). Let $A_k = H^{n+1}_{p^k}(F) \oplus H^n_{p^k}(F)$, $B_k = H^{n+1}_{p^k}(F((t)))$, $C_k = H^{n+1}_{p^k}(F^{\mathrm{sep}}((t)))$. In view of (6.2), we have a commutative diagram with exact rows:

$$\begin{array}{ccccccccc} & & 0 & & 0 & & 0 & & \\ & & \downarrow & & \downarrow & & \downarrow & & \\ 0 & \to & A_{k-1} & \to & A_k & \to & A_1 & \to & 0 \\ & & \downarrow & & \downarrow & & \downarrow & & \\ 0 & \to & B_{k-1} & \to & B_k & \to & B_1 & \to & 0 \\ & & \downarrow & & \downarrow & & \downarrow & & \\ 0 & \to & C_{k-1} & \to & C_k & \to & C_1 & \to & 0 \end{array}$$

The left and the right column are exact by the inductive hypothesis. Hence the middle one is also exact. Therefore
$$H^{n+1}_{p^k}(F) \oplus H^n_{p^k}(F) = A_k \cong \ker(B_k \to C_k) \cong H^{n+1}_{p^k,\mathrm{ur}}(F((t))). \qquad \square$$

6.7. COROLLARY. *Let L be a complete discrete valuation field. Then the sequences*

$$0 \to H^{n+1}_{p^k,\mathrm{ur}}(L) \xrightarrow{r} H^{n+1}_{p^{k+l},\mathrm{ur}}(L) \xrightarrow{\mu} H^{n+1}_{p^l,\mathrm{ur}}(L) \to 0,$$

$$0 \to \widetilde{H}^{n+1}_{p^k}(L) \xrightarrow{r} \widetilde{H}^{n+1}_{p^{k+l}}(L) \xrightarrow{\mu} \widetilde{H}^{n+1}_{p^l}(L) \to 0$$

are exact.

PROOF. Since $L \cong F((t))$, we have $H^{n+1}_{p^k,\mathrm{ur}}(L) \cong H^{n+1}_{p^k}(F) \oplus H^n_{p^k}(F)$. Hence the first sequence is exact in view of Proposition 6.6 and Lemma 6.2. The second one is exact since $\widetilde{H}^{n+1}_{p^k}(L) = H^{n+1}_{p^k}(L)/H^{n+1}_{p^k,\mathrm{ur}}(L)$ and because the sequence of Lemma 6.2 is exact for L. □

6.8. PROPOSITION. *Let L be a complete discrete valuation field of characteristic $p > 0$ with residue field F. Then*

(1) *There is a well-defined homomorphism*

$$i \colon H^{n+1}_{p^k}(F) \to H^{n+1}_{p^k,\mathrm{ur}}(L) :$$

$$(\bar{a}_1, \ldots, \bar{a}_k) \otimes \bar{b}_1 \otimes \cdots \otimes \bar{b}_n \mapsto (a_1, \ldots, a_k) \otimes b_1 \otimes \cdots \otimes b_n,$$

where $a_1, \ldots, a_n \in \mathcal{O}_L$, $b_1, \ldots, b_n \in U_L$.

(2) *The group $H^{n+1}_{p^k,\mathrm{ur}}(L)$ is generated by elements of the form*

$$(a_1, \ldots, a_k) \otimes \pi \otimes b_1 \otimes \cdots \otimes b_{n-1},$$

where $a_1, \ldots, a_n \in \mathcal{O}_L$, $b_1, \ldots, b_{n-1} \in U_L$, and $\pi \in L$ is a prime element. The homomorphism

$$\partial \colon H^{n+1}_{p^k,\mathrm{ur}}(L) \to H^n_{p^k}(F) :$$

$$(a_1, \ldots, a_k) \otimes \pi \otimes b_1 \otimes \cdots \otimes b_{n-1} \mapsto (\bar{a}_1, \ldots, \bar{a}_k) \otimes \bar{b}_1 \otimes \cdots \otimes \bar{b}_{n-1}$$

is well defined.

(3) *The sequence $0 \to H^{n+1}_{p^k}(F) \xrightarrow{i} H^{n+1}_{p^k,\mathrm{ur}}(L) \xrightarrow{\partial} H^n_{p^k}(F) \to 0$ is exact.*

PROOF. Since $L \cong F((t))$, we have $H^{n+1}_{p^k,\mathrm{ur}}(L) \cong H^{n+1}_{p^k,\mathrm{ur}}(F((t))) \cong H^{n+1}_{p^k}(F) \oplus H^n_{p^k}(F)$. Hence there is an exact sequence

$$0 \to H^{n+1}_{p^k}(F) \xrightarrow{\mathrm{in}_1} H^{n+1}_{p^k,\mathrm{ur}}(L) \xrightarrow{\mathrm{pr}_2} H^n_{p^k}(F) \to 0.$$

It is easy to verify that $i = \mathrm{in}_1$ and $\partial = \mathrm{pr}_2$. □

6.9. DEFINITION. Let v be a discrete valuation on $F(t)$. Let

$$i_v \colon H^n_{m,\mathrm{sep}}(F(t)) \to H^n_{m,\mathrm{ur}}(F(t)_v) \quad \text{and} \quad i_v \colon \widetilde{H}^n_m(F(t)) \to \widetilde{H}^n_m(F(t)_v)$$

be the inclusion homomorphisms.

Denote by $\partial_v \colon H^n_{m,\mathrm{sep}}(F(t)) \to H^{n-1}_m(F_v)$ the composition

$$H^n_{m,\mathrm{sep}}(F(t)) \xrightarrow{i_v} H^n_{m,\mathrm{ur}}(F(t)_v) \xrightarrow{\partial} H^{n-1}_m(F_v).$$

Let $s \colon H^n_{m,\mathrm{sep}}(F(t)) \to H^n_m(F)$ be the composition

$$H^n_{m,\mathrm{sep}}(F(t)) \to H^n_{m,\mathrm{ur}}(F((t))) \cong H^n_m(F) \oplus H^{n-1}_m(F) \xrightarrow{\mathrm{pr}_1} H^n_m(F).$$

6.10. THEOREM. (i) *The map*

$$s \oplus \coprod_{v \neq \infty} \partial_v \colon H^n_{m,\mathrm{sep}}(F(t)) \to H^n_m(F) \oplus \coprod_{v \neq \infty} H^{n-1}(F_v)$$

is bijective.

(ii) *The map*

$$\coprod i_v \colon \widetilde{H}^n_m(F(t)) \to \coprod_v \widetilde{H}^n_m(F(t)_v)$$

is bijective.

PROOF. In view of [Se, Chapter 2, Annexe, §4] it is sufficient to consider the case $\mathrm{char}(F) = p$, $m = p^k$. Let

$$R_k = R_k(F) = \ker\left(H^n_{p^k}(F(t)) \xrightarrow{\coprod i_v} \coprod_v \widetilde{H}^n_{p^k}(F(t)_v) \right).$$

By the definitions of $\widetilde{H}^n_{p^k}(F(t)_v)$ and $R_k(F)$, we see that $i_v(R_k(F)) \subset H^n_{p^k,\mathrm{ur}}(F(t)_v)$. Hence we can define the map $\partial_v \colon R_k(F) \to H^{n-1}_{p^k}(F_v)$ as the composition

$$R_k(F) \xrightarrow{i_v} H^n_{p^k,\mathrm{ur}}(F(t)_v) \xrightarrow{\partial} H^{n-1}_{p^k}(F_v)$$

and the map $s \colon R_k(F) \to H^n_{p^k}(F)$ as the composition

$$R_k(F) \xrightarrow{i_t} H^n_{p^k,\mathrm{ur}}(F(t)_{(t)}) = H^n_{p^k,\mathrm{ur}}(F((t))) \cong H^n_{p^k}(F) \oplus H^{n-1}_{p^k}(F) \xrightarrow{\mathrm{pr}_1} H^n_{p^k}(F).$$

Set

$$A_k = A_k(F) = H^n_{p^k}(F) \oplus \coprod_{v \neq \infty} H^{n-1}_{p^k}(F_v),$$

$$B_k = B_k(F) = H^n_{p^k}(F(t)), \qquad C_k = C_k(F) = \coprod_v \widetilde{H}^n_{p^k}(F(t)_v).$$

Consider the homomorphisms $\alpha = s \oplus \coprod \partial_v \colon R_k \to A_k$ and $\beta = \coprod_v i_v \colon B_k \to C_k$.

In view of (5.11) and the definition of $R_k(F)$, the sequence $0 \to R_1 \xrightarrow{\alpha} B_1 \xrightarrow{\beta} C_1 \to 0$ is exact.

The following commutative diagram with exact rows and columns reduces the proof of the exactness of the sequence $0 \to R_k \xrightarrow{\alpha} B_k \xrightarrow{\beta} C_k \to 0$ to the case $k = 1$ considered above.

$$\begin{array}{ccccccccc}
& & R_{k-1} & & R_k & & R_1 & & \\
& & \uparrow & & \uparrow & & \uparrow & & \\
0 & \to & B_{k-1} & \to & B_k & \to & B_1 & \to & 0 \\
& & \downarrow & & \downarrow & & \downarrow & & \\
0 & \to & C_{k-1} & \to & C_k & \to & C_1 & \to & 0
\end{array}$$

Using this diagram we see that the sequence $0 \to R_{k-1} \to R_k \to R_1 \to 0$ is exact. It follows from Proposition 5.8(i) that $R_1 \xrightarrow{\alpha} A_1$ is bijective. Applying the five lemma to the commutative diagram

$$\begin{array}{ccccccccc} 0 & \longrightarrow & R_{k-1} & \longrightarrow & R_k & \longrightarrow & R_1 & \longrightarrow & 0 \\ & & \downarrow & & \downarrow & & \downarrow & & \\ 0 & \longrightarrow & A_{k-1} & \longrightarrow & A_k & \longrightarrow & A_1 & \longrightarrow & 0 \end{array}$$

with exact rows, we see that the homomorphism $R_k \to A_k$ is bijective for any $k \geqslant 0$.

If F is a separably closed field, then

$$A_k(F) = H^n_{p^k}(F) \oplus \coprod_{v \neq \infty} H^{n-1}_{p^k}(F_v) = 0.$$

Hence $R_k(F^{\text{sep}}) \cong A_k(F^{\text{sep}}) = 0$. Therefore

$$R_k(F) \subset \ker(H^n_{p^k}(F(t)) \to H^n_{p^k}(F^{\text{sep}}(t))) = H^n_{p^k,\text{sep}}(F(t)).$$

On the other hand,

$$H^n_{p^k,\text{sep}}(F(t)) \subset \ker(H^n_{p^k}(F(t)) \to \coprod_v \widetilde{H}(F(t)_v)) = R_k(F).$$

Hence $R_k(F) = H^n_{p^k,\text{sep}}(F(t))$. Finally we have

$$H^n_{p^k,\text{sep}}(F(t)) = R_k \cong A_k = H^n_{p^k}(F) \oplus \coprod_{v \neq \infty} H^{n-1}_{p^k}(F_v),$$

$$\widetilde{H}^n_{p^k,\text{sep}}(F(t)) = H^n_{p^k}(F(t))/H^n_{p^k,\text{sep}}(F(t)) = B_k/R_k \cong C_k = \coprod_v \widetilde{H}^n_{p^k}(F(t)_v).$$

6.11. COROLLARY. (i) *If F is a perfect field, then*

$$H^n_m(F(t)) \cong H^n_m(F) \oplus \coprod_{v \neq \infty} H^{n-1}_m(F_v).$$

(ii) *If F is a separably closed field, then* $H^n_m(F(t)) \cong \coprod_v H^n_m(F(t)_v)$. □

6.12. COROLLARY. *The homomorphism $H^n_m(F(t)) \to \coprod_v H^n_m(F(t)_v)$ is injective.* □

References

[BK] S. Bloch and K. Kato, *p-adic étale cohomology*, Inst. Hautes Études Sci. Publ. Math. **63** (1986), 107–152.

[BT] H. Bass and J. Tate, *The Milnor ring of a global field*, Lecture Notes Math., vol. 342, Springer-Verlag, Berlin, 1973, pp. 349–446.

[Fa] D. K. Faddeev, *Simple algebras over a field of algebraic function of one variable*, Trudy Mat. Inst. Steklov. **38** (1951), 321–344; English transl., Amer. Math. Soc. Transl. (2) **3** (1956), 15–38.

[K1] K. Kato, *A generalization of local class field theory by using K-groups*, J. Fac. Sci. Univ. Tokyo Sect. IA Math. **26** (1079), no. 2, 303–376; **27** (1980), no. 3, 603–683.

[K2] _____, *Galois cohomology of complete discrete valuation fields*, Algebraic K-theory, Lecture Notes in Math., vol. 967, Springer-Verlag, 1982, pp. 215–237.

[IL] L. Illusie, *Complexe de Rham–Witt et cohomologie cristalline*, Ann. Sci. École Norm. Sup. (4) **12** (1979), 501–661.

[Bl] S. Bloch, *Algebraic K-theory and crystalline cohomology*, Inst. Hautes Études Sci. Publ. Math. **47** (1977), 187–268.

[Se] J.-P. Serre, *Cohomologie Galoisienne*, Lecture Notes Math., vol. 5, Springer-Verlag, Berlin and Heidelberg, 1994.

Translated by THE AUTHOR

ST. PETERSBURG STATE UNIVERSITY, RUSSIA MATH. & MECH. DEPT., BIBLIOTECHNAYA PL. 2, STARY PETERGOF, ST. PETERSBURG, 198904, RUSSIA

On Topological Filtration for Severi–Brauer Varieties II

N. A. Karpenko

Dedicated to Boarding School #45 on the occasion of its 30th anniversary

ABSTRACT. The topological filtration on K_0 of a Severi–Brauer variety is computed when the ratio of the index and the exponent is a squarefree number, and for each prime p dividing this ratio, the p-primary component of the corresponding division algebra is decomposable. In particular, this gives a description of Ch^2 for such varieties.

Let D be a central simple algebra over the field F and $X = SB(D)$ be the Severi–Brauer variety of D. In [3] the topological filtration on the Grothendieck group $K(X)$ has been computed provided that $\operatorname{ind} D = \exp D$. The topic of these notes is the case when the ratio $\operatorname{ind} D / \exp D$ is any squarefree number but with one more additional restriction on D: for each prime $p \mathrel{\vdots} \operatorname{ind} D / \exp D$, the p-primary component of the corresponding division algebra must be decomposable (i.e., isomorphic to the tensor product $D_1 \otimes_F D_2$ with $D_j \neq F$ for both j).

In addition to the notation introduced above, we fix the following:

The notation relative to the Grothendieck group as introduced in [3]. In particular $G^i K(X)$ is the quotient group of topological filtration of codimension i; $\overline{G^i K(X)}$ is the image of the homomorphism $G^i K(X) \to G^i K(\bar{X}) = \mathbb{Z}$, where $\bar{X}$ is the variety X over the algebraic closure of F.

For a prime p, v_p is the p-adic valuation on $\mathbb{Q}$; $\binom{n}{k}$ is the binomial coefficient; $(\,,\,)$ is the greatest common divisor.

I owe to A. S. Merkurjev the idea that the cycle $SB(D_1) \times SB(D_2)$ on the variety $SB(D_1 \otimes D_2)$ may be an interesting one.

THEOREM. *Let D be a central simple algebra with $\operatorname{ind} D = r$, $\exp D = e$ and let $X = SB(D)$.*

If r/e is a squarefree number and for each prime $p \mathrel{\vdots} r/e$ the p-primary component of the division algebra similar to D is decomposable, then the map

$$G^i K(X) \to G^i K(\bar{X}) \qquad (0 \leqslant i \leqslant \dim X)$$

1991 *Mathematics Subject Classification.* 14C15, 14C25.
Key words and phrases. Severi–Brauer variety, Chow groups, K-groups.

1) *is injective;*
2) *has the image*
$$\frac{r}{(i,r)\prod p} \cdot \mathbb{Z},$$
where the product $\prod p$ is taken over all prime $p \, \vdots \, r/e$ such that $0 < v_p(i+p-1) < v_p(r)$.

EXAMPLE. Let D be a division algebra of index p^2 and exponent p, and let $X = SB(D)$. If D is decomposable, then for $1 \leq i \leq p^2 - 1$ we have
$$G^i K(X) = \begin{cases} p\mathbb{Z}, & \text{if } i \,\vdots\, p \text{ or } i-1 \,\vdots\, p \text{ without } i = p^2 - p + 1, \\ p^2\mathbb{Z}, & \text{otherwise.} \end{cases}$$

PROOF OF THE THEOREM. It suffices to consider only the case when $r = p^n$ for a prime p. Then e equals to p^n or p^{n-1}. The first case was considered in [3]. Below we assume that $e = p^{n-1}$.

The proof consists of several lemmas.

First consider the case when D is a division algebra.

LEMMA 1. *If $D = D_1 \otimes_F D_2$ is a nontrivial decomposition of division algebra D with $\text{ind } D = p^n$ and $\exp D = p^{n-1}$, then the index and exponent of one of the factors equal p, the index and exponent of the other one equal p^{n-1}.*

PROOF. Put $\text{ind } D_j = p^{k_j}$, $\exp D_j = p^{l_j}$. Then
$$k_1 + k_2 = n \quad \text{and} \quad n-1 \geq \max\{k_1, k_2\} \geq \max\{l_1, l_2\} \geq n-1.$$
So, all the inequalities are equalities, which implies the statement.

Let us fix a decomposition $D = D_1 \otimes_F D_2$ with $\text{ind } D_1 = \exp D_1 = p$ and $\text{ind } D_2 = \exp D_2 = p^{n-1}$.

LEMMA 2. *For $X = SB(D)$, where $D = D_1 \otimes_F D_2$ is division algebra with $\text{ind } D_1 = \exp D_1 = p$ and $\text{ind } D_2 = \exp D_2 = p^{n-1}$, we have*
$$\log_p |K(\bar{X})/K(X)| = n \cdot p^n - \alpha, \quad \text{where } \alpha = v_p(p^n!) + p^{n-1} - 1.$$

PROOF. It is known from [5] that
$$|K(\bar{X})/K(X)| = \prod_{i=1}^{p^n} \text{ind } D^{\otimes i}.$$
Put $\log_p \text{ind } D^{\otimes i} = n - \alpha_i$. If $v_p(i) = 0$, then $\alpha_i = 0$. If $v_p(i) > 0$, then
$$\text{ind } D^{\otimes i} = \text{ind } D_2^{\otimes i} = \text{ind } D_2/(i, \text{ind } D_2)$$
(see [3] for the last equality), so $\alpha_i = \min\{v_p(i) + 1, n\}$. Consequently
$$\alpha = \sum_{i=1}^{p^n} \alpha_i = v_p(p^n!) + p^{n-1} - 1.$$

Before we deal with Lemma 3, let us formulate a fact from [4], which will be needed below.

PROPOSITION [4]. *For $N \geq 1$ denote by X^N the variety $SB(M_N(D))$, where $M_N(D)$ is the F-algebra of $N \times N$-matrices over an arbitrary central simple algebra D of degree d. Then*

$$\mathrm{Ch}_*(X^N) = \mathrm{Ch}_*(X) \oplus \mathrm{Ch}_{*-d}(X) \oplus \cdots \oplus \mathrm{Ch}_{*-(N-1)d}(X),$$

where Ch_* denotes the Chow group graded by dimensions of cycles (so, $\mathrm{Ch}(X^N) = (\mathrm{Ch}(X))^N$).

Now we come to the main point in proving the theorem.

LEMMA 3. *In the notation introduced before the proposition, D is a decomposed division algebra of index p^n and exponent p^{n-1}, while $X = SB(D)$. If $i - 1 \vdots p$, where $0 \leq i \leq p^n - 1$ and $i \neq p^n - p + 1$, then $\overline{G^i K}(X) \ni p^{n-1}$.*

PROOF. Replace the quotient groups of the topological filtration by Chow groups indexed by dimension. What we need to show is that $\overline{\mathrm{Ch}}_i(X) \ni p^{n-1}$, where $\overline{\mathrm{Ch}}_i(X) = \mathrm{Im}(\mathrm{Ch}_i(X) \to \mathrm{Ch}_i(\bar{X}) = \mathbb{Z})$ if $i = kp - 2$ and $2 \leq k \leq p^{n-1}$. Fix a large number N (it suffices to take any $N \geq p$) and consider the closed embedding $X_1 \times X_2^N \hookrightarrow X^N$, where $X_j = SB(D_j)$ is induced by the tensor product of ideals [1]. According to the proposition, it suffices to show that $p^{n-1} \in \overline{\mathrm{Ch}}_i(X^N)$ if $i = kp - 2$ (and $2 \leq k \leq p^{n-1}$ as above).

Consider the commutative diagram:

$$\begin{array}{ccccc}
\mathrm{Ch}_{p-1}(\bar{X}_1) \otimes \mathrm{Ch}_{(k-1)p-1}(\bar{X}_2^N) & \longrightarrow & \mathrm{Ch}_i(\bar{X}_1 \times \bar{X}_2^N) & \longrightarrow & \mathrm{Ch}_i(\bar{X}^N) \\
\uparrow & & \uparrow & & \uparrow \\
\mathrm{Ch}_{p-1}(X_1) \otimes \mathrm{Ch}_{(k-1)p-1}(X_2^N) & \longrightarrow & \mathrm{Ch}_i(X_1 \times X_2^N) & \longrightarrow & \mathrm{Ch}_i(X^N)
\end{array}$$

The morphism $\bar{X}_1 \times \bar{X}_2^N \hookrightarrow \bar{X}^N$ is a Segre embedding [1]. Now we need several standard statements on Segre embeddings.

SUBLEMMA [2]. *The image of the Segre embedding $\mathbb{P}^{a_1} \times \mathbb{P}^{a_2} \hookrightarrow \mathbb{P}^a$, where $a = a_1 a_2 + a_1 + a_2$, has degree $\binom{a_1 + a_2}{a_1}$. If $\mathbb{P}^{b_j} \subset \mathbb{P}^{a_j}$ $(j = 1, 2)$ are linear subvarieties, then the image of $\mathbb{P}^{b_1} \times \mathbb{P}^{b_2}$ is contained in some linear subspace $\mathbb{P}^b \subset \mathbb{P}^a$ with $b = b_1 b_2 + b_1 + b_2$ and the induced morphism $\mathbb{P}^{b_1} \times \mathbb{P}^{b_2} \hookrightarrow \mathbb{P}^b$ is Segre embedding as well.*

As a corollary, we see by looking at the diagram that the image of $1 \otimes 1$ from the upper left corner in the upper right corner equals $\binom{i}{p-1}$.

Further, $\overline{\mathrm{Ch}}_{p-1}(X_1) \ni 1$ in a trivial way and $\overline{\mathrm{Ch}}_{(k-1)p-1}(X_2^N) \ni p^{n-m-2}$ with $m = v_p(k-1)$ according to [3], since

$$X_2^N = SB(M_N(D_2)) \quad \text{and} \quad \mathrm{ind}(M_N(D_2)) = \exp(M_N(D_2)) = p^{n-1}.$$

Thus, $\overline{\mathrm{Ch}}_i(X^N) \ni \binom{i}{p-1} \cdot p^{n-m-2}$ and the last observation we need is that $v_p(\binom{i}{p-1}) = m + 1$.

END OF THE PROOF OF THE THEOREM. Put $\log_p |G^* K(\bar{X}) / \overline{G^* K}(X)| = n \cdot p^n - \beta$. We claim that $\beta \geq \alpha$ for any α from Lemma 1 (compare with [3]). For each i

$(0 \leqslant i \leqslant p^n - 1)$, put

$$\beta_i = \begin{cases} n, & \text{if } i = 0, \\ v_p(i), & \text{if } i \vdots p \text{ and } i \neq 0, \\ 1, & \text{if } i - 1 \vdots p \text{ and } i \neq p^n - p + 1, \\ 0 & \text{otherwise.} \end{cases}$$

Then $\log_p |G^i K(\bar{X})/\overline{G^i K}(X)| \leqslant n - \beta_i$. Indeed, for each i we have $p^n/(i, p^n) \in \overline{G^i K}(X)$ [3], and

$$p^{n-1} \in \overline{G^i K}(X) \text{ if } i - 1 \vdots p, i \neq p^n - p + 1$$

by Lemma 3, which implies the required inequality.

Adding all the inequalities together, we obtain

$$\beta \geqslant \sum_{i=0}^{p^n - 1} \beta_i.$$

It is easy to see that the last sum equals α. Consequently, by the proposition from [3], $G^* K(X) \to G^* K(\bar{X})$ is injective and $p^{n-\beta_i}$ generates $\overline{G^i K}(X)$.

To complete the proof, we must consider the case when D is arbitrary, not necessarily a division algebra. Write $D \simeq M_s(D')$ with a skewfield D' and put $X' = SB(D')$. Then

$$|G^* K(\bar{X}')/\overline{G^* K}(X')| = |K(\bar{X}')/K(X')|$$

as shown above. Since

$$|G^* K(\bar{X})/\overline{G^* K}(X)| = s|G^* K(\bar{X}')/\overline{G^* K}(X')|$$

by the Proposition and

$$|K(\bar{X})/K(X)| = s|K(\bar{X}')/K(X')|$$

(see [5]), the same equality holds for X. It implies the injectivity of $G^* K(X) \to G^* K(\bar{X})$ and the second assertion of the theorem in the same way as above.

COROLLARY. *If X is a Severi–Brauer variety satisfying the assumption of the theorem, then*

$$\text{Ch}^2(X) = \frac{r}{(2, r)} \cdot \mathbb{Z}.$$

References

1. M. Artin, *Brauer–Severi varieties*, Lecture Notes in Math., vol. 917, Springer-Verlag, Heidelberg, 1982, pp. 194–210.
2. R. Hartshorne, *Algebraic geometry*, Springer-Verlag, Heidelberg, 1977.
3. N. A. Karpenko, *On topological filtration for Severi–Brauer varieties*, K-Theory and Algebraic Geometry: Connections with Quadratic Forms and Division Algebras Proc. Symp. Pure Math., vol. 58; Part 2, Amer. Math. Soc., Providence, RI, 1995, pp. 275–277 (to appear).
4. A. S. Merkurjev, *Certain K-cohomology of Severi–Brauer varieties*, Preprint (1992).
5. D. Quillen, *Higher algebraic K-theory*, I, Lecture Notes Math., vol. 341, Springer-Verlag, Heidelberg, 1973, pp. 85–147.

Translated by THE AUTHOR

MATHEMATISCHES INSTITUT WESTFÄLISCHE WILHELMS-UNIVERSITÄT, EINSTEINSTR. 62 D–48149 MÜNSTER, GERMANY

On the Norm Residue Homomorphism for Fields

A. S. Merkur'ev

Dedicated to Boarding School #45 on the occasion of its 30th anniversary

Let F be a field and m a natural number prime to the characteristic of F. The interaction between Milnor K-theory and Galois cohomology of the field F is presented by the *norm residue homomorphism* of degree n modulo m:

$$h_{n,m,F}\colon K_nF/mK_nF \to H^n(F,\mu_m^{\otimes n}).$$

There is a well-known conjecture saying that $h_{n,m,F}$ is always an isomorphism. This is clear if $n = 0$, simply follows from Hilbert Theorem 90 for $n = 1$, and is known to be true if $n = 2$ for arbitrary m [8]. It is also known to be true if m is a power of 2 and $n = 3$ [9, 14] or $n = 4$ (Rost, unpublished) for an arbitrary field F.

In the present paper we consider the case of arbitrary n, $m = 2$ and a field F of characteristic different from 2. It is easy to see (Lemma 2.1) that the surjectivity of the norm residue homomorphism $h_n = h_{n,2,F}$ implies the triviality of the Bockstein connecting map of degree n,

$$B_n = B_{n,F}\colon H^n(F,\mu_2^{\otimes n}) \to H^{n+1}(F,\mu_2^{\otimes n})$$

given by the short exact sequence

$$1 \to \mu_2^{\otimes n} \to \mu_4^{\otimes n} \to \mu_2^{\otimes n} \to 1.$$

We prove that, conversely, the purely cohomological property of the Bockstein map to be trivial implies the bijectivity of the norm residue homomorphism h_n. More precisely, our main result is the following

THEOREM. *Suppose that for some $n \geqslant 0$ the Bockstein map B_n is trivial for all fields. Then the norm residue homomorphism h_n is an isomorphism for all fields.*

COROLLARY. *The surjectivity of h_n for all fields implies the bijectivity of h_n also for all fields.*

1991 *Mathematics Subject Classification.* Primary 12G05, 19D45.
Key words and phrases. Galois cohomology, Milnor K-theory, étale cohomology.

§1. Notation and preliminary results

Milnor K-theory. Let F be a field. The Milnor graded ring K_*F of the field F is the factor ring of the tensor $\mathbb{Z}$-algebra $T(F^\times)$ modulo the ideal generated by the tensors $a \otimes (1-a)$ for all $a \neq 0, 1$ in F [13]. Hence, $K_*F = \coprod_{n\geq 0} K_nF$, $K_0F = \mathbb{Z}$, $K_1F = F^\times$.

For any field extension E/F, there is a natural ring homomorphism $i_{E/F}: K_*F \to K_*E$ induced by the inclusion $F \hookrightarrow E$.

If E/F is a finite extension, then the norm homomorphism $n_{E/F}: K_*E \to K_*F$ is defined [5]. The *projection formula* $n_{L/F}(u \cdot i_{L/F}(v)) = n_{L/F}(u) \cdot v$ holds for any $u \in K_*E$, $v \in K_*F$. In particular, the composition $n_{E/F} \circ i_{E/F}$ is the multiplication by the degree $[E:F]$ of the extension E/F.

Throughout the paper we use the notation k_*F for $K_*F/2K_*F$. The image of the tensor $a_1 \otimes \cdots \otimes a_n$ in k_*F will be denoted by $\{a_1, \ldots, a_n\}$ and called a *symbol*.

If E/F is a quadratic extension, then k_nE is generated by symbols of the type $u = \{x, a_1, \ldots a_{n-1}\}$, where $x \in E^\times$, $a_i \in F^\times$ [2] and $n_{E/F}(u) = \{n_{E/F}(x), a_1, \ldots, a_{n-1}\}$ by the projection formula.

Let X be an algebraic variety defined over F. For any $i \geq 0$ denote by $X^{(i)}$ the subset of points in X of codimension i. The group

$$C_n^i(X) = \coprod_{x \in X^{(i)}} k_nF(x)$$

is called the group of k_n-*cycles of codimension i on X*. We have the complex [6]:

$$\cdots \to C_{n+1}^{i-1}(X) \to C_n^i(X) \to C_{n-1}^{i+1}(X) \to \cdots.$$

We denote by $H^i(X, k_{n+i})$ the homology group of this complex in the term $C_n^i(X)$.

If $X = \mathbb{A}_F^1$ is the affine line, then $H^0(X, k_n) = k_nF$ and $H^i(X, k_n) = 0$ if $i > 0$. In other words, there is an exact sequence [13]:

$$0 \to k_nF \to k_nF(t) \to \coprod_{x \in \mathbb{A}^1} k_{n-1}F(x) \to 0.$$

Cohomology. Denote by F_s a separable closure of the field F, and let $\Gamma = \mathrm{Gal}(F_s/F)$. For any discrete Γ-module M by $H^*(F, M)$ we denote the cohomology groups $H^*(\Gamma, M)$ [16].

For any field extension E/F there is a natural *restriction* homomorphism:

$$i_{E/F}: H^*(F, M) \to H^*(E, M).$$

If E/F is a finite extension, then we have the *norm homomorphism (corestriction)*:

$$n_{E/F}: H^*(E, M) \to H^*(F, M).$$

The composition $n_{E/F} \circ i_{E/F}$ is the multiplication by $[E:F]$.

The group $H^*(F, \mathbb{Z}/2\mathbb{Z})$ will be simply denoted by H^*F. For a quadratic extension $E = F(\sqrt{a})/F$, char $F \neq 2$, there is an infinite exact sequence [1]:

$$\cdots \to H^{n-1}F \xrightarrow{?\cup(a)} H^nF \xrightarrow{i_{L/F}} H^nL \xrightarrow{n_{L/F}} H^nF \to \cdots.$$

Let X be an algebraic variety defined over F.

The group
$$Z_n^i(X) = \coprod_{x \in X^{(i)}} H^n F(x)$$
is called the group of H^n-*cycles of codimension i on X*. There is the complex [3, 12] (char $F \neq 2$):
$$\cdots \to Z_{n+1}^{i-1}(X) \to Z_n^i(X) \to Z_{n-1}^{i+1}(X) \to \cdots.$$
We denote by $H^i(X, H^{n+i})$ the cohomology groups of this complex in the term $Z_n^i(X)$.

If $X = \mathbb{A}_F^1$, then $H^0(X, H^n) = H^n F$ and $H^i(X, H^n) = 0$ for $i \geq 1$. In other words, there exists an exact sequence [4]:
$$0 \to H^n F \to H^n F(t) \to \coprod_{x \in \mathbb{A}^1} H^{n-1} F(x) \to 0.$$

If $(m, \text{char } F) = 1$, then the group $\mu_m \subset F_s^\times$ of mth roots of unity is a Γ-module. By $\mu_m^{\otimes n}$ we denote the nth tensor power $\underbrace{\mu_m \otimes \cdots \otimes \mu_m}_{n}$ of μ_m. The exact sequence of Γ-modules $1 \to \mu_m \to F_s^\times \xrightarrow{m} F_s^\times \to 1$ induces the connecting homomorphism
$$F^\times / F^{\times m} \to H^1(F, \mu_m), \qquad a \cdot F^{\times m} \mapsto (a),$$
which is an isomorphism in view of the Hilbert Theorem 90 [16].

The cup-product in cohomology theory induces the homomorphism
$$\underbrace{F^\times \otimes \cdots \otimes F^\times}_{n} \to H^n(F, \mu_m^{\otimes n}),$$
$$a_1 \otimes \cdots \otimes a_n \mapsto (a_1) \cup \cdots \cup (a_n) \stackrel{\text{def}}{=} (a_1, \ldots, a_n),$$
which factors through the *norm residue homomorphism of degree n modulo m* [18]:
$$h_{n,m,F} : K_n F / m K_n F \to H^n(F, \mu_m^{\otimes n}).$$
If $m = 2$, then the norm residue homomorphism is simply denoted by
$$h_n = h_{n,F} : k_n F \to H^n F.$$

Comparing the exact sequences of Milnor groups and Galois cohomology for an affine line $\mathbb{A}^1$, one deduces that if h_{n-1} is an isomorphism for any field, then $\ker h_{n,F} \simeq \ker h_{n,F(t)}$, $\operatorname{coker} h_{n,F} \simeq \operatorname{coker} h_{n,F(t)}$, i.e., kernel and cokernel of the norm residue homomorphism do not change under purely transcendental extensions.

The norm residue homomorphism commutes with norms: for a finite field extension E/F, the diagram
$$\begin{array}{ccc} k_n E & \xrightarrow{h_{n,E}} & H^n E \\ {\scriptstyle n_{E/F}} \downarrow & & \downarrow {\scriptstyle n_{E/F}} \\ k_n F & \xrightarrow{h_{n,F}} & H^n F \end{array}$$
commutes [17].

If F is a finite, local or global field, then $h_{n,F}$ is an isomorphism [13, 18].

If X is an algebraic variety over F, then for an étale sheaf M on $X_{\text{ét}}$ by $H^*(X, M)$ we denote étale cohomology groups [12]. We shall denote by $H^* X$ the group $H^*(X, \mathbb{Z}/2\mathbb{Z})$.

The degree $\deg x$ of a closed point $x \in X$ is the degree of the extension $F(x)/F$.

By $[1, n]$ we denote the set $\{1, \ldots, n\}$.

In §2 we prove certain properties of the Bockstein map. The next section is devoted to the study of étale cohomology of conic curves (projective and affine). In §4 we express the fact that a given element in $k_n F$ is trivial in terms of algebraic equations. This statement generalizes Lemma 4 from [7]. In §5, we develop a specialization technique, part of which was already used in [7]. The next section is devoted to the computation of K-cohomology groups of affine conic curves. In §7 we prove that under certain conditions the surjectivity of $h_{n,F}$ implies the surjectivity of $h_{n,F(X)}$ for a conic curve X over F. The proof of the theorem is presented in §8. We follow the approach of [7] in the case $n = 2$.

§2. Bockstein map

Let F be a field of characteristic different from 2. The map $B_n = B_{n,F}$ in the exact sequence

$$H^n(F, \mu_4^{\otimes n}) \to H^n F \xrightarrow{B_n} H^{n+1} F \to H^{n+1}(F, \mu_4^{\otimes n})$$

associated to the short exact sequence

$$1 \to \mu_2^{\otimes n} \to \mu_4^{\otimes n} \to \mu_2^{\otimes n} \to 1$$

will be called the *Bockstein map*.

LEMMA 2.1. *The surjectivity of the norm residue homomorphism $h_{n,F}$ implies $B_{n,F} = 0$.*

PROOF. This follows from the commutative diagram

$$\begin{array}{ccc} K_n F & =\!=\!= & K_n F \\ {\scriptstyle h_{4,F}} \downarrow & & {\scriptstyle h_{2,F}} \downarrow \\ H^n(F, \mu_4^{\otimes n}) & \longrightarrow H^n F \xrightarrow{B_n} & H^{n+1} F. \end{array} \qquad \square$$

Later we shall need the following two statements:

LEMMA 2.2. *If $B_{n,F} = 0$, then the restriction of the natural map*

$$H^{n+1}(F, \mu_4^{\otimes n}) \to H^{n+1}(F, \mu_2^{\otimes n}) = H^{n+1} F$$

to the subgroup of elements of exponent 2 is trivial.

PROOF. This follows from the fact that the composition

$$H^{n+1}(F, \mu_4^{\otimes n}) \to H^{n+1} F \to H^{n+1}(F, \mu_4^{\otimes n})$$

is the multiplication by 2 and the second map is injective since $B_{n,F} = 0$. $\square$

LEMMA 2.3. *If B_n is trivial for all fields, then $B_{n-1} = 0$ also for all fields.*

PROOF. Let $E = F((t))$ be the field of formal power series over F. The statement follows from the following diagram with surjective rows [13]:

$$\begin{array}{ccc} H^n E & \xrightarrow{\partial} & H^{n-1} F \\ \downarrow{B_{n,E}} & & \downarrow{B_{n-1,F}} \\ H^{n+1} E & \xrightarrow{\partial} & H^n F. \end{array} \quad \square$$

§3. Étale cohomology of conic curves

Let C be a smooth curve over a field F, S be a finite set of closet points in C, $U = C - S$. Let m be a natural number prime to char F.

Consider the Gysin exact sequence [12]

$$\cdots \to \coprod_{c \in S} H^{n-2}(F(c), \mu_m^{\otimes(j-1)}) \to H^n(C, \mu_m^{\otimes j}) \to H^n(U, \mu_m^{\otimes j}) \to \cdots.$$

Passing to the inductive limit with respect to S, we get the localization sequence

$$\cdots \to \coprod_{c \in C} H^{n-2}(F(c), \mu_m^{\otimes(j-1)}) \to H^n(C, \mu_m^{\otimes j}) \to H^n(F(C), \mu_m^{\otimes j}) \to \cdots.$$

The arrows in this sequence are compatible with the multiplication on $H^*(F, \mu_m^{\otimes *})$ up to sign.

In the special case $n = 2$, $j = 1$ we get a monomorphism

$$\text{Pic } C/m \text{ Pic } C \to H^2(C, \mu_m).$$

Consider the Hochschild–Serre spectral sequence [12]:

$$E_2^{p,q} = H^p(F, H^q(C_s, \mu_m^{\otimes j})) \Rightarrow H^{p+q}(C, \mu_m^{\otimes j})$$

where $C_s = C \otimes_F F_s$. To compute the E_2-term, one must calculate the groups

$$H^q(C_s, \mu_m^{\otimes j}) = H^q(C_s, \mu_m) \otimes \mu_m^{\otimes(j-1)}.$$

It is known that $H^0(C_s, \mu_m^{\otimes j}) = \mu_m^{\otimes j}$ and $H^q(C_s, \mu_m^{\otimes j}) = 0$ if $q \geqslant 3$ [12]. The exact sequence

$$1 \to \mu_m \to \mathbb{G}_m \xrightarrow{m} \mathbb{G}_m \to 1$$

implies that $H^2(C_s, \mu_m) \simeq \text{Pic } C_s/m \text{ Pic } C_s$ since $H^1(C_s, \mathbb{G}_m) \simeq \text{Pic } C_S$ and m-torsion in $H^2(C_s, \mathbb{G}_m)$ is trivial [12], and we also obtain the exact sequence

$$1 \to F_s[C_s]^\times / F_s[C_s]^{\times m} \to H^1(C_s, \mu_m) \to \text{Pic } C_s \xrightarrow{m} \text{Pic } C_s.$$

Let Y be the projective conic curve given in $\mathbb{P}_F^2$ by the equation $aU^2 + bV^2 = W^2$, where $a, b \in F^\times$ (we assume that char $F \neq 2$). The following conditions are clearly equivalent:
1. Y has a F-rational point.
2. Y is isomorphic to $\mathbb{P}_F^1$.
3. The quaternion algebra $\left(\frac{a,b}{F}\right)$ splits, i.e., $(a,b) = 0 \in H^2 F$.

If Y satisfies these conditions, then it is called a *split* conic curve.

If Y is not split, then the degree of each closed point is even. The exact sequence [15]
$$0 \to \operatorname{Pic} Y \to H^0(F, \operatorname{Pic} Y_s) \to \operatorname{Br} F \to \operatorname{Br} F(Y)$$
shows that the kernel of $\operatorname{Br} F \to \operatorname{Br} F(Y)$ is generated by $(a, b) \in H^2 F$, while $\operatorname{Pic} Y = \mathbb{Z}$ is generated by the class of any F-rational point if Y is split, and $\operatorname{Pic} Y = 2\mathbb{Z}$ is generated by the class of an arbitrary closed point of degree 2.

Since $Y_s \simeq \mathbb{P}^1_s$, we have $F_s[Y_s] = F_s$ and $\operatorname{Pic} Y_s = \mathbb{Z}$. Applying the above computation one gets:
$$H^q(Y_s, \mu_m^{\otimes j}) = \begin{cases} \mu_m^{\otimes j}, & \text{if } q = 0, \\ \mu_m^{\otimes (j-1)}, & \text{if } q = 2, \\ 0, & \text{if } q = 1 \text{ or } q \geq 3. \end{cases}$$

Therefore, we have the commutative diagram

$$\cdots \xrightarrow{\partial} \coprod_{y \in Y} H^{n-2}(F(y), \mu_m^{\otimes(j-1)}) \to H^n(Y, \mu_m^{\otimes j}) \to H^n(F(Y), \mu_m^{\otimes j}) \to \cdots$$

with $H^n(F, \mu_m^{\otimes j})$ mapping via i to $H^n(F(Y), \mu_m^{\otimes j})$, and $\pi: H^n(Y, \mu_m^{\otimes j}) \to H^{n-2}(F, \mu_m^{\otimes(j-1)})$, j from the coproduct to $H^{n-2}(F, \mu_m^{\otimes(j-1)})$,

where i is the natural map.

LEMMA 3.1. *The homomorphism j is the sum of the norm homomorphisms.*

PROOF. First consider the case where $n = 2$, $j = 1$, and $y \in Y$ is an F-rational point. The statement follows from the commutative diagram

$$\begin{array}{ccc} H^0(F(y), \mathbb{Z}/m\mathbb{Z}) & \longrightarrow & H^2(Y, \mu_m) \\ \| & & \downarrow \\ \operatorname{Pic} Y_s / m \operatorname{Pic} Y_s \xrightarrow{\sim} \mathbb{Z}/m\mathbb{Z} & \xrightarrow{\sim} & H^2(Y_s, \mu_m) \end{array}$$

The case of a rational point $y \in Y$ (with arbitrary n and j) follows from the above and the fact that the homomorphisms in the diagram are compatible up to sign with the multiplication on $H^*(F, \mu_m^{\otimes *})$.

For the general case, let $E = F(y)$ and y be an E-rational point of Y_E over y, $E(y) = F(y)$. The statement follows from the commutative diagram

$$\begin{array}{ccccc} H^{n-2}(E(y'), \mu_m^{\otimes(j-1)}) & \longrightarrow & H^n(Y_E, \mu_m^{\otimes j}) & \xrightarrow{\pi_E} & H^{n-2}(E, \mu_m^{\otimes(j-1)}) \\ \| & & \downarrow n_{E/F} & & \downarrow n_{E/F} \\ H^{n-2}(F(y), \mu_m^{\otimes(j-1)}) & \longrightarrow & H^n(Y, \mu_m^{\otimes j}) & \xrightarrow{\pi_F} & H^{n-2}(F, \mu_m^{\otimes(j-1)}). \end{array} \quad \square$$

It follows from Lemma 3.1 that the norm maps induce a well-defined norm map
$$N \colon H^1(Y, H^{n-1}) \to H^{n-2} F.$$

In the special case $n = m = 2$, we have the monomorphism $\operatorname{Pic} Y/2 \operatorname{Pic} Y \to H^2 Y$.

LEMMA 3.2. *If Y is not split, then the image of the canonical generator of $\operatorname{Pic} Y$ in $H^2 Y$ equals $(a, b)_Y$.*

PROOF. The group $\operatorname{Pic} Y$ is generated by the class $[y]$ of arbitrary point $y \in Y$ of degree 2. Since $[y] \notin \operatorname{im} \partial$ and $j([y]) = 0$, we see that the image of $[y]$ in $H^2 Y$ comes from a nontrivial element of the kernel of $H^2 F \to H^2 F(Y)$. But the only nontrivial element in this kernel is (a, b). $\square$

Let X be an open subvariety in Y defined by $w \neq 0$, so that X is an affine conic curve: $u^2 - av^2 = b$, where $u = U/W$, $v = V/W$. The complement $Y - X$ is isomorphic to $\operatorname{Spec} L$, where $L = F(\sqrt{a})$. There is only one "infinite" point $\infty \in Y$ of degree 2 if $a \notin F^{\times 2}$ and the union of two points if $a \in F^{\times 2}$. If X is not split, then $\operatorname{Pic} X = \operatorname{Pic} Y / \langle \infty \rangle = 0$ since $\operatorname{Pic} Y$ is generated by the class of the point ∞ of degree 2.

If X is split but $a \notin F^{\times 2}$, then the group $\operatorname{Pic} X \simeq \mathbb{Z}/2\mathbb{Z}$ is generated by the class of any rational point. If $a \in F^{\times 2}$, then $\operatorname{Pic} X = 0$ since $X \simeq \mathbb{A}^1 - pt$.

Assume that $(m, \operatorname{char} F) = 1$. It follows from the computation above that the group $H^1(X_s, \mu_m)$ is isomorphic to $F_s[X]^\times / F_s^\times \simeq \mathbb{Z}/m\mathbb{Z}$ and is generated by the class of the invertible regular function $u + v\sqrt{a}$ on X. The Galois group Γ acts through the action of $\operatorname{Gal}(L/F)$ whose generator takes $k \in \mathbb{Z}/m\mathbb{Z}$ to $-k$. Denote the group $\mathbb{Z}/m\mathbb{Z}$ with this Γ-action by $\widetilde{\mathbb{Z}/m\mathbb{Z}}$ and put $\tilde{\mu}_m^{\otimes j} = \widetilde{\mathbb{Z}/m\mathbb{Z}} \otimes \mu_m^{\otimes j}$. Hence we have:

$$H^q(X_s, \mu_m^{\otimes j}) = \begin{cases} \mu_m^{\otimes j} & \text{if } q = 0, \\ \tilde{\mu}_m^{\otimes (j-1)} & \text{if } q = 1, \\ 0 & \text{if } q \geq 2. \end{cases}$$

The Hochschild–Serre spectral sequence induces the exact sequence

$$\cdots \to H^n(F, \mu_m^{\otimes j}) \to H^n(X, \mu_m^{\otimes j}) \to H^{n-1}(F, \tilde{\mu}_m^{\otimes (j-1)}) \xrightarrow{\Delta} H^{n+1}(F, \mu_m^{\otimes j}).$$

LEMMA 3.3. *If $m = 2$, then the differential map $\Delta: H^{n-1} F \to H^{n+1} F$ is the multiplication by $(a, b) \in H^2 F$.*

PROOF. Since Δ commutes with the multiplication on $H^* F$, it follows that Δ is the multiplication by $u = \Delta(1 + 2\mathbb{Z}) \in H^2 F$.

The commutative diagram

$$\begin{array}{ccccccc}
& & & & \operatorname{Pic} X & & \\
& & & & \downarrow & \searrow^{j} & \\
H^0 F & \xrightarrow{\Delta} & H^2 F & \to & H^2 X & \to & H^1 F \\
& & & \searrow^{i} & \downarrow & & \\
& & & & H^2 F(X) & &
\end{array}$$

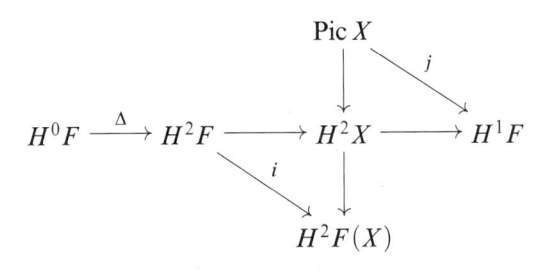

shows that $u \in \ker i$. If X is split, then $u = 0 = (a, b)$. If X is not split, then $\operatorname{Pic} X = 0$ and therefore u generates $\ker i$, i.e., $u = (a, b)$. $\square$

Consider the commutative diagram

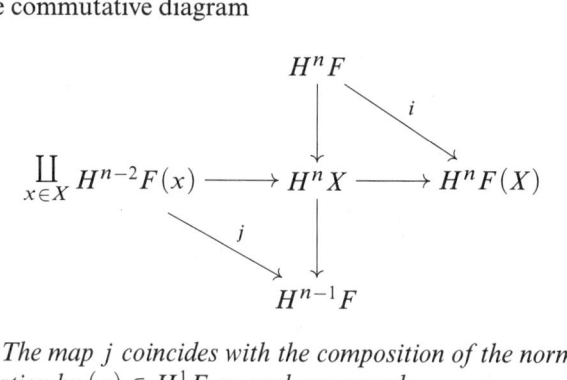

$$\begin{array}{ccc}
 & & H^n F \\
 & & \downarrow \quad \searrow i \\
\coprod_{x \in X} H^{n-2}F(x) \longrightarrow & H^n X & \longrightarrow H^n F(X) \\
 & \searrow j \quad \downarrow & \\
 & H^{n-1}F &
\end{array}$$

LEMMA 3.4. *The map j coincides with the composition of the norm homomorphism and the multiplication by $(a) \in H^1 F$ on each summand.*

PROOF. As in the proof of Lemma 3.1, it is sufficient to consider the case of an F-rational point $x \in X$ (X is therefore split) and $n = 2$, so we must show that $j(1 + 2\mathbb{Z}) = (a) \in H^1 F$. If $(a) = 0$, i.e., $a \in F^{\times 2}$, then $\operatorname{Pic} X = 0$, hence $1 + 2\mathbb{Z}$ belongs to the image of $H^1 F(X) \to \coprod_{x \in X} H^0 F(x)$ and $j(1 + 2\mathbb{Z}) = 0 = (a)$.

If $(a) \neq 0$, then the group $\operatorname{Pic} X = \mathbb{Z}/2\mathbb{Z}$ is generated by the class of x. Since the map i is injective, the diagram in the proof of Lemma 3.3 shows that $j(1 + 2\mathbb{Z}) \neq 0$.

Since $(a)_L = 0 \in H^1 L$, by considering X over L and the case described above, one shows that $j(1 + 2\mathbb{Z})_L = 0$, hence $j(1 + 2\mathbb{Z}) \in \ker(H^1 F \to H^1 L)$ and $j(1 + 2\mathbb{Z}) = (a)$. □

COROLLARY 3.5. *The sum of the norm homomorphisms induces a well-defined norm map*

$$N: H^1(X, H^{n-1}) \to H^{n-2}F/n_{L/F}(H^{n-2}L).$$

PROOF. The image of the map j belongs to

$$\ker(H^{n-1}F \to H^{n-1}L) \simeq H^{n-2}F/n_{L/F}(H^{n-2}L). \quad \square$$

Consider the following sequence:

$$0 \to H^1(X, H^n) \xrightarrow{N} H^{n-1}F/n_{L/F}(H^{n-1}L) \xrightarrow{\cup(a,b)} H^{n+1}F \to H^{n+1}F(X).$$

The map in the middle is the multiplication by (a, b); it is well defined since

$$n_{L/F}(H^{n-1}L) \cup (a) = 0.$$

LEMMA 3.6. *The sequence*

$$0 \to H^1(X, H^n) \xrightarrow{N} H^{n-1}F/n_{L/F}(H^{n-1}L) \xrightarrow{\cup(a,b)} H^{n+1}F \to H^{n+1}F(X)$$

is a complex.

PROOF. Since $(a, b)_{F(X)} = 0$, the composition of the last two maps is zero. The equality $(a, b) \circ N = 0$ follows from the fact that N commutes with the multiplication by (a, b) and $(a, b)_{F(x)} = 0$ for any $x \in X$. □

REMARK 3.7. The homology of the complex

$$H^1(X, H^n) \xrightarrow{N} H^{n-1}F/n_{L/F}(H^{n-1}L) \xrightarrow{\cup (a,b)} H^{n+1}F$$

is naturally isomorphic to the homology of the complex

$$H^1(Y, H^n) \xrightarrow{N} H^{n-1}F \xrightarrow{(a,b)} H^{n+1}F.$$

LEMMA 3.8. *For any $x \in X$ the composition*

$$H^{n-1}F(x) \to H^{n+1}X \to H^{n+1}(X, \mu_4^{\otimes n})$$

is trivial.

PROOF. As in the proof of Lemma 3.1, we may assume that $n = 1$ and x is an F-rational point of X. But in this case the statement follows from the fact that $2 \cdot \operatorname{Pic} X = 0$ and the commutativity of the diagram

$$\begin{array}{ccccc} H^0 F(x) & \longrightarrow & \operatorname{Pic} X & \longrightarrow & H^2 X \\ & & \downarrow 2 & & \downarrow \\ & & \operatorname{Pic} X & \longrightarrow & H^2(X, \mu_4). \end{array} \quad \square$$

LEMMA 3.9. *Assume that the Bockstein map $B_{n-1,F}$ is trivial. Then the image of the composition*

$$H^n(X, \mu_4^{\otimes n}) \to H^n X \to H^{n-1}F$$

is contained in $n_{L/F}(H^{n-1}L)$.

PROOF. It suffices to show that the map ∂ in the diagram

$$\begin{array}{ccc} H^n(X, \mu_4^{\otimes n}) & \longrightarrow & H^n X \\ \downarrow & & \downarrow \\ H^{n-1}(F, \tilde{\mu}_4^{\otimes(n-1)}) & \longrightarrow & H^{n-1}F \xrightarrow{\partial} H^n F \end{array}$$

coincides with the multiplication by $(a) \in H^1 F$. The map δ is the connected map relative to the short exact sequence

$$0 \to \mathbb{Z}/2\mathbb{Z} = \mu_2^{\otimes(n-1)} \to \tilde{\mu}_4^{\otimes(n-1)} \to \mu_2^{\otimes(n-1)} = \mathbb{Z}/2\mathbb{Z} \to 0.$$

The class of this sequence in $\operatorname{Ext}^1_\Gamma(\mathbb{Z}/2\mathbb{Z}, \mathbb{Z}/2\mathbb{Z})$ is the sum of classes of the following sequences: $0 \to \mathbb{Z}/2\mathbb{Z} \to \mathbb{Z}/2\mathbb{Z}[G] \to \mathbb{Z}/2\mathbb{Z} \to 0$, where $G = \operatorname{Gal}(L/F)$ and $0 \to \mathbb{Z}/2\mathbb{Z} \to \mu_4^{\otimes(n-1)} \to \mathbb{Z}/2\mathbb{Z} \to 0$. The connecting homomorphism for the first sequence is the multiplication by (a) [1] and for the second sequence is equal to $B_{n-1,F} = 0$. $\square$

Consider the following diagram:

$$\begin{array}{ccc}
& \coprod_{x \in X} H^{n-1}F(x) \longrightarrow & H^{n+1}X \\
& \downarrow & \downarrow \\
H_4^n X \longrightarrow H_4^n F(X) \longrightarrow & \coprod_{x \in X} H_4^{n-1}F(x) \longrightarrow & H_4^{n+1}(X, \mu_4^{\otimes n}) \\
\downarrow \quad\quad\quad \downarrow p & \downarrow & \\
\coprod_{x \in X} H^{n-2}F(x) \longrightarrow H^n X \longrightarrow H^n F(X) \longrightarrow & \coprod_{x \in X} H^{n-1}F(x) & \\
\downarrow & & \\
H^{n-1} F & & \\
\downarrow (a,b) & & \\
H^{n+1} F & &
\end{array}$$

where for brevity we write $H_4^n(-)$ for $H^n(-, \mu_4^{\otimes n})$.

A diagram chase and Lemmas 3.8 and 3.9 show that there is a well-defined map

$$\theta: \coprod_{x \in X} H^{n-1}F(x) \to H^{n-1}F/n_{L/F}(H^{n-1}L).$$

LEMMA 3.10. *The homomorphism θ is induced by the norm map*

$$N: H^1(X, H^n) \to H^{n-1}F/n_{L/F}(H^{n-1}L).$$

PROOF. We may assume that $(a) \neq 0 \in H^1 F$. As in Lemma 3.1, it suffices to consider the case in which $n = 1$ and X is split. Let x be an F-rational point in X. We must show that a nonzero element in $H^0 F(x)$ corresponds to a nonzero element in $H^0 F$. Choose a function $f \in F(X)^\times$ such that $\operatorname{div} f = 2x$. Since $\operatorname{div} f \vdots 2$, we have $(f) = g_{F(X)}$ for some $g \in H^1 X$. The image of g in $H^0 F$ is nontrivial. For if it is trivial, then $g \in \operatorname{im}(H^1 F \to H^1 X)$ and therefore $f \in F^\times \cdot F(X)^{\times 2}$, but this is impossible since $0 \neq [x] \in \operatorname{Pic} X \simeq \mathbb{Z}/2\mathbb{Z}$. On the other hand, the image of f in $H^0(F(x), \mathbb{Z}/4\mathbb{Z}) = \mathbb{Z}/4\mathbb{Z}$ equals 2 mod $4\mathbb{Z}$ and hence is nontrivial. □

LEMMA 3.11. *Let $B_{n-1,F}$ and $B_{n,F(x)}$ be trivial. Then the sequence*

$$H^1(X, H^n) \xrightarrow{N} H^{n-1}F/n_{L/F}(H^{n-1}L) \xrightarrow{\cup(a,b)} H^{n+1}F$$

is exact.

PROOF. Diagram chase using the surjectivity of p and Lemma 3.10. □

LEMMA 3.12. *Assume that $B_{n,F} = 0$ and h_{n-1} is surjective for all fields. Suppose*

$$u = (u_y) \in \coprod_{y \in Y} H^{n-1}F(y), \quad \text{where } \sum n_y(u_y) = 0 \in H^{n-1}F.$$

Then the image of u in $H^{n+1}Y$ is equal to $(a,b) \cdot v_Y$ for some $v \in H^{n-1}F$.

PROOF. Choose $w_y \in H^{n-1}(F(y), \mu_4^{\otimes(n-1)}) \mapsto u_y \in H^{n-1}F(y)$ and assume first that
$$\sum n_y(w_y) = 0 \in H^{n-1}(F, \mu_4^{\otimes(n-1)}).$$
Then the image of w in $H^{n+1}(Y, \mu_4^{\otimes n})$ equals t_Y for some $t \in H^{n+1}(F, \mu_4^{\otimes n})$. In particular, $t_{F(Y)} = 0$.

Since L splits Y, the map
$$H^{n+1}(L, \mu_4^{\otimes n}) \to H^{n+1}(L(Y), \mu_4^{\otimes n})$$
is injective, and hence $t_L = 0$. Applying the norm map we obtain: $2t = n_{L/F}(t_L) = 0 \in H^{n+1}(F, \mu_4^{\otimes n})$. By Lemma 2.2, the image of t in $H^{n+1}F$ is trivial, therefore, the image of u in $H^{n+1}Y$ is trivial.

In general, we have $\sum n_y(w_y) = k(s)_y$ where $s \in H^{n-1}F$ and $k: H^{n-1}F \to H^{n-1}(F, \mu_4^{\otimes(n-1)})$. Choose $s' \in H^{n-1}(F, \mu_4^{\otimes(n-1)}) \mapsto s \in H^{n-1}F$. Then $k(s) = 2s' \in H^{n-1}(F, \mu_4^{\otimes(n-1)})$ and by the first part of the proof we can replace w by w' with the support at one point ∞ of degree 2:
$$w'_y = \begin{cases} s'_L \in H^{n-1}(L, \mu_4^{\otimes(n-1)}) & \text{if } y = \infty, \\ 0 & \text{if } y \neq \infty. \end{cases}$$

Therefore, we can replace u by u':
$$u'_y = \begin{cases} s_L \in H^{n-1}L & \text{if } y = \infty, \\ 0 & \text{if } y \neq \infty. \end{cases}$$

Finally, using the multiplicative property, we may assume that $n = 1$ and apply Lemma 3.2. $\square$

COROLLARY 3.13. *In the conditions of Lemma* 3.12, *the homomorphism*
$$H^1(X, H^n) \xrightarrow{N} H^{n-1}F/n_{L/F}(H^{n-1}L)$$
is injective.

PROOF. Let $(u_x) \in \coprod H^{n-1}F(x)$ and $\sum n_x(u_x) = n_{L/F}(w)$ for some $w \in H^{n-1}L$. Consider $u \in \coprod_{y \in Y} H^{n-1}F(y)$ such that
$$u_y = \begin{cases} u_x & \text{if } y = x \in X, \\ w & \text{if } y = \infty. \end{cases}$$

Then $\sum n_y(u_y) = 0$ and by Lemma 3.12 the image of u in $H^{n+1}Y$ equals $(a, b) \cdot v_Y$ for some $v \in H^{n-1}F$. Hence the image of (u_x) in $H^{n+1}X$ equals zero and we apply the exact sequence
$$H^n F(X) \to \coprod_{x \in X} H^{n-1}F(x) \to H^{n+1}X. \quad \square$$

LEMMA 3.14. *The homology of the complex*
$$0 \to H^1(X, H^n) \xrightarrow{N} H^{n-1}F/n_{L/F}(H^{n-1}L) \xrightarrow{\cup (a,b)} H^{n+1}F \to H^{n+1}F(X)$$
at the first and third terms are isomorphic.

PROOF. This follows from the commutative diagram:

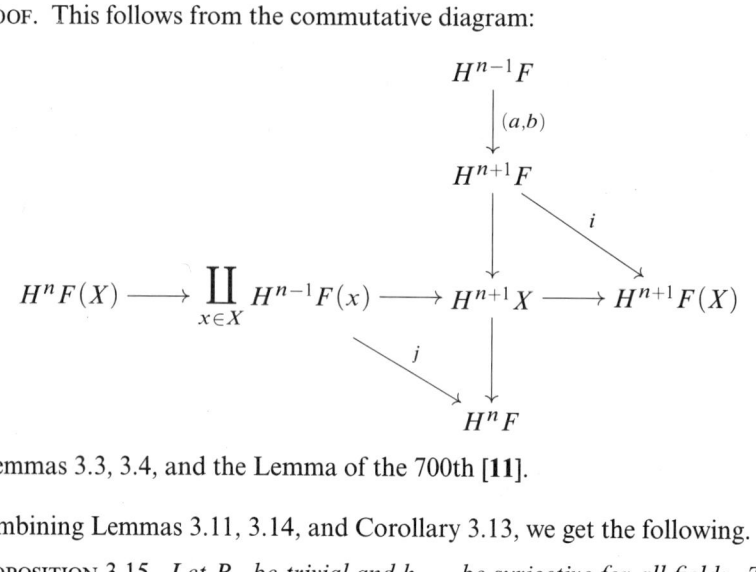

$$
\begin{array}{ccccccc}
 & & & & H^{n-1}F & & \\
 & & & & \downarrow (a,b) & & \\
 & & & & H^{n+1}F & & \\
 & & & & & \searrow^{i} & \\
H^n F(X) & \longrightarrow & \coprod_{x \in X} H^{n-1}F(x) & \longrightarrow & H^{n+1}X & \longrightarrow & H^{n+1}F(X) \\
 & & & \searrow^{j} & \downarrow & & \\
 & & & & H^n F & &
\end{array}
$$

from Lemmas 3.3, 3.4, and the Lemma of the 700th [11]. $\square$

Combining Lemmas 3.11, 3.14, and Corollary 3.13, we get the following.

PROPOSITION 3.15. *Let B_n be trivial and h_{n-1} be surjective for all fields. Then the sequence*

$$0 \to H^1(X, H^n) \xrightarrow{N} H^{n-1}F/n_{L/F}(H^{n-1}L) \xrightarrow{\cup (a,b)} H^{n+1}F \to H^{n+1}F(X)$$

is exact. $\square$

§4. When an element in k_n equals zero

In this section we generalize Lemma 4 from [7].

Let $n, m \in \mathbb{N}$, $a_1, \ldots, a_m \in F^\times$ and $b_\alpha \in F^\times$ for any map $\alpha: [1, n-1] \to [1, m]$. Put $\underline{a}_\alpha = \{a_{\alpha(1)}, \ldots, a_{\alpha(n-1)}\} \in k_{n-1}F$.

PROPOSITION 4.1. *Let the images of $a_1, \ldots, a_m$ be linearly independent in $F^\times/F^{\times 2}$ (regarded as a vector space over $\mathbb{Z}/2\mathbb{Z}$). The following conditions are equivalent:*

(1) $\sum_\alpha b_\alpha \cdot \underline{a}_\alpha = 0 \in k_n F$, *where the sum is taken over all maps* $\alpha: [1, n-1] \to [1, m]$.

(2) *There exists an integer $r \geq m$ and elements $a_{m+1}, \ldots, a_r \in F^\times$, $c_{s,j}, d_{s,j} \in F$ for $j = 1, \ldots, n-1$ and all maps $s: [1, n-1] \to \{nonempty\ subsets\ in\ [1, r]\}$ such that*

$$b_\alpha = \prod_s \left(\prod_{j=1}^{n-1} (c_{s,j}^2 - a_{s(j)} \cdot d_{s,j}^2) \right)$$

for any map $\alpha: [1, n-1] \to [1, r]$, where the product is taken over all s such that $s(k) \ni \alpha(k)$ for any $k = 1, \ldots, n-1$ (also $a_T = \prod_{i \in T} a_i$ for any subset $T \subset [1, r]$; $b_\alpha = 1$, if $\mathrm{im}(\alpha) \not\subset [1, m]$).

PROOF. $(2) \Longrightarrow (1)$:

$$\sum_\alpha b_\alpha \cdot \underline{a}_\alpha = \sum_{j=1}^{n-1} \left(\sum_s (c_{s,j}^2 - a_{s(j)} \cdot d_{s,j}^2) \cdot \sum_{\alpha(k) \in s(k)} \underline{a}_\alpha \right)$$

$$= \sum_{j=1}^{n} \left(\sum_s \{c_{s,j}^2 - a_{s(j)} \cdot d_{s,j}^2, a_{s(1)}, \ldots, a_{s(j)}, \ldots, a_{s(n-1)}\} \right) = 0$$

since $\{c^2 - ad^2, a\} = 0 \in k_2 F$ for $a, c, d \in F$. Note that here we do not need a_i to be linearly independent in $F^\times/F^{\times 2}$.

(1) $\Longrightarrow$ (2): The group $k_n F$ is isomorphic to the factor group of $(F^\times/F^{\times 2})^{\otimes n}$ by the subgroup generated by $u \otimes v_1 \otimes \cdots \otimes v_{n-1}$, where u is a norm in the quadratic extension $F(\sqrt{v_j})/F$ for some $j = 1, 2, \ldots, n-1$. Therefore, the element $\theta = \sum_\alpha b_\alpha \otimes a_{\alpha(1)} \otimes \cdots \otimes a_{\alpha(n-1)}$ is the sum of elements of the type described above. Consider elements $a_{m+1}, \ldots, a_r$ such that the images of $a_1, \ldots, a_r$ in $F^\times/F^{\times 2}$ form a base of the vector space $V \subset F^\times/F^{\times 2}$ such that the above relation holds in $F^\times \otimes V \otimes \cdots \otimes V$. Then

$$\theta = \sum_{j=1}^{n-1} \left(\sum_s (c_{s,j}^2 - a_{s(j)} d_{s,j}^2) \otimes a_{s(1)} \otimes \cdots \otimes a_{s(j)} \otimes \cdots \otimes a_{s(n-1)} \right)$$

$$= \sum_{j=1}^{n-1} \sum_\alpha \prod_s (c_{s,j}^2 - a_{s(j)} d_{s,j}^2) \otimes a_{\alpha(1)} \otimes \cdots \otimes a_{\alpha(n-1)},$$

where the product is taken over all s such that $s(k) \ni \alpha(k)$ for any $k = 1, \ldots, n-1$. Hence the desired relations hold modulo $F^{\times 2}$. But squares can be included to $c_{s,1}$ and $d_{s,1}$ for s with $s(k) = \{\alpha(k)\}, k = 1, \ldots, n-1$. $\square$

§5. Specialization

We develop the specialization technique which will be used later many times.

Let X be an irreducible variety over a field F and $x \in X$ be a closed point. Assume that for any $n \geqslant 0$ and any point $y \in X$ such that $x \in \overline{\{y\}}$ there is an abelian group $V_n(y)$ such that:

(1) For any points $y, z \in X$ such that $x \in \overline{\{y\}}$ and y is a regular point of codimension 1 in $\overline{\{z\}}$ there is a homomorphism

$$\partial_{y,z}^n \colon V_n(z) \to V_{n-1}(y)$$

called the *residue map*, for any $n \geqslant 1$.

(2) For any $z \in X$ such that $x \in \overline{\{z\}}$ there is a multiplication

$$V_n(z) \otimes F(z)^\times \to V_{n+1}(z)$$

for any $n \geqslant 0$.

Let $y, z \in X$ be the two points as in (1), $\pi \in F(z)^\times$ be a prime element of the discrete valuation on the field $F(z)$ defined by the point y. Consider the composition

$$s_{y,z}^n \colon V_n(z) \to V_{n+1}(z) \xrightarrow{\partial_{y,z}^n} V_n(y),$$

where the first map is the multiplication by π.

Assume that x is a regular point of X. Then there exists a sequence of points in X: $x = x_0, x_1, \ldots, x_m$ such that $x_m = \xi$ is the generic point of X and for any $i = 0, 1, \ldots, m-1$ x_i is a regular point in $\overline{\{x_{i+1}\}}$ of codimension 1. Choose prime elements $\pi_{i+1} \in F(x_{i+1})$ with respect to the discrete valuation defined by x_i on $F(x_{i+1})$. The composition

$$s^n = s_{x_0,x_1}^n \circ s_{x_1,x_2}^n \circ \cdots \circ s_{x_{m-1},x_m}^n \colon V_n(\xi) \to V_n(x)$$

is called the *specialization map*. It depends on the choice of the sequence of points x_i and prime elements π_i.

EXAMPLES. 1) $V_n(y) = k_n F(y)$. The residue map $\partial^n_{y,z}$ in (1) is the usual residue map (see [2, 13]) and the multiplication is obvious. We therefore have the specialization map
$$s^n: k_n(F(X)) \to k_n F(x).$$
If the functions $f_1, \ldots, f_n \in F(X)^\times$ are defined and nonzero in x, then
$$s^n(\{f_1, \ldots, f_n\}) = \{f_1(x), \ldots, f_n(x)\}.$$

2) Let $\pi: V \to X$ be a flat dominating morphism. Fix some integer $i \geq 0$. For any $y \in X$, define
$$V_n(y) = H^i(V_y, k_n), \quad \text{where } V_y = \pi^{-1}(y).$$

Let $y, z \in X$ be as in (1) and $\mathcal{O}$ be the discrete valuation ring $\mathcal{O}_{y,z}$ with fraction field $F(z)$ and residue field $F(y)$. Consider $V_\mathcal{O} = V \times_X \operatorname{Spec} \mathcal{O}$. Since $V_\mathcal{O}^{(i)} = V_z^{(i)} \cup V_y^{(i-1)}$, we have the exact sequence of complexes:

$$0 \to \coprod_{s \in V_y^{(*-1)}} K_* F(s) \to \coprod_{s \in V_\mathcal{O}^{(*)}} K_* F(s) \to \coprod_{s \in V_z^{(*)}} K_* F(s) \to 0.$$

Taking the exact sequence of homology groups, we get the residue map $\partial^n_{y,z}: V_n(z) \to V_{n-1}(y)$.

The multiplication map clearly exists. We therefore have the specialization map
$$s^n: H^i(V_\xi, k_n) \to H^i(V_x, k_n).$$

§6. K-cohomology of a conic curve

We keep the notation of §3. Since the conic curve Y is projective, the norm maps $k_n F(y) \to k_n F$ for each $y \in Y$ determine the norm map [6]:
$$\overline{N}: H^1(Y, k_n) \to k_{n-1} F.$$
Therefore, the norm map $N: H^1(X, k_n) \to k_{n-1} F / n_{L/F}(k_{n-1} L)$ for an affine conic curve X is determined by the commutative diagram

$$\begin{array}{ccccc} k_{n-1} F(\infty) & \longrightarrow & H^1(Y, k_n) & \longrightarrow\!\!\!\rightarrow & H^1(X, k_n) \\ \Big\| & & \overline{N} \Big\downarrow & & N \Big\downarrow \\ k_{n-1} L & \xrightarrow{n_{L/F}} & k_{n-1} F & \longrightarrow & k_{n-1} F / n_{L/F} k_{n-1} L \end{array}$$

We shall show in this section that under certain conditions N is injective.

Assume that X is not split. For any $m \geq 0$, we consider the subspace $L_m \subset F[X]$ consisting of the functions $f(u, v)$, where f is a polynomial in two variables of degree $\leq m$ and u, v are regular functions on X such that $u^2 - av^2 = b$.

It is clear that $\dim L_m = 2m + 1$. Any function $g \in L_m$ has a pole at infinity of order at most m, therefore, the degree of the divisors of poles and zeros of g is at most $2m$. Hence for any $x \in X$ such that $g(x) = 0$ we have $\deg(x) \leq 2m$. Therefore for any $x \in X$ of degree $2m$, the map $L_{m-1} \xrightarrow{\rho} F(x)$, $g \mapsto g(x)$, is injective.

LEMMA 6.1. *For any $x \in X$ of degree $2m$ and any $w \in F(x)^\times$ there exist functions $g \in L_{m-1}, h \in L_1$ such that $w = (gh^{-1})(x)$.*

PROOF. Let A be the image of the map ρ above, $\dim F(x)/A = 1$. Consider the map
$$L_1 \to F(x)/A, \qquad h \mapsto h(x) \cdot w \mod A.$$
Since $\dim L_1 = 3$, there exists a nonzero function $h \in L_1$ such that $0 \neq h(x) \cdot w \in A$, i.e., $h(x) \cdot w = g(x)$ for some function $g \in L_{m-1}$. □

For any $m \geq 1$, we denote by A_m the image of the natural map
$$\coprod_{\deg x \leq 2m} k_{n-1} F(x) \to H^1(X, k_n).$$
It is clear that $A_1 \subset A_2 \subset \cdots \subset H^1(X, k_n) = \bigcup_m A_m$.

LEMMA 6.2. $H^1(X, k_n) = A_1$, i.e., any element in $H^1(X, k_n)$ can be represented by some k_{n-1}-cycle with support in the set of points of degree 2.

PROOF. We shall show that $A_1 = A_m$ by induction on m. Let $x \in X$ be a point of degree $2m > 2$ and $w = \{w_1, \ldots, w_{n-1}\} \in k_{n-1}F(x)$. By Lemma 6.1, we can find functions $g_i \in L_{m-1}$, $h_i \in L_1$ such that $w_i = (g_i h_i^{-1})(x)$ for any $i = 1, \ldots, n-1$. Since $\operatorname{Pic} X = 0$, there exists a function $f \in F(X)^\times$ such that $\operatorname{div} f = x$. Consider the symbol
$$t = \{g_1 h_1^{-1}, \ldots, g_{n-1} h_{n-1}^{-1}, f\} \in k_n F(X).$$
It is clear that $\partial(t) = u \cdot x + (k_{n-1}\text{-cycle with support at points of degree} \leq 2m-2) \in \coprod_{z \in X} k_{n-1} F(z)$. Therefore, $u \cdot x \in A_{m-1} = A_1$ by the inductive hypothesis. □

LEMMA 6.3. If $x \in X$ satisfies $F(x) \simeq L$, then the image of $k_{n-1}F(x)$ in $H^1(X, k_n)$ is trivial.

PROOF. The image of $k_{n-1}F(x)$ in $H^1(X, k_n)$ belongs to the image of the norm map $n_{L/F}: H^1(X_L, k_n) \to H^1(X, k_n)$. But $H^1(X_L, k_n) = 0$ since $X_L = \mathbb{A}_L^1 - pt$. □

COROLLARY 6.4. Any element in the kernel of the norm map
$$N: H^1(X, k_n) \to k_{n-1}F / n_{L/F}(k_{n-1}L)$$
can be represented by a k_{n-1}-cycle $\sum w_x \cdot x$, $w_x \in K_{n-1}F(x)$, with support in the set of points of degree 2, and such that $\sum n_x(w_x) = 0 \in k_{n-1}F$.

PROOF. By Lemma 6.2, we can find the representation we need, but $\sum n_x(w_x) = n_{L/F}(s)$ for some $s \in k_{n-1}L$. Choose a point $x_0 \in X$ such that $F(x_0) \simeq L$. By Lemma 6.3, we can replace w_x by
$$w'_x = \begin{cases} w_x & \text{if } x \neq x_0, \\ w_{x_0} + s & \text{if } x = x_0. \end{cases}$$
□

Let $D = \left(\frac{a,b}{F}\right)$ be the quaternion algebra. We have a canonical map [10]:
$$p: D^\times - F^\times \to \{\text{points of degree 2 in } Y\}$$

such that for any $w \in D^\times - F^\times$ there is a natural isomorphism of the (maximal) quadratic subfield $F[w] \subset D$ and the residue field of the point $p(w) \in Y$. Therefore, we have a map

$$D^\times - F^\times \to \{K_1\text{-cycles on } Y\}, \qquad w \mapsto w \cdot p(w).$$

We extend this map to $D^\times$ taking $c \in F^\times$ to $c \cdot \infty$. For any $w \in D^\times$, we simply denote the corresponding K_1-cycle on Y by $[w]$. It is proved in [10] that the composition $D^\times \to \{K_1\text{-cycles on } Y\} \to H^1(Y, K_2)$ is a group homomorphism. Using this property, Corollary 6.4 and also the fact that for any quadratic extension E/F the group $K_m E$ is generated by the symbols $\{x, a_1, \ldots, a_{m-1}\}$, where $x \in E^\times$, $a_i \in F^\times$, one proves the following

LEMMA 6.5. *Any element in the kernel of the map*

$$H^1(X, k_n) \to k_{n-1}F/n_{L/F}(k_{n-1}L)$$

can be represented by a k_{n-1}-cycle of the type $\sum_\alpha [u_\alpha] \cdot \underline{a}_\alpha$, where the sum is taken over all the maps $\alpha \colon [1, n-2] \to [1, m]$ for some $m > 0$, $u_\alpha \in D^\times$, $\underline{a}_\alpha = \{a_{\alpha(1)}, \ldots, a_{\alpha(n-2)}\} \in k_{n-2}F$, where the images of $a_1, \ldots, a_m \in F^\times$ are linearly independent in $F^\times/F^{\times 2}$ and

$$\sum_\alpha (\mathrm{Nrd}\, u_\alpha) \cdot \underline{a}_\alpha = 0 \in k_{n-1}F. \qquad \square$$

Let θ belong to the kernel of the map $H^1(X, k_n) \to k_{n-1}F/n_{L/F}(k_{n-1}L)$. We write $\theta = \sum_\alpha [u_\alpha] \cdot \underline{a}_\alpha$ as in Lemma 6.5. By Proposition 4.1, there exist $a_{m+1}, \ldots, a_r \in F^\times$, $c_{s,j}, d_{s,j} \in F$ such that

$$\mathrm{Nrd}(u_\alpha) = \prod_s \left(\prod_{j=1}^{n-2} (c_{s,j}^2 - a_s(j) \cdot d_{s,j}^2) \right).$$

Let $u_\alpha = s_\alpha + t_\alpha \cdot i + v_\alpha \cdot j + w_\alpha \cdot ij$, where $s_\alpha, t_\alpha, v_\alpha, w_\alpha \in F$ and i, j are generators of the quaternion algebra D such that $i^2 = a$, $j^2 = b$, $ij = -ji$. We have: $\mathrm{Nrd}\, u_\alpha = s_\alpha^2 - a t_\alpha^2 - b v_\alpha^2 + a b w_\alpha^2$.

Consider the indeterminates $A, B, P, Q, S_\alpha, T_\alpha, V_\alpha, W_\alpha$ for all $\alpha \colon [1, n-2] \to [1, r]$ and the affine variety V over F defined by the equations:

$$\begin{cases} S_\alpha^2 - A \cdot T_\alpha^2 - B \cdot V_\alpha^2 + AB \cdot W_\alpha^2 = \mathrm{Nrd}(u_\alpha), \\ P^2 - A \cdot Q^2 = B, \end{cases}$$

and also the variety W defined in the coordinates $A, B, S_\alpha, T_\alpha, V_\alpha, W_\alpha$ by the equations of the first type.

There is a natural dominating flat morphism $\pi \colon V \to W$. The point $x \in X$ with coordinates $A = a$, $B = b$, $S_\alpha = s_\alpha$, $T_\alpha = t_\alpha$, $V_\alpha = v_\alpha$, $W_\alpha = w_\alpha$ is nonsingular. The fiber Y_x is isomorphic to the affine conic X.

Let ξ be a generic point of W. Choose any specialization map

$$s \colon H^1(V_\xi, k_n) \to H^1(V_x, k_n) = H^1(X, k_n).$$

Let $\left(\begin{smallmatrix} A, B \\ F(W) \end{smallmatrix} \right)$ be the quaternion algebra over $F(W)$ with the anticommutative generators I and J such that $I^2 = A$, $J^2 = B$. Consider the element

$$\Theta = \sum_\alpha [S_\alpha + T_\alpha \cdot I + V_\alpha \cdot J + W_\alpha \cdot IJ] \cdot \underline{a}_\alpha \in H^1(V_\xi, k_n).$$

Since $N(\Theta) = \sum_\alpha (\mathrm{Nrd}\, u_\alpha) \cdot \underline{a}_\alpha = 0$, we have
$$\Theta \in \ker[H^1(V_\xi, k_n) \to k_{n-1}F(W)/n(k_{n-1}L(W))].$$
It is clear that $s(\Theta) = \theta$.

Consider the natural dominating projection $V \to \mathbb{A}^3$ with respect to the coordinates A, P, and Q. The generic fiber of this morphism is the product of several affine quadrics corresponding to the isotropic quadratic form $\langle 1, -A, -B, AB \rangle$ over $F(\mathbb{A}^3)$, and, therefore is a rational variety over $F(\mathbb{A}^3)$. Hence V is a rational variety over F.

LEMMA 6.6. *Assume that h_{n-1} is an isomorphism for all fields. Then for any purely transcendental extension E/F, the homomorphism $k_n E/k_n F \to H^n E/H^n F$ induced by the map $h_{n,E}$ is an isomorphism.*

PROOF. We may assume that $E = F(\mathbb{A}^1)$. In this case the homomorphism in question is isomorphic to
$$\coprod h_{n-1,F(x)}: \coprod_{x \in \mathbb{A}^1} k_{n-1}F(x) \to \coprod_{x \in \mathbb{A}^1} H^{n-1}F(x). \qquad \square$$

COROLLARY 6.7. *The natural homomorphism $H^1(V_\xi, k_n) \to H^1(V_\xi, H^n)$ is an isomorphism.*

PROOF. The field $F(\xi)(V_\xi) = F(V)$ is purely transcendental over F. $\qquad \square$

PROPOSITION 6.8. *If B_n is trivial and h_{n-1} is an isomorphism for all fields, then the norm map $N: H^1(X, k_n) \to k_{n-1}F/n_{L/F}(k_{n-1}L)$ is surjective.*

PROOF. We may clearly assume that X is not split. Let θ be in the kernel. It follows from Proposition 3.15, Corollary 6.7, and the commutative diagram:

$$\begin{array}{ccc} H^1(V_\xi, k_n) & \longrightarrow & H^1(V_\xi, H^n) \\ \downarrow N & & \downarrow N \\ k_{n-1}F(W)/n(k_{n-1}L(W)) & \xrightarrow{\sim} & H^{n-1}F(W)/n(H^{n-1}L(W)) \end{array}$$

that the left vertical map is injective, hence $\Theta = 0 \in H^1(V_\xi, k_n)$. Applying the specialization map s, we get: $\theta = s(\Theta) = 0$. $\qquad \square$

§7. Lifting property for a conic curve

Assume that X is an affine conic curve as in §3. Let us denote the groups $H^0(X, k_n)/k_n F$ and $H^0(X, H^n)/H^n F$ by $\overline{H}^0(X, k_n)$ and $\overline{H}^0(X, H^n)$ respectively.

LEMMA 7.1. *Let B_n be trivial and h_{n-1} be an isomorphism for all fields. Then the map*
$$\coprod_{E/F \text{ splits } X} \overline{H}^0(X_E, H^n) \xrightarrow{\sum n_{E/F}} \overline{H}^0(X, H^n)$$
is surjective.

PROOF. It follows from the commutative diagram:

$$\begin{array}{ccccccc} H^n F & \longrightarrow & H^n X & \longrightarrow & H^{n-1} F & \xrightarrow{(a,b)} & H^{n+1} F \\ & \searrow & \downarrow & & & & \\ & & H^n F(X) & & & & \\ & & \downarrow & & & & \\ & & \coprod\limits_{x \in X} H^{n-1} F(x) & & & & \end{array}$$

and Proposition 3.15 that there is a canonical surjection (see Remark 3.7)

$$H^1(Y, H^n) \twoheadrightarrow (\ker(H^{n-1} F \xrightarrow{(a,b)} H^{n+1} F)) \twoheadrightarrow \overline{H}^0(X, H^n).$$

The statement of the lemma follows from the commutative diagram:

$$\begin{array}{ccc} \coprod\limits_{E \text{ splits } X} H^1(Y_E, H^n) & \longrightarrow & \coprod\limits_{E \text{ splits } X} \overline{H}^0(X_E, H^n) \\ \downarrow & & \downarrow \\ H^1(Y, H^n) & \twoheadrightarrow & \overline{H}^0(X, H^n) \end{array}$$

and the surjectivity of the left vertical map. □

PROPOSITION 7.2. *Let B_n be trivial and h_{n-1} be an isomorphism for all fields. Then:*
(1) $\overline{H}^0(X, k_n) \to \overline{H}^0(X, H^n)$ *is surjective map,*
(2) $H^1(X, k_n) \to H^1(X, H^n)$ *is injective map.*

PROOF. (1) If X is split, then by Lemma 6.6 we have $k_n F(X) / k_n F \simeq H^n F(X) / H^n F$, hence the first map is surjective. In general, statement (1) follows from Lemma 7.1 and the commutative diagram:

$$\begin{array}{ccc} \coprod\limits_{E \text{ splits } X} \overline{H}^0(X_E, k_n) & \xrightarrow{\sim} & \coprod\limits_{E \text{ splits } X} \overline{H}^0(X_E, H^n) \\ \downarrow & & \downarrow \\ \overline{H}^0(X, k_n) & \longrightarrow & \overline{H}^0(X, H^n) \end{array}$$

(2) This follows from the commutative diagram:

$$\begin{array}{ccc} H^1(X, k_n) & \longrightarrow & H^1(X, H^n) \\ \uparrow & & \downarrow \\ k_{n-1} F / n_{L/F}(k_{n-1} L) & \xrightarrow{\sim} & H^{n-1} F / n_{L/F}(H^{n-1} L) \end{array}$$

and Proposition 6.8. □

COROLLARY 7.3. *Under the conditions of Proposition 7.2, assume that $h_{n,F}$ is surjective. Then $h_{n,F(X)}$ is also surjective.*

PROOF. This follows from the commutative diagram:

$$\begin{array}{ccccccccc}
k_nF & \longrightarrow & k_nF(X) & \longrightarrow & \coprod_{x\in X} k_{n-1}F(x) & \longrightarrow & H^1(X,k_n) & \longrightarrow & 0 \\
\downarrow h_{n,F} & & \downarrow h_{n,F(X)} & & \downarrow ? & & \downarrow & & \\
H^nH & \longrightarrow & H^nF(X) & \longrightarrow & \coprod_{x\in X} H^{n-1}F(x) & \longrightarrow & H^1(X,H^n) & \longrightarrow & 0. \quad \square
\end{array}$$

§8. Proof of the theorem

We follow the approach given in [7] in the case $n = 2$. Consider any quadratic extension $L = F(\sqrt{a})/F$.

PROPOSITION 8.1. *If B_n is trivial and $h_{n-1,F}$ is an isomorphism for all fields, then the sequence*

$$k_nF \to k_nL \xrightarrow{n_{L/F}} k_nF$$

is exact.

PROOF. Let $u \in k_nL$ and $n_{L/F}(u) = 0 \in k_nF$. We can write u in the form $u = \sum_\alpha x_\alpha \cdot \underline{a}_\alpha$. Here the sum is taken over all maps $\alpha: [1, n-1] \to [1, m]$ and we have

$$x_\alpha \in L^\times, \quad \underline{a}_\alpha = \{a_{\alpha(1)}, \ldots, a_{\alpha(n-1)}\} \in k_{n-1}F, \quad a_1, \ldots, a_m \in F^\times,$$

where the elements $a_1, \ldots, a_m$ are independent in $F^\times/F^{\times 2}$.

Since $0 = n_{L/F}(u) = \sum_\alpha n_{L/F}(x_\alpha) \cdot \underline{a}_\alpha \in k_nF$, it follows from Proposition 4.1 that there exists $a_{m+1}, \ldots, a_r \in F^\times$, $c_{s,j}, d_{s,j} \in F^\times$ such that

$$n_{L/F}(x_\alpha) = \prod_s \left(\prod_{j=1}^{n-1} (c_{s,j}^2 - a_{s(j)} d_{s,j}^2) \right)$$

for all $\alpha: [1, n-1] \to [1, r]$.

Consider the indeterminates A_i ($i = 1, \ldots, r$), X'_α, X''_α ($\alpha: [1, n-1] \to [1, r]$), $C_{s,j}$, $D_{s,j}$. Denote $\prod_{i \in T} A_i$ by A_T for $T \subset [1, r]$. We consider the affine variety X defined over the field $F_1 = F_0(a)$, where F_0 is the prime subfield in F, by the equations:

$$X'^2_\alpha - aX''^2_\alpha = \prod_s \left(\prod_{j=1}^{n-1} (C_{s,j}^2 - A_{s(j)} D_{s,j}^2) \right)$$

for all $\alpha: [1, n-1] \to [1, r]$.

Let $L_1 = F_1(\sqrt{a})$; consider the element

$$U = \sum_\alpha (X'_\alpha + X''_\alpha \sqrt{a}) \cdot \underline{A}_\alpha \in k_nL_1(X),$$

where $\underline{A}_\alpha = \{A_{\alpha(1)}, \ldots, A_{\alpha(n-1)}\} \in k_{n-1}F_1(X)$. By Proposition 4.1, we have $n_{L_1(X)/F_1(X)}(U) = 0$.

Let E be a purely transcendental extension $F_1(A_i, C_{s,j}, D_{s,j})$ of F_1. The field F_1 is either finite, or global, or isomorphic to $\mathbb{Q}(t)$. In all the cases h_{n,F_1} is an isomorphism, hence $h_{n,E}$ is an isomorphism.

The function field $F_1(X)$ is isomorphic over E to the function field of the product of several conics, therefore, by Corollary 7.3, $h_{n,F_1(X)}$ is surjective.

The conics mentioned above are split by the field $E(\sqrt{a})$. Hence $L_1(X)$ is a purely transcendental extension of L_1 and since h_{n,L_1} is an isomorphism, $h_{n,L_1(X)}$ is also an isomorphism. It follows from the commutative diagram

$$\begin{array}{ccccc} k_nF_1(X) & \xrightarrow{i} & k_nL_1(X) & \xrightarrow{n_{L/F}} & k_nF_1(X) \\ \downarrow & & \downarrow \wr & & \downarrow \\ H^nF_1(X) & \longrightarrow & H^nL_1(X) & \longrightarrow & H^nF_1(X) \end{array}$$

that $U \in \operatorname{im}(i)$.

There is a nonsingular point $x \in X$ and specialization maps s and s' fitting to the commutative diagram

$$\begin{array}{ccc} k_nF_1(X) & \xrightarrow{i} & k_nK_1(X) \\ s'\downarrow & & \downarrow s \\ k_nF & \xrightarrow{j} & k_nL \end{array}$$

such that $s(U) = u$, hence $u = s(U) \in \operatorname{im} j$. □

COROLLARY 8.2. *If α_F is injective, then α_L is also injective.*

PROOF. This follows from the commutative diagram:

$$\begin{array}{ccccccc} k_{n-1}F & \xrightarrow{\{a\}} & k_nF & \longrightarrow & k_nL & \xrightarrow{n_{L/F}} & k_nF \\ \downarrow \wr & & \downarrow & & \downarrow & & \uparrow \\ H^{n-1}F & \xrightarrow{(a)} & H^nF & \longrightarrow & H^nL & \xrightarrow{n_{L/F}} & H^nF. \end{array}$$ □

PROPOSITION 8.3. *Assume that B_n is trivial and h_{n-1} is an isomorphism for all fields. Then the sequence*

$$k_{n-1}F \xrightarrow{\{a\}} k_nF \to k_nL$$

is exact.

PROOF. Let $u \in k_nF$ and $u_L = 0$. We can write u in the form $u = \sum_\alpha b_\alpha \cdot \underline{a}_\alpha$ where the sum is taken over all maps $\alpha: [1, n-1] \to [1, m]$, $b_\alpha \in F^\times$, $\underline{a}_\alpha = \{a_{\alpha(1)}, \dots, a_{\alpha(n-1)}\} \in k_{n-1}F$, $a_1, \dots, a_m \in F^\times$ and we may assume that a_i are linearly independent in $L^\times/L^{\times 2}$. By Proposition 4.1, there exists $a_{m+1}, \dots, a_r \in L^\times$, $c_{s,j}, d_{s,j} \in L$ such that

$$b_\alpha = \prod_s \left(\prod_{j=1}^{n-1} (c_{s,j}^2 - a_{s(j)} d_{s,j}^2) \right)$$

for all maps $\alpha: [1, n-1] \to [1, r]$.

Consider the indeterminates

$$A_i \ (i = 1, \dots, m),\ A'_i, A''_i \ (i = m+1, \dots, r),$$
$$B_\alpha(\alpha: [1, n-1] \to [1, r]),\ C'_{s,j}, C''_{s,j}, D'_{s,j}, D''_{s,j}.$$

Put $A_i = A'_i + A''_i \cdot \sqrt{a}$ for $i = m+1,\ldots,r$ and $A_T = \prod_{i \in T} A_i$ for any $T \subset [1,r]$.
We consider the affine variety Y over the field $F_1 = F_0(a)$ given by the equations:

$$B_\alpha = \prod_s \left[\prod_{j=1}^{n-1} ((C'_{s,j} + C''_{s,j} \sqrt{a})^2 - A_{s(j)} \cdot (D'_{s,j} + D''_{s,j} \sqrt{a})^2) \right].$$

Each equation is considered as a system of two equations over F_1.

Consider the element $U = \sum_\alpha B_\alpha \cdot \underline{A}_\alpha \in k_n F_1(Y)$, where $\underline{A}_\alpha = \{A_{\alpha(1)},\ldots, A_{\alpha(n-1)}\} \in k_{n-1} F_1(X)$. By Proposition 4.1,

$$U \in \ker(k_n F_1(X) \to k_n L_1(X)),$$

where $L_1 = F_1(\sqrt{a})$.

There exists a nonsingular F-point x in X and a specialization map $s\colon k_n F_1(Y) \to k_n F$ such that $s(U) = u$.

Consider the purely transcendental extension E of F obtained by joining all indeterminates except $C'_{s,1}$, $C''_{s,1}$, $D'_{s,1}$, $D''_{s,1}$, where $s(k)$ consists of one element for all $k = 1,\ldots,n-1$. The function field $F_1(Y)$ is isomorphic over E to the function field of the product of varieties of the type $Z = R_{E(\sqrt{a})/E}(Q)$, where Q is an affine conic curve over $E(\sqrt{a})$ and R is the Weil transfer [16].

LEMMA 8.4. *There are quadratic extensions $E(Z) \supset E_1 \supset E_2$ such that E_2 is a purely transcendental extension of E.*

PROOF. Let $Q \to \mathbb{A}^1_{E(\sqrt{a})}$ be any double cover. Then it is easy to see that the extension $E(Z)$ of degree 4 over $R_{E(\sqrt{a})/E}(\mathbb{A}^1_{E(\sqrt{a})})$ contains an intermediate subfield. □

As in the proof of the Proposition 8.1, $h_{n,E}$ is an isomorphism. By Corollary 8.2 and Lemma 8.4, $\alpha_{F_1(Y)}$ is injective. It follows from the commutative diagram

$$\begin{array}{ccccc}
k_{n-1} F_1(Y) & \xrightarrow{\{a\}} & k_n F_1(Y) & \longrightarrow & k_n L_1(Y) \\
\downarrow{\wr} & & \uparrow & & \downarrow \\
H^{n-1} F_1(Y) & \xrightarrow{(a)} & H^n F_1(Y) & \longrightarrow & H^n L_1(Y)
\end{array}$$

that the top row is exact, hence U is divisible by $\{a\}$. Applying the specialization map s in the commutative diagram

$$\begin{array}{ccc}
k_{n-1} F_1(Y) & \xrightarrow{\{a\}} & k_n F_1(Y) \\
\downarrow{s'} & & \downarrow{s} \\
k_{n-1} F & \xrightarrow{\{a\}} & k_n F
\end{array}$$

we deduce that u is in the image of the bottom map. □

PROOF OF THE THEOREM. Using Lemma 2.3, by induction on n we may assume that h_{n-1} is an isomorphism for all fields.

Injectivity of $h_{n,F}$. Let $u \in k_n F$ and $h_n(u) = 0 \in H^n F$. Assume that u is the sum of k symbols in $k_n F$. We prove by induction on k that $u = 0$. One can find a quadratic extension $L = F(\sqrt{a})/F$ such that u_L is the sum of $k - 1$ symbols, hence by inductive hypothesis $u_L = 0$. It follows from the commutative diagram:

$$\begin{array}{ccccccc} k_{n-1}L & \xrightarrow{n_{L/F}} & k_{n-1}F & \xrightarrow{\{a\}} & k_n F & \longrightarrow & k_n L \\ \downarrow \imath & & \downarrow \imath & & \downarrow h_n & & \downarrow \\ H^{n-1} & \xrightarrow{n_{L/F}} & H^{n-1} F & \xrightarrow{(a)} & H^n F & \longrightarrow & H^n L \end{array}$$

and Proposition 8.3 that $u = 0$.

Surjectivity of $h_{n,F}$. Assume first that F has no odd degree extensions. Let $u \in H^n F$ and E be a finite extension of F such that $u_E = 0$. We prove by induction on the degree $[E : F]$ that $u \in \mathrm{im}\, h_{n,F}$. By the assumption, one can find an intermediate field L, $F \subset L \subset E$ such that $[L : F] = 2$. By the inductive hypothesis $u_L \in \mathrm{im}\, h_{n,L}$. The inclusion $u \in \mathrm{im}\, h_{n,F}$ follows from the injectivity of h_n (which is already proved), Proposition 8.1 and the diagram in the proof of Corollary 8.2.

In general, using the first case, one can find an odd degree extension E/F such that $u_E \in \mathrm{im}\, h_{n,E}$. Then $u = n_{E/F}(u_E) \in \mathrm{im}\, h_{n,F}$ since h_n commutes with norms. $\square$

COROLLARY. *The surjectivity of h_n for all fields implies the bijectivity of h_n also for all fields.*

PROOF. This follows from the theorem and Lemma 2.1. $\square$

References

1. J. K. Arason, *Kohomologische Invarianten quadratischer Formen*, J. Algebra **36** (1975), 448–491.
2. H. Bass and J. Tate, *The Milnor ring of a global field*, Lecture Notes in Math., vol. 342, Springer-Verlag, Berlin, 1973, pp. 349–446.
3. S. Bloch and A. Ogus, *Gersten's conjecture and the homology of schemes*, Ann. Sci. École Norm. Sup. (4) **7** (1974), 181–202.
4. R. Elman, *On Arason's theory of Galois cohomology*, Comm. Algebra **10** (**13**) (1982), 1449–1474.
5. K. Kato, *A generalization of local class field theory by using K-groups*, J. Fac. Sci. Univ. Tokyo Sect. IA Math. **26** (1079), no. 2, 303–376; **27** (1980), no. 3, 603–683.
6. _____, *Milnor K-theory and the Chow group of zero cycles*, Contemp. Math **55** (1986), 241–254.
7. A. S. Merkurjev, *On the norm residue homomorphism of degree* 2, Dokl. Akad. Nauk SSSR **261** (1981), 542–547; English transl. in Soviet Math. Dokl. **24** (1981).
8. A. S. Merkurjev and A. A. Suslin, *K-cohomology of Severi–Brauer varieties and norm residue homomorphism*, Izv. Akad. Nauk SSSR Ser. Mat. **46** (1982), no. 5, 1011–1061; English transl. in Math. USSR-Izv. **21** (1983).
9. _____, *The norm residue homomorphism of degree* 3, Izv. Akad. Nauk SSSR Ser. Mat. **54** (1990), no. 2, 339–356; English transl. in Math. USSR-Izv. **36** (1991).
10. _____, *The group of K_1-zero-cycles on Severi–Brauer varieties*, Nova J. Algebra Geometry **1** (1992), no. 3, 297–315.
11. A. S. Merkurjev and J. P. Tignol, *Galois cohomology of biquadratic extensions*, Comment. Math. Helv. **68** (1993), 138–169.
12. J. S. Milne, *Étale cohomology*, Princeton Math. Series, vol. 33, Princeton, NJ, 1980.
13. J. Milnor, *Algebraic K-theory and quadratic forms*, Invent. Math. **9** (1970), 318–344.
14. M. Rost, *Hilbert 90 for K_3 for degree-two extensions*, Preprint Regensburg (1986).
15. J.-J. Sansuc, *Group de Brauer et arithmétique des groupes algébriques linéaires sur un corps de nombres*, J. Reine Angew. Math. **327** (1981), 12–80.

16. J.-P. Serre, *Cohomologie Galoisienne, 5-éme édition*, Lecture Notes in Math., vol. 5, Springer-Verlag, Heidelberg, 1994.
17. A. A. Suslin, *Algebraic K-theory and the norm-residue homomorphism*, Itogi Nauki i Tekhniki. Sovremennye Problemy Mathematiki, vol. 25, VINITI, Moscow, 1984, pp. 115–208; English transl. in J. Soviet Math. **30** (1985), no. 2.
18. J. Tate, *Relation between K_2 and Galois cohomology*, Invent. Math. **36** (1976), 257–274.

Translated by THE AUTHOR

St. Petersburg State University, Russia Math. @ Mech. Dept., Bibliotechnaya pl. 2, Stary Petergof, St. Petersburg, 198904, Russia

Concerning Hall's Theorem

A. G. Moshonkin

Dedicated to my mathematics teacher Boris Bekker

§1. Statement of Hall's theorem

We begin with the statement of a finite version of the theorem concerning systems of distinct Hall representatives.

Let A be a finite set, and let $\{A_i\}$, $i = 1, \ldots, n$ be a family of subsets of A. We say that $\{x_i\}$ is a *system of distinct representatives* (SDR) for $\{A_i\}$ if $x_i \in A_i$ for all $i = 1, \ldots, n$ and $x_i \neq x_j$ for all $i \neq j$.

We shall study the existence of SDR for a given family $\{A_i\}$. The following statement provides a trivial necessary condition.

CONDITION A. *For each natural number k, the union of any k distinct sets A_i contains at least k elements.*

In [H] P. Hall proved that Condition A is also sufficient for the existence of an SDR. Thus the following statement is valid.

THEOREM 1. *For a family $\{A_i\}$ of subsets of a finite set A, an SDR exists if and only if Condition A holds for the $\{A_i\}$.*

It is not easy to check Condition A explicitly in particular cases. In the sequel, we present an algorithm of polynomial complexity (with respect to n) which, for any family of n sets either produces a SDR or indicates a counterexample to Condition A.

It is more or less obvious that the theorem remains valid for the case in which the number of sets or some of the sets is infinite. However, in the present paper, we do not discuss these infinite versions of the theorem.

§2. An analog of Hall's theorem for vector spaces

Let L be a finite-dimensional vector space over a field l. Let $\{L_i\}$, $i = 1, \ldots, n$, be a family of subspaces of L. We say that $\{l_i\}$ is a *system of linearly independent representatives* $(SLIR)$ for all $\{L_i\}$ if $l_i \in L_i$ for all $i = 1, \ldots, n$, and the vectors l_i are linearly independent.

1991 *Mathematics Subject Classification.* Primary 05A18.

©1996, American Mathematical Society

As above, we have the following trivial necessary condition for the existence of a SLIR for $\{L_i\}$.

CONDITION B. *For each natural number k, the linear span of any k distinct subspaces L_i has dimension at least k.*

Now we state and prove a theorem similar to Theorem 1.

THEOREM 2. *Let L be a finite-dimensional vector space over a field l. For a family $\{L_i\}$ of subspaces of L a SLIR exists if and only if Condition B holds for $\{L_i\}$.*

PROOF. (In the case in which l is the field of real numbers, the proof of this theorem was communicated to the author by S. Traile.)

We need only prove the sufficiency of Condition B. Let Condition B be fulfilled.

Case 1. The field l is infinite. (As will be seen from what follows, in this case the vectors $x_i \in L_i$, $i = 1, \ldots, n$, "selected randomly" "almost surely" form a SLIR).

For each $i = 1, \ldots, n$, we choose an arbitrary $x_i \in L_i$. If the system $\{x_i\}$ is linearly independent, we consider an arbitrary minimal (with respect to inclusion) linearly dependent subsystem of the system $\{x_i\}$. We may assume that this subsystem is $x_1 \in L_1, \ldots, x_m \in L_m$. Since $x_1, \ldots, x_m$ is minimal, the linear span $\langle x_1, \ldots, x_m \rangle$ has dimension $m - 1$. Owing to Condition B, this span does not contain at least one of the spaces $L_1, \ldots, L_m$. We may assume that $L_1 \not\subset \langle x_1, \ldots, x_m \rangle$. Consider a vector $y_1 \subset L_1$ such that $y \neq \langle x_1, \ldots, x_m \rangle$, and therefore, $\alpha x_1 + y_1 \notin \langle x_1, \ldots, x_m \rangle$ for all $\alpha \in l$. This means that $\alpha x_1 + y_1, x_2, \ldots, x_m$ are linearly independent. Now, in the system $x_1, \ldots, x_n$, we replace $x_1 \in L$ by $\alpha x_1 + y_1 \in L$, where α is an arbitrary parameter. The new system of representatives for $\{L_i\}$ has at least one subsystem which remains linearly independent after this change. Of course, it may happen that some other linearly independent subsystem becomes linearly dependent after this change. However, for a specific subsystem, this may happen for a single value of α. Since the field l is infinite, and the number of subsystems of $x_1, \ldots, x_m$ is finite, we can choose an α such that, for the new system of the representatives $\alpha x_1 + y_1, x_2, \ldots, x_m$, the number of linearly independent subsystems is strictly less than that of $x_1, \ldots, x_n$. After a finite number of such steps, we come to a SLIR.

Case 2. The field l is finite. In this case we need the following statement.

BASIC LEMMA. *Let L_i be a family of subspaces of L. In each L_i, we fix a basis. If $\{L_i\}$ has a SLIR, then it has a SLIR such that each L_i is represented by an element of its basis.*

PROOF. Let $x_1 \in L_1, \ldots, x_n \in L_n$ be a SLIR. Then the vectors $x_2, \ldots, x_n$ are linearly independent, and $x_1 \notin \langle x_2, \ldots, x_n \rangle$. Hence there is an element of the basis of L_1 which does not belong to $\langle x_2, \ldots, x_n \rangle$, and, therefore, x_1 can be replaced by this element. The same can be done with the other x_i. □

Now it is clear how to prove Theorem 2 by means of the basic lemma in the case of finite field l. We fix a basis in each L_i. Next, we extend the field l to an infinite field l'. In this case, we have the standard construction of the canonical embedding of sets $\phi: L \to L'$, where L' is a vector space over l' and $\dim(L') = L$. We set $L'_i = \langle \phi(L_i) \rangle$, $i = 1, \ldots, n$. From Condition B applied to $\{L_i\}$, it follows that Condition B is valid for the family $\{L'_i\}$ of subspaces of L'. Thus, there is a SLIR for $\{L'_i\}$. For each i, we can lift the basis of L_i to a basis of L'_i. By the Basic Lemma, each L'_i can be represented, in some *SLIR*, by an element of this basis. This *SLIR* can be restricted to one for $\{L_i\}$. □

§3. Deduction of Theorem 1 from Theorem 2

Theorem 1 can be deduced from Theorem 2. Indeed, let us regard the set A in the statement of Theorem 1 as the basis of a vector space over a field. To each subset A_i of A we assign the span of A_i, and A_i is the canonical basis of this span. Condition A for sets implies Condition B for the corresponding subspaces. Finally, the Basic Lemma implies the existence of a *SLIR* that consists of elements of the corresponding bases. This system is the desired *SDR*. □

§4. Concerning a certain class of matrices

Let A be an $n \times n$ matrix. We shall consider operations that consist in replacing certain matrix elements by zeros. It is natural to interpret each such operation as an $n \times n$ matrix F whose entries are zeros and ones, and the replacement operation itself can be written as $A \times F$, when $\times$ denotes the element by element multiplication, i.e., we obtain the matrix B for which $b_{ij} = a_{ij} \cdot f_{ij}$. Square matrices whose entries are zeros and ones will be called *replacements*. A replacement F will be called *admissible* if the permanent of F is nonzero, in other words, if there exists a permutation σ on n elements such that all elements $f_{i,\sigma(i)}$ are equal to 1. We shall say that the matrix A *survives* the replacement F if the matrix $A \times F$ is nondegenerate. It is clear that any matrix can survive only after an admissible replacement. Let us call a matrix *strongly nondegenerate* if it survives all admissible replacements.

For any n, one can indicate an $n \times n$ strongly nondegenerate matrix whose entries are natural numbers. An important problem, whose solution is not known to the author, is to find a polynomial time algorithm that would produce, given n, a strongly nondegenerate $n \times N$ matrix whose entries are natural numbers increasing no faster than the exponential in degrees of a polynomial in n.

If such an algorithm existed, it could be used to construct a sequence of strongly nondegenerate matrices $A_1, A_2, \ldots$ such that the $(i \times i)$ matrix A_i consists of entries that are natural numbers increasing no faster than an exponential function in the degrees of a polynomial in i. Let us call such sequences *slowly increasing sequences of strongly nondegenerate matrices* (SISSNM). The absence of any SISSNM would mean that the algorithm described above does not exist.

In this section we prove the existence of SISSNM's.

THEOREM. *There exist slowly increasing sequences of strongly nondegenerate matrices.*

For the proof, we introduce certain notions and give some definitions. Let V be an n-dimensional vector space with fixed basis. The linear span of any subset of the set of basis elements will be called a *coordinate subspace* in V. The coordinate subspaces are in one-to-one correspondence with all the subsets of the set of basis vectors or, equivalently, with all vectors of length n with coordinates zero or one.

Now with each replacement $F(n \times n)$ we can associate a family of subspaces $V_1, \ldots, V_n$ corresponding to the column vectors of the matrix F. It is easy to see that the admissibility of the replacement F means precisely that we have Condition B for the subspaces $V_1, \ldots, V_n$ corresponding to this replacement. If the matrix A survives the replacement F, then, as pointed out above, this means that the matrix $B = A \times F$

is nondegenerate. It is easy to see how the matrix $B = A \times F$ is constituted in terms of the subspaces $V_1, \ldots, V_n$ corresponding to the replacement F. Each row vector b_i of the matrix B is the projection of the corresponding row vector a_i of the matrix A on the coordinate subspace V_i assigned to the appropriate row vector f_i of the replacement F. The strong nondegeneracy of the matrix A means that for any family of subspaces $V_1, \ldots, V_n$ satisfying Condition B, the projections of the row vectors of the matrix A on the corresponding subspaces of this family are linearly independent.

Now consider the family of subspaces $E_1, \ldots, E_n$ of the space E and the family of vectors $e_1 \in E_1, \ldots, e_k \in E_k$, $k \leq n$. We shall say that this family of vectors is *extensible* if there exists a *SLIR* for the subspaces $E_1, \ldots, E_n$ in which the subspaces $E_1, \ldots, E_k$ are represented by the vectors $e_1, \ldots, e_k$ respectively. It is easy to see that the vectors $e_{k+1} \in E_{k+1}$ for which the sequence $e_1 \in E_1, \ldots, e_k \in E_{k+1}$ is not extensible form a (proper) subset of E_{k+1}, hence, if we want to extend the sequence of vectors $e_1 \in E_1, \ldots, e_k \in E_k$ to an extensible sequence $e_1 \in E_1, \ldots, e_k \in E_k, e_{k+1} \in E_{k+1}$, any vector in E_{k+1} will do, as long as it does not belong to a certain proper subspace.

Finally we introduce one last notion needed for the proof of the theorem. Consider a set of vectors from the n-dimensional vector space E. We shall say that the vectors of this set are in *general position* if any n of them are linearly independent. Clearly, if a set of vectors is in general position, then so are any of its subsets.

PROOF OF THE THEOREM. Consider the sequence $e(\lambda)$, $\lambda = 1, 2, \ldots$ of vectors from the space E with coordinates $(1, \lambda, \lambda^2, \ldots, \lambda^{n-1})$. It follows from the invariance of the Vandermonde matrix that the vectors of this sequence are in general position. Moreover, the projection of the vectors of this sequence on any coordinate subspace will also be in general position with respect to this subspace. Let us present a brief proof of this fact, which, as can easily be seen is equivalent to the nondegeneracy of the following square matrix C,

$$C = \begin{pmatrix} \lambda_1^{i_1} & \lambda_1^{i_2} & \ldots & \lambda_1^{i_k} \\ \lambda_2^{i_1} & \lambda_2^{i_2} & \ldots & \lambda_2^{i_k} \\ \vdots & \vdots & \ddots & \vdots \\ \lambda_k^{i_1} & \lambda_k^{i_1} & \ldots & \lambda_k^{i_1} \end{pmatrix},$$

where all the numbers λ_i are positive and pairwise distinct, while the i_s are nonnegative pairwise distinct integers. If the matrix C were degenerate, this would mean that $C \cdot b = 0$ for some vector $b = (b_1, \ldots, b_k)$, i.e., the polynomial $b_1 x^{i_1} + \cdots + b_k x^{i_k}$ would have k real positive distinct roots, but a polynomial with only k nonzero coefficients cannot have more than $k - 1$ real positive roots. Indeed, if it had k distinct positive real roots, we could divide by a smaller degree of the variable, take the derivative, and thus obtain a polynomial with $k - 1$ nonzero coefficients having $k - 1$ distinct positive real roots. This ensures the induction step. The base of the induction is trivial.

Now let us choose a positive integer n and number all the admissible $n \times n$ replacements $F_1, F_2, \ldots$. Their total number is certainly less than 2^{n^2}.

Let us show that there exists a strongly nondegenerate $n \times n$ matrix A constituted by row vectors $e(\lambda)$ for which the value of the parameter λ is less than $(n-1)2^{n^2} + 1$.

Consider the first replacement F_1. Suppose the corresponding family of coordinate subspaces is $E_1^1, \ldots, E_n^1$; as noted above, they possess property C_1. Hence all the vectors e_1 from E_1, except those in a subspace $\widehat{E}_1^1$, constitute a sequence of length 1 which is extensible for the family $E_1, \ldots, E_n$. Since the projections of the vectors $e(\lambda)$

on E_1 are in general position, it follows that no more than $n-1$ of them can be in the subspace $\widehat{E}_1$, so that among the parameters λ there are no more than $n-1$ for which the projection of the vector $e(\lambda)$ on E_1^1 constitutes a nonextensible sequence of length 1 for the family $E_1^1, \ldots, E_n^1$. Now consider the second replacement F_2 and let $E_1^2, \ldots, E_n^2$ be the corresponding family of coordinate subspaces. As before, there exists no more than $n-1$ values of the parameter λ for which the projection of the vector $e(\lambda)$ on the subspace E_1^2 constitutes a nonextensible sequence of length 1 for this family. According to the Dirichlet principle, among the vectors $e(\lambda)$, $1 \leqslant \lambda \leqslant (n-1)2^{n^2}+1$, there is a vector $e(\lambda_1)$ such that for any admissible replacement F_i whose family of corresponding subspaces is $E_1^i, \ldots, E_n^i$, the projection of this vector on the space E_1^i constitutes an extensible sequence of length 1 for the family $E_1^i, \ldots, E_n^i$. This vector is good for the first row of the matrix A.

Now exactly in the same way we can prove the existence of a vector $e(\lambda_2)$, $1 \leqslant \lambda_2 \leqslant (n-1)2^{n^2}+1$, which is good for the second row of the matrix A. Indeed, for any admissible replacement F_i for which the corresponding family of coordinate subspaces is $E_1^i, \ldots, E_n^i$, the projection of $e(\lambda_1)$ on E_1^i constitutes an extensible sequence for this family, and, as we indicated above, in order to extend this sequence by one term we may use any vector from E_2^i except for a proper subspace of E_2^i, e.g., the projections of the vectors $e(\lambda)$ on the subspace E_2^i for all values of λ, except $n-1$ values or less of this parameter. Since there are less than 2^{n^2} admissible replacements, the vector $e(\lambda)$ for some value $1 \leqslant \lambda \leqslant (n-1)2^{n^2}+1$ is good for the second row of the matrix A. Repeating these arguments, we finally establish that there exists a strongly nondegenerate $n \times n$ matrix A whose rows are vectors $e(\lambda)$ with certain nonnegative integer values of λ not greater than $(n-1)2^{n^2}+1$. The theorem is proved.

References

[H] P. Hall, *On representatives of subsets*, J. London Math. Soc. **10** (1935), 26–30.

Translated by A. SOSSINSKY

Simplicial Determinant Maps and the Second Term of Weight Filtrations

A. Yu. Nenashev

To the school that I remember with love

Introduction

The notion of determinant occurs twice in the algebraic K-theory of a scheme X: first, we have the map $\det: K_0 X \to \operatorname{Pic} X$ that takes a vector bundle on X to its highest exterior power (we assume X is irreducible); second, we have the map $\det: K_1 X \to \Gamma(X, \mathcal{O}_X^*)$ induced by the usual determinant map $\operatorname{GL}(R) \to R^*$ in the affine case $X = \operatorname{Spec}(R)$.

Let W^1 be the union of components of rank zero in the G-construction of Gillet and Grayson [GG] associated with the category $\mathcal{P}_X$ of vector bundles on X. We can regard W^1 as the first term of the weight filtration, since

$$\pi_0 W^1 \cong F_\gamma^1 K_0 X = \ker(\operatorname{rank}: K_0 X \to \mathbb{Z}),$$
$$\pi_m W^1 \cong F_\gamma^1 K_m X \cong K_m X \quad \text{for } m \geq 1.$$

In the present paper, we define a simplicial set T such that

$$\pi_0 T \cong \operatorname{Pic} X, \quad \pi_1 T \cong \Gamma(X, \mathcal{O}_X^*), \quad \pi_m T = 0 \text{ for } m \geq 2,$$

and a simplicial map $\det: W^1 \to T$ that yields the above two determinant maps on the homotopy groups:

$$F_\gamma^1 K_0 X = \ker(\operatorname{rank}: K_0 X \to \mathbb{Z}) \cong \pi_0 W^1 \xrightarrow{\det} \pi_0 T \cong \operatorname{Pic} X,$$
$$K_1 X \cong \pi_1 W^1 \xrightarrow{\det} \pi_1 T \cong \Gamma(X, \mathcal{O}_X^*).$$

We also describe the homotopy fiber of the map $\det: W^1 \to T$ as a simplicial set W^2. A vertex in W^2 is a triple $(P, P'; \psi)$, where P and P' are vector bundles on X such that $\operatorname{rank} P = \operatorname{rank} P'$ and $\psi: \det P \xrightarrow{\sim} \det P'$ is an isomorphism. An

---

1991 *Mathematics Subject Classification.* Primary 14C35, 19D06, 19E08.

©1996, American Mathematical Society

edge in W^2 connecting $(P_0, P_0'; \psi_0)$ to $(P_1, P_1'; \psi_1)$ is a pair of short exact sequences $(P_0 \to P_1 \to P_{1/0}; P_0' \to P_1' \to P_{1/0})$ such that the diagram

$$\begin{array}{ccc} \det P_1 & \xrightarrow{\sim} & \det P_0 \otimes \det P_{1/0} \\ \psi_1 \downarrow \wr & & \wr \downarrow \psi_0 \otimes 1 \\ \det P_1' & \xrightarrow{\sim} & \det P_0' \otimes \det P_{1/0} \end{array}$$

commutes; here the horizontal isomorphisms are naturally induced by these short exact sequences. Higher-dimensional simplices in W^2 are defined in a similar way.

The long exact sequence associated with $W^2 \to W^1 \to T$ yields

$$\pi_0 W^2 \cong \ker((\text{rank}, \det) \colon K_0 X \to \mathbb{Z} \oplus \text{Pic } X),$$
$$\pi_1 W^2 \cong \ker(\det \colon K_1 X \to \Gamma(X, \mathcal{O}_X^*)),$$
$$\pi_m W^2 \cong K_m X \quad \text{for } m \geq 2.$$

Thus W^2 provides the groups $SK_m X$ as homotopy groups for all $m \geq 0$. The SK-groups can be defined for $m \geq 1$ as the homotopy groups of $BSL^+(R)$ in the affine case $X = \text{Spec}(R)$ and by means of the generalized cohomology of the sheafification of BSL^+ in the general case (cf. [**Sou**, p. 524]).

In a future paper, we hope to define λ- and γ-operations on W^2 as simplicial maps and prove on the simplicial level that the map $\gamma^1 + \gamma^2 + \cdots$ is contractible. This would directly imply that $F_\gamma^2 K_m X = SK_m X$ for each $m \geq 0$.

§1. Definitions

Let X be an irreducible scheme.

We denote by $\mathcal{P} = \mathcal{P}_X$ the category of vector bundles on X. Suppose we are given a choice of the tensor product $P_1 \otimes \cdots \otimes P_k$ for each collection of objects $(P_1, \ldots, P_k)$ in $\mathcal{P}$ and a choice of the exterior product $P_1 \wedge \cdots \wedge P_k$ for each admissible filtration $P_1 \rightarrowtail \cdots \rightarrowtail P_k$ (by definition, the latter is isomorphic to the image of $P_1 \otimes P_2 \otimes \cdots \otimes P_k$ in $\bigwedge^k P_k$). These operations satisfy the usual functoriality and compatibility conditions (cf. [**Gr**, §7]).

Let $\mathcal{L} = \mathcal{L}_X$ be the category of linear bundles on X and their isomorphisms. We set $\det P = \bigwedge^{\text{rank } P} P$ for every P in $\mathcal{P}$, where $\bigwedge^k P$ now stands precisely for the exterior product $P \wedge \cdots \wedge P$ associated with $P \xrightarrow{1} \cdots \xrightarrow{1} P$ (k copies). Thus we obtain the map

$$\det \colon Ob\mathcal{P} \to Ob\mathcal{L}.$$

Let $I = \mathcal{O}_X$ be the trivial linear bundle. We assume that $\det 0 = I$ for any zero object 0 in $\mathcal{P}$. We also assume that an object $L^{-1} \cong \text{Hom}(L, I)$ is chosen for each L in $\mathcal{L}$.

PROPOSITION 1.1. (i) *Any exact sequence*

(1.1) $$0 \to P_0 \to P_1 \to P_{1/0} \to 0$$

in $\mathcal{P}$ gives rise to an isomorphism $\delta = \delta_{0,1} \colon \det P_1 \xrightarrow{\sim} \det P_0 \otimes \det P_{1/0}$ *in a natural way.*

(ii) *Given a commutative diagram of the form*

(1.2)
$$\begin{array}{ccc} & & P_{2/1} \\ & & \uparrow \\ & P_{1/0} \longrightarrow & P_{2/0} \\ & \uparrow & \uparrow \\ P_0 \longrightarrow P_1 \longrightarrow & P_2 \end{array}$$

such that the sequences $0 \to P_i \to P_j \to P_{j/i} \to 0$, *with* $0 \leqslant i < j \leqslant 2$, *and* $0 \to P_{1/0} \to P_{2/0} \to P_{2/1} \to 0$ *are exact, the diagram*

(1.3)
$$\begin{array}{ccc} \det P_2 & \xrightarrow{\delta_{0,2}} & \det P_0 \otimes \det P_{2/0} \\ \delta_{1,2} \downarrow & & \downarrow 1 \otimes \delta_{1/0, 2/0} \\ \det P_1 \otimes \det P_{2/1} & \xrightarrow{\delta_{0,1} \otimes 1} & \det P_0 \otimes \det P_{1/0} \otimes \det P_{2/1} \end{array}$$

commutes.

PROOF. For any $m > 0$, we have the Grothendieck filtration

$$P_0 \wedge \cdots \wedge P_0 \rightarrowtail P_0 \wedge \cdots \wedge P_0 \wedge P_1 \rightarrowtail \cdots \rightarrowtail P_0 \wedge P_1 \wedge \cdots \wedge P_1 \rightarrowtail P_1 \wedge \cdots \wedge P_1$$

associated with the left arrow in (1.1) in which all products contain m factors. The successive quotients are the products

$$\underbrace{P_0 \wedge \cdots \wedge P_0}_{r} \otimes \underbrace{P_{1/0} \wedge \cdots \wedge P_{1/0}}_{s}$$

with $r + s = m$, the quotient maps being induced by the right arrow in (1.1). In particular, if $m = \operatorname{rank} P_1$, the only nonvanishing quotient corresponds to the pair $r = \operatorname{rank} P_0$, $s = \operatorname{rank} P_{1/0}$, and we obtain the isomorphism

$$\begin{array}{ccc} \underbrace{P_0 \wedge \cdots \wedge P_0}_{r} \wedge \underbrace{P_1 \wedge \cdots \wedge P_1}_{s} & \longrightarrow & \underbrace{P_1 \wedge \cdots \wedge P_1}_{m} \\ \downarrow & \swarrow \delta & \\ \underbrace{P_0 \wedge \cdots \wedge P_0}_{r} \otimes \underbrace{P_{1/0} \wedge \cdots \wedge P_{1/0}}_{s} & & \end{array}$$

Now the desired isomorphism $\delta_{0,1} \colon \det P_1 \xrightarrow{\sim} \det P_0 \otimes \det P_{1/0}$ can be defined from the above diagram.

Let rank $P_i = r_i$ and rank $P_{j/i} = r_{j/i}$ in (1.2). We have natural commutative diagrams

$$\Lambda^{r_2}P_2 \longleftarrow \Lambda^{r_0}P_0 \wedge \Lambda^{r_{2/0}}P_2 \longleftarrow \Lambda^{r_0}P_0 \wedge \Lambda^{r_{1/0}}P_1 \wedge \Lambda^{r_{2/1}}P_2$$

$$\searrow_{\delta_{0,2}} \quad \downarrow \quad \downarrow$$

$$\Lambda^{r_0}P_0 \otimes \Lambda^{r_{2/0}}P_{2/0} \longleftarrow \Lambda^{r_0}P_0 \otimes \Lambda^{r_{1/0}}P_{1/0} \wedge \Lambda^{r_{2/1}}P_{2/0}$$

$$\searrow_{1 \otimes \delta_{1/0, 2/0}} \quad \downarrow$$

$$\Lambda^{r_0}P_0 \otimes \Lambda^{r_{1/0}}P_{1/0} \otimes \Lambda^{r_{2/1}}P_{2/1}$$

and

$$\Lambda^{r_2}P_2 \longleftarrow \Lambda^{r_1}P_1 \wedge \Lambda^{r_{2/1}}P_2 \longleftarrow \Lambda^{r_0}P_0 \wedge \Lambda^{r_{1/0}}P_1 \wedge \Lambda^{r_{2/1}}P_2$$

$$\searrow_{\delta_{1,2}} \quad \downarrow \quad \downarrow$$

$$\Lambda^{r_1}P_1 \otimes \Lambda^{r_{2/1}}P_{2/1} \longleftarrow \Lambda^{r_0}P_0 \wedge \Lambda^{r_{1/0}}P_1 \otimes \Lambda^{r_{2/1}}P_{2/1}$$

$$\searrow_{\delta_{0,1} \otimes 1} \quad \downarrow$$

$$\Lambda^{r_0}P_0 \otimes \Lambda^{r_{1/0}}P_{1/0} \otimes \Lambda^{r_{2/1}}P_{2/1}$$

in which all the arrows are obviously isomorphisms (where for brevity we write

$$\Lambda^a A \wedge \Lambda^b B \wedge \Lambda^c C = \underbrace{A \wedge \cdots \wedge A}_{a} \wedge \underbrace{B \wedge \cdots \wedge B}_{b} \wedge \underbrace{C \wedge \cdots \wedge C}_{c}$$

whenever we are given an admissible filtration $A \rightarrowtail B \rightarrowtail C$, etc.). The desired commutativity in (1.3) is now equivalent to the commutativity of the diagram

$$\begin{array}{ccc}
\Lambda^a P_0 \wedge \Lambda^b P_1 \wedge \Lambda^c P_2 & \longrightarrow & \Lambda^a P_0 \otimes \Lambda^b P_{1/0} \wedge \Lambda^c P_{2/0} \\
\downarrow & & \downarrow \\
\Lambda^a P_0 \wedge \Lambda^b P_1 \otimes \Lambda^c P_{2/1} & \longrightarrow & \Lambda^a P_0 \otimes \Lambda^b P_{1/0} \otimes \Lambda^c P_{2/1}
\end{array}$$

The latter is obvious (cf. (E2) in [**Gr**, §7]).

DEFINITION. Let A be a partially ordered set. Let $\mathrm{Ar}(A)$ denote the set $\{j/i \mid i, j \in A, i \leqslant j\}$. By a multiplicative map on $\mathrm{Ar}(A)$ with values in $\mathcal{L}$, we mean a map $D \colon \mathrm{Ar}(A) \to \mathcal{L}$ endowed with a collection of isomorphisms $\delta_{i,j,k} \colon D(k/i) \xrightarrow{\sim} D(j/i) \otimes D(k/j)$ for every $i \leqslant j \leqslant k$ in A such that

(1.4)(i) $D(i/i) = I$ for every $i \in A$;

(1.4)(ii) for every $i \leqslant j$ in A, $\delta_{i,i,j}$ and $\delta_{i,j,j}$ are the natural isomorphisms
$D(j/i) \xrightarrow{\sim} I \otimes D(j/i)$ and $D(j/i) \xrightarrow{\sim} D(j/i) \otimes I$, respectively;

(1.4)(iii) for every $i \leqslant j \leqslant k \leqslant l$ in A, the diagram

$$\begin{array}{ccc}
D(l/i) & \xrightarrow{\delta_{i,j,l}} & D(j/i) \otimes D(l/j) \\
{\scriptstyle \delta_{i,k,l}} \downarrow & & \downarrow {\scriptstyle 1 \otimes \delta_{j,k,l}} \\
D(k/i) \otimes D(l/k) & \xrightarrow{\delta_{i,j,k} \otimes 1} & D(j/i) \otimes D(k/j) \otimes D(l/k)
\end{array}$$

commutes.

We let $\mathrm{Mult}(\mathrm{Ar}(A), \mathcal{L}) = \{(D; \delta_{i,j,k})\}$ denote the set of all $\mathcal{L}$-valued multiplicative maps on $\mathrm{Ar}(A)$.

DEFINITION. We define the simplicial set $Z = Z.\mathcal{L}$ by

$$Z(A) = \mathrm{Mult}(\mathrm{Ar}(A), \mathcal{L}), \quad A \in \Delta,$$

where as usual Δ denotes the category of finite nonempty totally ordered sets and nondecreasing maps.

By definition, there is a unique 0-simplex $*$ in Z. A 1-simplex in Z is an object $L = D(1/0)$ of $\mathcal{L}$. A 2-simplex in Z is a tuple $(L_1, L_2, L_{2/1}; \delta)$, where in the above notation $L_1 = D(1/0)$, $L_2 = D(2/0)$, and $L_{2/1} = D(2/1)$ are objects of $\mathcal{L}$ and $\delta = \delta_{0,1,2} \colon L_2 \xrightarrow{\sim} L_1 \otimes L_{2/1}$ is an isomorphism. Thus, Z looks in a sense like the classifying space of the Picard group, and it is easy to see that $\pi_1 Z \cong \mathrm{Pic}\, X$. However, Z is not homotopy equivalent to $B\,\mathrm{Pic}\, X$. In fact, we have $\pi_2 Z \cong \mathrm{Aut}(I) \cong \Gamma(X, \mathcal{O}_X^*)$ and $\pi_m Z \cong 0$ for $m \geqslant 3$ (cf. Proposition 2.1 and Theorem 3.1).

Given a partially ordered set A, we regard the set $\mathrm{Ar}(A)$ as a category in which the set of morphisms $\mathrm{Mor}(j/i, j'/i')$ consists of a unique morphism if $i \leqslant i'$ and $j \leqslant j'$, and is empty otherwise. We say that a functor $F \colon \mathrm{Ar}(A) \to \mathcal{P}$ is *exact* if
  (i) $F(i/i) = 0$ for every $i \in A$, where 0 denotes a distinguished zero object in $\mathcal{P}$;
  (ii) the sequence $0 \to F(j/i) \to F(k/i) \to F(k/j) \to 0$ is exact for every $i \leqslant j \leqslant k$ in A.

Recall that the S-construction of Waldhausen associated with the category $\mathcal{P}$ is the simplicial set $S = S.\mathcal{P}$ given by

$$S(A) = \mathrm{Exact}(\mathrm{Ar}(A), \mathcal{P}), \quad A \in \Delta,$$

where Exact refers to the set of exact functors.

Proposition 1.1 obviously implies the following

PROPOSITION 1.2. *Let A be a partially ordered set and $F \colon \mathrm{Ar}(A) \to \mathcal{P}$ an exact functor. Consider the map $D = \det \circ F \colon \mathrm{Ar}(A) \to \mathcal{L}$.*
  (i) *For every $i \leqslant j \leqslant k$ in A, the exact sequence $0 \to F(j/i) \to F(k/i) \to F(k/j) \to 0$ gives rise to an isomorphism $\delta_{i,j,k} \colon D(k/i) \xrightarrow{\sim} D(j/i) \otimes D(k/j)$ in a natural way.*
  (ii) *For every $i \leqslant j \leqslant k \leqslant l$ in A, the diagram (1.4)(iii) commutes, i.e., $(D; \delta_{i,j,k})$ is a multiplicative map ((1.4)(i) and (ii) obviously hold).*
  (iii) *It gives rise to a simplicial map*

(1.5) $$\det \colon S.\mathcal{P} \to Z.\mathcal{L}.$$

§2. Applying the loop space functor

Let $F \colon X \to Y$ be a simplicial map, $A \in \Delta$ and $y_0 \in Y(A)$. Following [GG, §1], we define the *right fiber* over y_0 to be the simplicial set $y_0 | F$ given by

$$(y_0 | F)(B) = \varprojlim \begin{pmatrix} & & X(B) \\ & & \downarrow \\ & Y(AB) & \to Y(B) \\ & \downarrow & \\ \{y_0\} & \hookrightarrow Y(A) & \end{pmatrix}, \quad B \in \Delta,$$

where AB denotes the concatenation of A and B, i.e., the disjoint union $A \coprod B$ ordered so $A < B$. By definition, a B-simplex in $y_0|F$ is a pair (y, x), where y is an AB-simplex in Y and x is a B-simplex in X such that the A-face of y is equal to y_0 and the B-face of y is equal to $F(x)$.

If $F = 1\colon Y \to Y$, we write $y|Y$ for $y|F$. It is easy to see that $y|Y$ is contractible for any Y and any $y \in Y(A)$ (cf. [GG, Lemma 1.4]). A B-simplex in $y|Y$ is an AB-simplex in Y whose A-face coincides with y.

Suppose Y has a distinguished vertex $*$, i.e., $* \in Y(\{b\})$, where $\{b\} \in \Delta$ is a one-element set. Let $\mathrm{Pr}\colon *|Y \to Y$ denote the natural projection. We define the (simplicial) loop space of Y at $*$ as the simplicial set $\Omega Y = *|\,\mathrm{Pr}$ or, equivalently, ΩY can be defined from the cartesian square

$$\begin{array}{ccc} \Omega Y & \longrightarrow & *|Y \\ \downarrow & & \downarrow \\ *|Y & \longrightarrow & Y \end{array}$$

(cf. [GG, §2]). By definition, a B-simplex in ΩY is a pair of $\{b\}B$-simplices in Y whose B-faces coincide and whose $\{b\}$-vertices are equal to $*$.

Recall that the G-construction of Gillet and Grayson associated with $\mathcal{P}$ is the simplicial set $G = G.\mathcal{P} = \Omega S.\mathcal{P}$. By [GG, Theorem 3.1], there is a homotopy equivalence $|G| \xrightarrow{\sim} \Omega |S|$. It follows that $\pi_m G \cong K_m X$ for $m \geqslant 0$.

For $A \in \Delta$, let $\gamma(A)$ denote the disjoint union $\{L, R\} \coprod A$ ordered so that the symbols L and R are not comparable, $L < a$ and $R < a$ for any $a \in A$, and A is an ordered subset in $\gamma(A)$. Let $\Gamma(A) = \mathrm{Ar}\,\gamma(A)$. It is easy to see that the G-construction can be described as follows:

(2.1) $$G(A) = \mathrm{Exact}(\Gamma(A), \mathcal{P}), \qquad A \in \Delta.$$

DEFINITION. We define the simplicial set $T = T.\mathcal{L}$ by $T.\mathcal{L} = \Omega Z.\mathcal{L}$. As in (2.1), we write

(2.2) $$T(A) = \mathrm{Mult}(\Gamma(A), \mathcal{L}), \qquad A \in \Delta.$$

Thus, a p-simplex in T is a collection of objects

$$\begin{pmatrix} & & & L_{p/p-1} \\ & & \cdots & \cdots \\ & L_{1/0} & \cdots & L_{p/0} \\ L_0 & L_1 & \cdots & L_p \\ L'_0 & L'_1 & \cdots & L'_p \end{pmatrix}$$

in $\mathcal{L}$ endowed with isomorphisms

$$\delta_{L,i,j}\colon L_j \xrightarrow{\sim} L_i \otimes L_{j/i} \quad \text{and} \quad \delta_{R,i,j}\colon L'_j \xrightarrow{\sim} L'_j \otimes L_{j/i}$$

for every $0 \leqslant i \leqslant j \leqslant p$ and

$$\delta_{i,j,k}\colon L_{k/i} \xrightarrow{\sim} L_{j/i} \otimes L_{k/j}$$

for every $0 \leqslant i \leqslant j \leqslant k \leqslant p$ satisfying (1.4)(iii) (here we write briefly L_i, L'_i, and $L_{j/i}$ for $D(i/L)$, $D(i/R)$, and $D(j/i)$, respectively). In particular, a vertex in T is a pair of objects $\begin{bmatrix} L \\ L' \end{bmatrix}$ in $\mathcal{L}$. An edge connecting $\begin{bmatrix} L_0 \\ L'_0 \end{bmatrix}$ to $\begin{bmatrix} L_1 \\ L'_1 \end{bmatrix}$ is a triple $(L_{1/0}; \delta, \delta')$, where $L_{1/0}$ is an object of $\mathcal{L}$ and $\delta \colon L_1 \xrightarrow{\sim} L_0 \otimes L_{1/0}$, $\delta' \colon L'_1 \xrightarrow{\sim} L'_1 \otimes L_{1/0}$ are isomorphisms.

PROPOSITION 2.1. $|T| \sim \Omega|Z|$.

PROOF. By [GG, Lemma 2.1], it suffices to show that the map $*|Z \to Z$ is fibered (see §4 for the definition of a fibered map). In fact, any simplicial map $X \to Z$ is fibered, since Z satisfies the condition (4.1) of Proposition 4.2. The verification of (4.1) for Z is similar to the proof of Proposition 4.3 and we omit it, because we will not use the homotopy equivalence $|T| \sim \Omega|Z|$ in the sequel.

Applying the loop space functor to the map (1.5), we obtain a simplicial map $G.\mathcal{P} \to T.\mathcal{L}$ which we will also denote by det. We set $W^0 = G$, and let W^1 be the union of components in G of rank zero, i.e., the components whose vertices $\begin{bmatrix} P \\ Q \end{bmatrix}$ satisfy rank P = rank Q. The restriction of the above map to W^1 yields the simplicial map

$$\det \colon W^1 \to T$$

which plays the central role in the paper. By definition, this map takes a simplex $F \in W^1(A) \subset \mathrm{Exact}(\Gamma(A), \mathcal{P})$ to the multiplicative map $D = \det \circ F \colon \Gamma(A) \to \mathcal{L}$ (cf. (2.2)) such that for every $i \leqslant j \leqslant k$ in $\gamma(A)$ the structural isomorphism

$$\delta_{i,j,k} \colon D(k/i) \xrightarrow{\sim} D(j/i) \otimes D(k/j)$$

is the isomorphism associated with the exact sequence

$$0 \to F(j/i) \to F(k/i) \to F(k/j) \to 0$$

as in Proposition 1.2.

§3. The homotopy groups of the simplicial set T

We make T into an H-space using tensor products in $\mathcal{L}$; i.e., for $D, D' \in T(A)$, we define $D \otimes D' \in T(A)$ by

$$(D \otimes D')(j/i) = D(j/i) \otimes D'(j/i) \quad \text{for } i \leqslant j \text{ in } \gamma(A)$$

and let the isomorphism

$$\delta_{i,j,k} \colon (D \otimes D')(k/i) \xrightarrow{\sim} (D \otimes D')(j/i) \otimes (D \otimes D')(k/j)$$

be the product map

$$D(k/i) \otimes D'(k/i) \xrightarrow{\sim} (D(j/i) \otimes D(k/j)) \otimes (D'(j/i) \otimes D'(k/j))$$
$$\xrightarrow{\sim} (D(j/i) \otimes D'(j/i)) \otimes (D(k/j) \otimes D'(k/j))$$

where the second arrow denotes the natural permutation map. The verification of (1.4) is trivial (we assume strictly $I \otimes I = I$). This H-space structure on T endows $\pi_0 T$ with the structure of a monoid. The vertex $\begin{bmatrix} I \\ I \end{bmatrix}$ is a strict identity in T, and therefore its component is the identity element of $\pi_0 T$.

THEOREM 3.1. (i) $\pi_0 T \cong \mathrm{Pic}\, X$;
(ii) $\pi_1 T \cong \Gamma(X, \mathcal{O}_X^*)$ and $\pi_m T \cong 0$ for $m \geqslant 2$.

PROOF. (i) For any two vertices $\begin{bmatrix} L \\ L' \end{bmatrix}$ and $\begin{bmatrix} M \\ M' \end{bmatrix}$ in T, there exists an edge connecting these vertices if and only if $L \otimes (L')^{-1} \cong M \otimes (M')^{-1}$. Thus the assignment $\begin{bmatrix} L \\ L' \end{bmatrix} \to \{L \otimes (L')^{-1}\}$ gives rise to a bijective map $\pi_0 T \to \operatorname{Pic} X$, and the operation on $\operatorname{Pic} X$ obviously agrees with the operation on $\pi_0 T$ induced by the H-space structure.

(ii) It follows from (i) that all the components of T are homotopy equivalent. Nevertheless, we shall construct a universal covering for an arbitrary component of T, which will enable us to compute its homotopy groups.

For $\{L\} \in \operatorname{Pic} X$, let T_L denote the component of the vertex $\begin{bmatrix} L \\ I \end{bmatrix}$ in T. We define the simplicial set $\widetilde{T}_L$ as follows. An A-simplex x in $\widetilde{T}_L$ is a tuple $x = (D; \varepsilon_i, i \in A)$, where $D \in T_L(A) \subset \operatorname{Mult}(\Gamma(A), \mathcal{L})$ and

$$\varepsilon_i \colon D(i/L) \xrightarrow{\sim} L \otimes D(i/R), \qquad i \in A,$$

are isomorphisms such that the diagram

$$\begin{array}{ccc} D(j/L) & \xrightarrow{\delta_{L,i,j}} & D(i/L) \otimes D(j/i) \\ \varepsilon_j \downarrow & & \downarrow \varepsilon_i \otimes 1 \\ L \otimes D(j/R) & \xrightarrow{1 \otimes \delta_{R,i,j}} & L \otimes D(i/R) \otimes D(j/i) \end{array}$$

commutes for every $i < j$ in A. Thus, a vertex in $\widetilde{T}_L$ is a pair of objects $\begin{bmatrix} L_0 \\ L'_0 \end{bmatrix}$ in $\mathcal{L}$ endowed with an isomorphism $\varepsilon_0 \colon L_0 \xrightarrow{\sim} L \otimes L'_0$. We have an obvious simplicial map $\widetilde{T}_L \to T_L$ which forgets the choice of ε_i.

LEMMA 3.2. *Let $D \in T_L(A)$ and $k \in A$. Then for any isomorphism $\varepsilon_k \colon D(k/L) \xrightarrow{\sim} L \otimes D(k/R)$, there exist uniquely determined isomorphisms $\varepsilon_i \colon D(i/L) \xrightarrow{\sim} L \otimes D(i/R)$ with $i \in A$, $i \neq k$, such that $x = (D; \varepsilon_i, i \in A) \in \widetilde{T}_L(A)$.*

PROOF. The uniqueness of ε_j for $j > k$ follows directly from diagram (3.1). For $i < k$, (3.1) implies that the isomorphism

$$\varepsilon_i \otimes 1 \colon D(i/L) \otimes D(k/i) \xrightarrow{\sim} L \otimes D(i/R) \otimes D(k/i)$$

is uniquely determined. But for any linear bundles L, L', and L'', with $\{L'\} = \{L''\}$ in $\operatorname{Pic} X$, the map $\operatorname{Iso}(L', L'') \to \operatorname{Iso}(L' \otimes L, L'' \otimes L)$ given by $\varepsilon \to \varepsilon \otimes 1$ is a bijection, since ε can be restored from the diagram

$$\begin{array}{ccc} L' & \xrightarrow{\varepsilon} & L'' \\ \uparrow & & \uparrow \\ L' \otimes L \otimes L^{-1} & \xrightarrow{\varepsilon \otimes 1 \otimes 1} & L'' \otimes L \otimes L^{-1} \end{array}$$

in which the vertical arrows are induced by the natural map $L \otimes L^{-1} \to I$ (in particular, $\operatorname{Aut}(L)$ is naturally isomorphic to $\operatorname{Aut}(I) \cong \Gamma(X, \mathcal{O}_X^*)$ for any $L \in \mathcal{L}$). Hence the isomorphisms ε_i, with $i < k$, are also uniquely determined. The commutativity of (3.1) for an arbitrary pair $i < j$ can be deduced from the commutativity for (i, k) and for (j, k) and the properties (1.4) of the isomorphisms δ.

Given a simplex $x = (D; \varepsilon_i) \in \widetilde{T}_L(A)$ and an element $\varepsilon \in \operatorname{Aut}(L)$, we define $\varepsilon(x)$ to be the simplex

$$\varepsilon(x) = (D; (\varepsilon \otimes 1_{D(i/R)}) \circ \varepsilon_i, \, i \in A) \in \widetilde{T}_L(A).$$

This is really a simplex in $\widetilde{T}_L(A)$, because the diagram

$$\begin{array}{ccccc}
D(j/L) & \xrightarrow{\delta_{L,i,j}} & D(i/L) \otimes D(j/i) & \xrightarrow{\varepsilon_i \otimes 1} & L \otimes D(i/R) \otimes D(j/i) \\
{\scriptstyle \varepsilon_j} \downarrow & & & & \downarrow {\scriptstyle \varepsilon \otimes 1 \otimes 1} \\
L \otimes D(j/R) & \xrightarrow{\varepsilon \otimes 1} & L \otimes D(j/R) & \xrightarrow{1 \otimes \delta_{R,i,j}} & L \otimes D(i/R) \otimes D(j/i)
\end{array}$$

obviously commutes for every $i < j$ in A. Thus we obtain a free left action of the group $\operatorname{Aut}(L)$ on the simplicial set $\widetilde{T}_L$, and it follows from Lemma 3.2 that the forgetful map $\widetilde{T}_L \to T_L$ is the quotient map associated with this action. Hence $|\widetilde{T}_L| \to |T_L|$ is a covering, and to complete the proof of Theorem 3.1, it now remains to establish the following

PROPOSITION 3.3. $\widetilde{T}_L$ *is contractible.*

PROOF. We shall define simplicial maps $f: *|Z \to \widetilde{T}_L$ and $g: \widetilde{T}_L \to *|Z$ such that $g \circ f = 1$, and $f \circ g$ admits a simplicial homotopy to the identity map of $\widetilde{T}_L$. This will suffice, since $*|Z$ is contractible (cf. [**GG**, Lemma 1.4]).

We can describe the simplicial set $*|Z$ as follows. For $A \in \Delta$, let $\sigma(A)$ denote the concatenation $\{b\}A$, where b is a symbol ("base element"), and let $\Sigma(A) = \operatorname{Ar}(\sigma(A))$. Then we can identify $(*|Z)(A)$ with the set $\operatorname{Mult}(\Sigma(A), \mathcal{L})$. Given $D \in \operatorname{Mult}(\Sigma(A), \mathcal{L})$, we define $f(D): \Gamma(A) \to \mathcal{L}$ by

$$\begin{aligned}
f(D)(j/i) &= D(j/i) & \text{for } i < j \text{ in } A, \\
f(D)(j/L) &= L \otimes D(j/b) & \text{for } j \in A, \\
f(D)(j/R) &= D(j/b) & \text{for } j \in A.
\end{aligned}$$

The isomorphisms δ for $f(D)$ are naturally induced by those for D, and we also define the map $\varepsilon_i \colon f(D)(i/L) \xrightarrow{\sim} L \otimes f(D)(i/R)$ to be the identity map for every $i \in A$. This makes $f(D)$ an A-simplex in $\widetilde{T}_L$. The definition obviously agrees with the face and degeneracy maps, and we obtain the simplicial map $f: *|Z \to \widetilde{T}_L$.

Given a simplex of $\widetilde{T}_L(A)$, i.e., a multiplicative map $D: \Gamma(A) \to \mathcal{L}$ endowed with isomorphisms ε_i satisfying (3.1), let $g(D)$ be the composite map

$$\Sigma(A) \hookrightarrow \Gamma(A) \xrightarrow{D} \mathcal{L},$$

where the inclusion $\Sigma(A) \hookrightarrow \Gamma(A)$ is the identity on $\operatorname{Ar}(A)$ and sends b to R. Then $g: \widetilde{T}_L \to *|Z$ is a simplicial map, and obviously we have $g \circ f = 1_{*|Z}$.

We now proceed to show that there is a simplicial homotopy connecting $f \circ g$ and $1_{\widetilde{T}_L}$. A p-simplex x in $\widetilde{T}_l$ is a collection of objects in $\mathcal{L}$ of the form

$$x = \begin{pmatrix} & & & L_{p/p-1} \\ & & \cdots & \cdots \\ & L_{1/0} & \cdots & L_{p/0} \\ L_0 & L_1 & \cdots & L_p \\ L'_0 & L'_1 & \cdots & L'_p \end{pmatrix}$$

endowed with isomorphisms δ (cf. (2.3)) and isomorphisms $\varepsilon_i \colon L_i \to L \otimes L'_i$ for $0 \leqslant i \leqslant p$. By definition

$$(f \circ g)(x) = \begin{pmatrix} & & & & L_{p/p-1} \\ & & & & \cdots \\ & & L_{1/0} & \cdots & L_{p/0} \\ L \otimes L'_0 & L \otimes L'_1 & \cdots & L \otimes L'_p \\ L'_0 & L'_1 & \cdots & L'_p \end{pmatrix},$$

where the isomorphisms $\varepsilon_i \colon L \otimes L'_i \xrightarrow{\sim} L \otimes L'_i$ are the identity maps.

A p-simplex in $\Delta[1]$ can be thought of as a representation of the set $[p] = \{0, 1, \ldots, p\}$ in the form of a concatenation $[p] = \{0, \ldots, n\}\{n+1, \ldots, p\}$, where $n \in \{-1, 0, \ldots, p\}$. For brevity, we shall denote this simplex by n. We define a homotopy $H \colon \widetilde{T}_L \times \Delta[1] \to \widetilde{T}_L$ by

$$H(x; n) = \begin{pmatrix} & & & & & & L_{p/p-1} \\ & & & & & L_{p-1/p-2} & L_{p/p-2} \\ & & \cdots & & \cdots & \cdots \\ & \cdots & \cdots & \cdots & \cdots & \cdots \\ & L_{1/0} & \cdots & L_{n/0} & L_{n+1/0} & \cdots & L_{p/0} \\ L_0 & \cdots & \cdots & L_n & L \otimes L'_{n+1} & \cdots & L \otimes L'_p \\ L'_0 & \cdots & \cdots & L'_n & L'_{n+1} & \cdots & L'_p \end{pmatrix},$$

where ε_i is as in x for $0 \leqslant i \leqslant n$ and $\varepsilon_i = 1_{L \otimes L'_i}$ for $n+1 \leqslant i \leqslant p$.

To make $H(x; n)$ a simplex of $\widetilde{T}_L$, it remains to define the isomorphisms δ. They will be the same as in x except for the case $\delta_{L,i,j} \colon L \otimes L'_j \xrightarrow{\sim} L_i \otimes L_{j/i}$, where $i \leqslant n$ and $j \geqslant n+1$. In this case we define $\delta \colon L \otimes L'_j \to L_i \otimes L_{j/i}$ to be the composite isomorphism in any of the two possible ways in the diagram

$$\begin{array}{ccc} L \otimes L'_j & \xrightarrow{1 \otimes \delta_{R,i,j}} & L \otimes L'_i \otimes L_{j/i} \\ \varepsilon_j^{-1} \downarrow \sim & & \downarrow (\varepsilon_i \otimes 1)^{-1} \\ L_j & \xrightarrow[\sim]{\delta_{L,i,j} \text{ of } x} & L_i \otimes L_{j/i} \end{array}$$

which is commutative by (3.1). One checks directly the required compatibility conditions (1.4)(iii) and (3.1) for $H(x; n)$, whence H is the desired simplicial homotopy. This completes the proof of Proposition 3.3 and of Theorem 3.1.

§4. The map $\det \colon W^1 \to T$ is fibered

Let $F \colon X \to Y$ be a map of simplicial sets, $A \in \Delta$, and $y_0 \in Y(A)$. Any map $f \colon A' \to A$ in Δ gives rise to the base change map $y_0|F \to (f^*y_0)|F$ which takes a B-simplex (y, x) to (f^*, x) where we write simply f^*y to denote the inverse image of y under the map $A'B \xrightarrow{f \amalg 1} AB$ (cf. §2). We say that F is *fibered* if $y|F \to (f^*y)|F$ is a homotopy equivalence for any $f \colon A' \to A$ in Δ and any $y \in Y(A)$.

THEOREM B′ [**GG**, p. 580]. *If $F: X \to Y$ is a fibered simplicial map, then for any $A \in \Delta$ and $y \in Y(A)$ the square*

$$\begin{array}{ccc} y|F & \longrightarrow & X \\ \downarrow & & \downarrow \\ y|Y & \longrightarrow & Y \end{array}$$

is homotopy Cartesian, and therefore $|y|F|$ can be regarded as homotopy fiber of the map $|F|: |X| \to |Y|$.

THEOREM 4.1. *The map $\det: W^1 \to T$ defined in §2 is fibered.*

We claim that in fact any simplicial map $X \to T$ is fibered. The latter follows from Propositions 4.2 and 4.3 below.

PROPOSITION 4.2. *Suppose that Y is a simplicial set such that*
(4.1) *for any map $f: \{a\} \to A$ in Δ and any simplex $y \in Y(A)$ there exists a simplicial map $\phi: (f^*y)|Y \to y|Y$ such that the diagram*

$$\begin{array}{ccc} (f^*y)|Y & \xrightarrow{\phi} & y|Y \\ \downarrow & & \downarrow \\ Y & \xrightarrow{1} & Y \end{array}$$

commutes, where the vertical arrows take $\{a\}B$ (resp. AB)-simplices to their B-faces, and

(4.1)(i) $(f^*y)|Y \xrightarrow{\phi} y|Y \xrightarrow{f^*} (f^*y)|Y$ *is the identity map;*

(4.1)(ii) *there exists a simplicial homotopy $h: (y|Y) \times \Delta[1] \to y|Y$ which connects the map $y|Y \xrightarrow{f^*} (f^*y)|Y \xrightarrow{\phi} y|Y$ with the identity and which is constant on the B-part (see the definition of $y|Y$ in §2), i.e., the diagram*

$$\begin{array}{ccc} (y|Y) \times \Delta[1] & \xrightarrow{h} & y|Y \\ \downarrow & & \downarrow \\ Y & \xrightarrow{1} & Y \end{array}$$

commutes.
Then any simplicial map $F: X \to Y$ is fibered.

PROOF. It suffices to prove that for any map $f: \{a\} \to A$ in Δ and any $y_0 \in Y(A)$, the map $y_0|F \to (f^*y_0)|F$ is a homotopy equivalence, for given a map $g: A' \to A$, we see that the base change maps $y_0|F \to (g_1^*g^*y_0)|F$ and $(g^*y_0)|F \to (g_1^*g^*y_0)|F$ are homotopy equivalences for any map $g_1: \{a\} \to A'$, the assertion for $y_0|F \to (g^*y_0)|F$ follows.

Let $\phi: (f^*y_0)|Y \to y_0|Y$ be the map of (4.1). We define a map

$$\Phi: (f^*y_0)|F \to y_0|F \text{ by } (y, x) \mapsto (\phi(y), x).$$

Then, by (4.1)(i), the composition $(f^*y_0)|F \xrightarrow{\Phi} y_0|F \xrightarrow{f^*} (f^*y_0)|F$ is the identity map. We define a homotopy $H: (y_0|F) \times \Delta[1] \to y_0|F$ which connects the map

$y_0|F \xrightarrow{f^*} (f^*y_0)|F \xrightarrow{\Phi} y_0|F$ with the identity map, by setting

$$H(y, x; n) = (h(y; n), x) \text{ for } (y, x) \in (y_0|F)(B) \text{ and } n \in \Delta[1](B).$$

PROPOSITION 4.3. T satisfies (4.1).

PROOF. Let $f: \{a\} \to A$ be the inclusion $\{t\} \hookrightarrow [p] = \{0, 1, \ldots, p\}$. We assume y_0 is the p-simplex in T given by (2.3). Then f^*y_0 is the vertex $\begin{bmatrix} L_t \\ L'_t \end{bmatrix}$. A q-simplex x in $(f^*y_0)|T$ is a collection of objects of $\mathcal{L}$ of the form

$$x = \begin{pmatrix} & & & & M_{q/q-1} \\ & & & \cdots & \cdots \\ & & M_{1/0} & \cdots & M_{q/0} \\ & M_{0,t} & \cdots & \cdots & M_{q,t} \\ L_t & M_0 & \cdots & \cdots & M_q \\ L'_t & M'_0 & \cdots & \cdots & M'_q \end{pmatrix}$$

together with isomorphisms δ satisfying (1.4). We set $\phi(x)$ to be the collection

$$\begin{pmatrix} & & & & & & & & M_{q/q-1} \\ & & & & & & & \cdots & \cdots \\ & & & & & & M_{1/0} & \cdots & M_{q/0} \\ & & & & & & \cdots & & \cdots \\ & & & & & & L_{j/t}^{-1} \otimes M_{0,t} & \cdots & L_{j/t}^{-1} \otimes M_{q,t} \\ & & \cdots & L_{j/t} & \cdots & \cdots & M_{0,t} & \cdots & M_{q,t} \\ & \cdots L_{t/i} & \cdots & L_{j/i} & \cdots & \cdots & L_{t/i} \otimes M_{0,t} & \cdots & L_{t/i} \otimes M_{q,t} \\ L_0 \cdots L_i & \cdots L_t & \cdots & L_j & \cdots & L_p & M_0 & \cdots & M_q \\ L'_0 \cdots L'_i & \cdots L'_t & \cdots & L'_j & \cdots & L'_p & M'_0 & \cdots & M'_q \end{pmatrix},$$

where $0 \leqslant i \leqslant t - 1$ and $t + 1 \leqslant j \leqslant p$ (recall that an inverse object L^{-1} is chosen for every L in $\mathcal{L}$). To make $\phi(x)$ a q-simplex in $y_0|T$, we must define the isomorphisms δ and verify (1.4). This amounts to the study of various locations of three (resp. six) objects in the above picture. In each case δ is naturally induced by the corresponding isomorphisms for x and y_0, and (1.4) for $\phi(x)$ follows easily from the same properties of x and y_0.

Thus we obtain a simplicial map $\phi: (f^*y_0)|T \to y_0|T$, and obviously $f^* \circ \phi$ is the identity map of $(f^*y_0)|T$. It remains to define a homotopy $h: (y_0|T) \times \Delta[1] \to y_0|T$ satisfying (4.1)(ii) which connects the map $\phi \circ f^*$ with $1_{y_0|T}$.

A q-simplex y in $y_0|T$ is a collection of objects of L

$$y = \begin{pmatrix} & & & & & & & M_{q/q-1} \\ & & & & & & \cdots & \cdots \\ & & & & & M_{1/0} & \cdots & M_{q/0} \\ & & & & M_{0,p} & \cdots & \cdots & M_{q,p} \\ & & & L_{p/p-1} & M_{0,p-1} & \cdots & \cdots & M_{q,p-1} \\ & & \cdots & \cdots & \cdots & \cdots & \cdots & \cdots \\ & L_{1/0} & \cdots & L_{p/0} & M_{0,0} & \cdots & \cdots & M_{q,0} \\ L_0 & L_1 & \cdots & L_p & M_0 & \cdots & \cdots & M_q \\ L'_0 & L'_1 & \cdots & L'_p & M'_0 & \cdots & \cdots & M'_q \end{pmatrix}$$

together with isomorphisms δ satisfying (1.4). Let $n \in \{-1, 0, \ldots, q\}$ denote a q-simplex in $\Delta[1]$. We set $h(y; x)$ to be the collection

$$\begin{pmatrix} & & & & & & & M_{q/q-1} \\ & & & & & & & \ldots \\ & & & M_{1/0} & \ldots & & \ldots & & \ldots & M_{q/0} \\ & & & \ldots & \ldots & \ldots & & \ldots & \\ & & M_{0,j} & \ldots & M_{n,j} & L_{j/t}^{-1} \otimes M_{n+1,t} & \ldots & L_{j/t}^{-1} \otimes M_{q,t} \\ \ldots & L_{p/t} & M_{0,t} & \ldots & M_{n,t} & M_{n+1,t} & \ldots & M_{q,t} \\ \ldots & L_{p/i} & M_{0,i} & \ldots & M_{n,i} & L_{t/i} \otimes M_{n+1,t} & \ldots & L_{t/i} \otimes M_{q,t} \\ L_0 & \ldots & L_p & M_0 & \ldots & M_n & M_{n+1} & \ldots & M_q \\ L_0' & \ldots & L_p' & M_0' & \ldots & M_n' & M_{n+1}' & \ldots & M_q' \end{pmatrix}.$$

Again we must consider various locations of objects in order to define the isomorphisms δ for $h(y; n)$ and check (1.4). We omit this trivial verification. This completes the proof of Proposition 4.3 and Theorem 4.1.

§5. The second term of the weight filtration

We define the simplicial set W^2 as follows. For $A \in \Delta$, an A-simplex in W^2 is a tuple $(F; \psi_i, i \in A)$, where $F : \Gamma(A) \to \mathcal{P}$ is an exact functor such that rank $F(i/L) =$ rank $F(i/R)$ for each $i \in A$ (i.e., $F \in W^1(A)$, cf. §2) and $\psi_i : \det F(i/L) \xrightarrow{\sim} \det F(i/R)$ are isomorphisms compatible with the isomorphisms δ in $\det \circ F$ (cf. Proposition 1.2), i.e., for every $i < j$ in A the diagram

$$\begin{array}{ccc} \det F(j/L) & \xrightarrow{\delta}_{\sim} & \det F(i/L) \otimes \det F(j/i) \\ \psi_j \downarrow & & \downarrow \psi_i \otimes 1 \\ \det F(j/R) & \xrightarrow{\delta}_{\sim} & \det F(i/R) \otimes \det F(j/i) \end{array}$$

commutes.

For brevity, set $P_i = F(i/L)$, $P_i' = F(i/R)$, and $P_{j/i} = F(j/i)$. We see, in particular, that a vertex in W^2 is a triple $(P, P'; \psi)$, where P and P' are objects of $\mathcal{P}$ such that rank P = rank P' and $\psi : \det P \xrightarrow{\sim} \det P'$ is an isomorphism. An edge in W^2 connecting $(P_0, P_0'; \psi_0)$ to $(P_1, P_1'; \psi_1)$ is a pair of short exact sequences

$$(0 \to P_0 \to P_1 \to P_{1/0} \to 0, 0 \to P_0' \to P_1' \to P_{1/0} \to 0)$$

such that the diagram

$$\begin{array}{ccc} \det P_1 & \xrightarrow{\delta}_{\sim} & \det P_0 \otimes \det P_{1/0} \\ \psi_1 \downarrow & & \downarrow \psi \otimes 1 \\ \det P_1' & \xrightarrow{\sim}_{\delta} & \det P_0' \otimes \det P_{1/0} \end{array}$$

commutes.

There is an obvious simplicial map $W^2 \to W^1$ which forgets the choice of the isomorphisms ψ_i.

THEOREM 5.1. $W^2 \to W^1 \xrightarrow{\det} T$ *is a homotopy fibration sequence.*

This assertion together with Theorem 3.1 yields a long exact sequence

$$\cdots \to 0 \to \pi_2 W^2 \xrightarrow{\sim} \pi_2 W^1 \to 0 \to \pi_1 W^2 \to K_1 X \xrightarrow{\text{epi}} \Gamma(X, \mathcal{O}_X^*)$$
$$\xrightarrow{0} \pi_0 W^2 \to \ker(\operatorname{rank}: K_0 X \to \mathbb{Z}) \to \operatorname{Pic} X \to 0.$$

COROLLARY 5.2. (i) $\pi_0 W^2 \cong \ker((\operatorname{rank}, \det): K_0 X \to \mathbb{Z} \oplus \operatorname{Pic} X)$,
(ii) $\pi_1 W^2 \cong \ker(\det: K_1 X \to \Gamma(X, \mathcal{O}_X^*))$,
(iii) $\pi_m W^2 \cong K_m X$ *for* $m \geq 2$.

PROOF OF THEOREM 5.1. Let $*$ denote the vertex $\begin{bmatrix} I \\ I \end{bmatrix}$ of T regarded as a $\{b\}$-simplex. By Theorem B' and Theorem 4.1, it suffices to construct homotopy inverse maps $f: *|\det \to W^2$ and $g: W^2 \to *|\det$.

A p-simplex in $*|\det$ is a pair (x, F), where

$$(5.2) \qquad x = \begin{pmatrix} & & & & L_{p/p-1} \\ & & & \cdots & \cdots \\ & & L_{1/0} & \cdots & L_{p/0} \\ & L_{0/b} & L_{1/b} & \cdots & L_{p/b} \\ L_b = I & L_0 & L_1 & \cdots & L_p \\ L_b' = I & L_0' & L_1' & \cdots & L_p' \end{pmatrix}$$

is a collection of objects of $\mathcal{L}$ endowed with isomorphisms δ (i.e., x is a $\{b\}[p]$-simplex in T whose $\{b\}$-vertex is $*$) and F is a p-simplex in W^1 such that $\det F$ is equal to the p-face of x.

We set $\psi_i = \delta_{R,b,i}^{-1} \circ \delta_{L,b,i}: L_i \xrightarrow{\sim} L_i'$, where $\delta_{L,b,i}: L_i \xrightarrow{\sim} I \otimes L_{i/b}$ and $\delta_{R,b,i}: L_i' \xrightarrow{\sim} I \otimes L_{i/b}$, and claim that $(F; \psi_i, 0 \leq i \leq p)$ is a p-simplex in W^2. To check that, it suffices to verify (5.1) for every $i < j$ in $[p]$. This follows from the diagram

$$\begin{array}{ccc}
L_j & \xrightarrow[\sim]{\delta_{L,i,j}} & L_i \otimes L_{j/i} \\
\delta_{L,b,j} \downarrow & & \downarrow \delta_{L,b,i} \otimes 1 \\
I \otimes L_{j/b} & \xrightarrow[\sim]{1 \otimes \delta_{b,i,j}} & I \otimes L_{i/b} \otimes L_{j/i} \\
\delta_{R,b,j} \uparrow & & \uparrow \delta_{R,b,i} \otimes 1 \\
L_j' & \xrightarrow{\delta_{R,i,j}} & L_i' \otimes L_{j/i}
\end{array}$$

in which both squares are commutative by (1.4)(iii) for x.

Thus we obtain a simplicial map $f: *|\det \to W^2$. We define a homotopy inverse map $g: W^2 \to *|\det$ as follows. Given a p-simplex $(F; \psi_i, 0 \leq i \leq p)$ in W^2, we set $L_i = \det F(i/L)$, $L_i' = L_{i/b} = \det F(i/R)$, and $L_{j/i} = \det F(j/i)$ for $0 \leq i < j \leq p$. We define a $\{b\}[p]$-simplex x by (5.2), where the isomorphisms δ in the p-face are the same as in $\det F$ (cf. Proposition 1.2). Further, we set

$$\delta_{L,b,i}: L_i = \det F(i/L) \xrightarrow{\psi_i} \det F(i/R) = L_{i/b} \to I \otimes L_{i/b},$$
$$\delta_{R,b,i}: L_i' = \det F(i/R) \xrightarrow{1} \det F(i/R) = L_{i/b} \to I \otimes L_{i/b}$$

where $L_{i/b} \to I \otimes L_{i/b}$ is the natural map (recall that $I = \mathcal{O}_X$), and

$$\delta_{b,i,j}: L_{j/b} = \det F(j/R) \xrightarrow{\delta_{R,i,j} \text{ of } \det F} \det F(i/R) \otimes \det F(j/i) = L_{i/b} \otimes L_{j/i};$$

the compatibility condition follows trivially. Thus (x, F) is a p-simplex in $*|\det$, which gives rise to a simplicial map g. Clearly, $f \circ g = 1_{W^2}$, and it is easy to define a simplicial homotopy which connects $g \circ f$ with $1_{*|\det}$. Theorem 5.1 is proved.

References

[GG] H. Gillet and D. Grayson, *The loop space of the S-construction*, Illinois J. Math. **31** (1987), no. 4, 574–597.
[Gr] D. Grayson, *Exterior power operations on higher K-theory*, K-theory **3** (1989), no. 3, 247–260.
[Sou] C. Soulé, *Opérations en K-théorie algébrique*, Canad. J. Math **37** (1985), no. 3, 488–550.

Translated by THE AUTHOR

POMI, Fontanka 27, St. Petersburg 191011, Russia
E-mail address: nenashev@lomi.spb.su

On Common Zeros of Two Quadratic Forms

A. S. Sivatsky

Dedicated to Boarding School #45 on the occasion of its 30th anniversary

This paper is devoted to some questions of the algebraic theory of quadratic forms. More specifically, we are interested in the conditions on the field under which any two quadratic forms in a fixed number of variables over this field have a common zero (of course, a nontrivial one, and we shall not mention this specially in the sequel).

Below we use the following notation: F is a field of characteristic $\neq 2$, $\overline{F}$ and F^q are its algebraic and quadratic closures respectively, F_f is the extension of F by means of the polynomial f, and $G = \text{Gal}(L/F)$ is the Galois group of the normal extension L/F. If L is an extension of F^q, then $\widetilde{L}$ is the normal closure of L over F^q. If $f \in K$, where K is a field and $f \neq 0$, then $\overline{f}$ denotes the image of f in K^*/K^{*2}. Let $I(F)$ be the ideal of even forms in the Witt ring $W(F)$ of the field F; then the symbol $\langle\!\langle a_1, \ldots, a_n \rangle\!\rangle = \langle 1, a_1 \rangle \otimes \cdots \otimes \langle 1, a_n \rangle$ denotes the corresponding Pfister form, $D(g)$ is the set of values of the quadratic form g, and $u(F)$ is the so-called u-invariant of the field F, i.e., the maximal number n such that there exists an anisotropic form of dimension n over F, $k_2(F) = K_2(F)/2K_2(F)$, where $K_2(F)$ is the Milnor K-group of F. If G is a group, then by $|G|$ we denote its order.

The simplest case of our investigation is that of algebraically closed fields, and the corresponding situation is well known: any two quadratic forms in 3 variables have a common zero (the proof uses methods of algebraic geometry). It is a little bit more difficult to solve a similar problem for fields with the trivial Sylow 2-subgroup of the Galois group. But it turns out that exactly the same result holds for such fields. To prove this and other facts of the paper, we use the following lemma, which is our basic tool.

LEMMA 1. *Suppose F is a field, f and g two quadratic forms of the same dimension over F, t an independent variable, and $F(t)$ the field of rational functions over F. Then the following two conditions are equivalent.*
1) *The form $f + tg$ is isotropic over $F(t)$.*
2) *The forms f and g have a common zero over F.*

1991 *Mathematics Subject Classification.* Priimary 15A63.
Key words and phrases. Quadratic forms, Galois group.

©1996, American Mathematical Society

The proof is similar to that of the Cassels–Pfister theorem and can be found in [**B**], but we present it here for completeness.

PROOF. 2) $\Longrightarrow$ 1). Let $f(x) = g(x) = 0$. Then $(f + tg)(x) = 0$.

1) $\Longrightarrow$ 2). Let f and g be forms on a vector space V. The form $f + tg$ may be considered as a form on the space $V(t) = V \otimes_F F(t)$. Taking an arbitrary basis in V, we define the *degree* of a vector in $V(t)$ as the maximal degree of its components. Let $\mathbf{p} = (p_1, \ldots, p_n)$ be a zero of $f + tg$, $\deg \mathbf{p} = n$. Obviously we may suppose that p_i belongs to $V \otimes_F F[t]$ and have no common denominator. Divide $\mathbf{p}$ by t^n: $\mathbf{p} = t^n \mathbf{q} + \mathbf{r}$, where $\mathbf{q}, \mathbf{r} \in V \otimes_F V[t]$, $\deg \mathbf{r} < n$, $\mathbf{r} \neq 0$. Let α be the bilinear form corresponding to the form $f + tg$. Consider the vector $\mathbf{s} = \alpha(\mathbf{q},\mathbf{q})\mathbf{p} - 2\alpha(\mathbf{p},\mathbf{q})\mathbf{q}$. We have

$$\alpha(\mathbf{s},\mathbf{s}) = \alpha(\alpha(\mathbf{q},\mathbf{q})\mathbf{p} - 2\alpha(\mathbf{p},\mathbf{q})\mathbf{q}, \alpha(\mathbf{q},\mathbf{q})\mathbf{p} - 2\alpha(\mathbf{p},\mathbf{q})\mathbf{q})$$
$$= \alpha(\mathbf{q},\mathbf{q})^2 \alpha(\mathbf{p},\mathbf{p}) - 4\alpha(\mathbf{q},\mathbf{q})\alpha(\mathbf{p},\mathbf{q})\alpha(\mathbf{p},\mathbf{q}) + 4\alpha(\mathbf{p},\mathbf{q})^2 \alpha(\mathbf{q},\mathbf{q})$$
$$= \alpha(\mathbf{q},\mathbf{q})^2 \alpha(\mathbf{p},\mathbf{p}) = 0,$$

since $\alpha(\mathbf{p},\mathbf{p}) = 0$.

If $\mathbf{p}$ and $\mathbf{q}$ are collinear, then $\alpha(\mathbf{r},\mathbf{r}) = 0$, otherwise $\mathbf{s} \neq 0$. Obviously

$$\mathbf{s} = \alpha\left(\frac{\mathbf{p}-\mathbf{r}}{t^n}, \frac{\mathbf{p}-\mathbf{r}}{t^n}\right)\mathbf{p} - 2\alpha\left(\mathbf{p}, \frac{\mathbf{p}-\mathbf{r}}{t^n}\right)\frac{\mathbf{p}-\mathbf{r}}{t^n} = \frac{1}{t^{2n}}(\alpha(\mathbf{r},\mathbf{r})\mathbf{p} - 2\alpha(\mathbf{p},\mathbf{r})\mathbf{r}).$$

Since $\deg \mathbf{r} \leqslant n-1$, $\deg \mathbf{p} = n$ and all the coefficients of $f + tg$ are linear, we have $\deg \mathbf{s} \leqslant n-1$. So, having an original zero, we can obtain another one of smaller degree. We continue until we have a zero of degree 0.

THEOREM 2. *Let F be a field with trivial Sylow 2-subgroup of its Galois group, and let f and g be two quadratic forms over F. If $\dim f = \dim g \leqslant 3$, then these forms have a common zero.*

PROOF. In view of Lemma 1, it is sufficient to check that $(f + tg)$ is isotropic over $F(t)$. But since any finite extension of F is of odd degree and an anisotropic form cannot become isotropic over such an extension, it is sufficient to prove that the form $(f + tg)$ is isotropic over $\overline{F}(t)$. It is well known that $\overline{F}(t)$ is a C_1-field [**P**] and so the proof is complete.

The next case that we want to study is that of a quadratically closed field. The following theorem holds.

THEOREM 3. *Let F be a field. The following conditions are equivalent.*
1) *For any field L such that $F \subset L \subset F^q$ there exists no irreducible polynomial over L of degree 4 or 8 with Galois group of order 12 or 24 (over L).*
2) *Any extension of F^q of degree 3 is a quadratically closed field.*
3) *Any two quadratic forms in 3 variables over F^q have a common zero.*

PROOF. 1) $\Longrightarrow$ 2). Let $[L : F^q] = 3$, $[E : L] = 2$. Then the extension L/F^q is separable. Indeed, otherwise we can consider the maximal separable subextension K of E over F^q. It is easy to see that $[K : F^q] = 2$, which is impossible.

Evidently $[\widetilde{L} : F^q] = 3$ or 6. The second case is impossible, because considering any Sylow 3-subgroup of $\mathrm{Gal}(\widetilde{L}/F^q)$, we see that F^q has some quadratic extension. Hence $[L : F^q] = 3$ and $L = \widetilde{L}$. Suppose $E = L(\sqrt{a})$, $L = F^q(\alpha)$, $a, \alpha \in L$, $\sigma_1, \sigma_2,$ and σ_3 are the different elements of $\mathrm{Gal}(\widetilde{L}/F^q)$. Then $\widetilde{E} = F^q(\alpha, \sqrt{\sigma_1 a}, \sqrt{\sigma_2 a}, \sqrt{\sigma_3 a})$

hence $[\widetilde{E} : F^q] = 6, 12,$ or 24. Considering any Sylow 3-subgroup of $\mathrm{Gal}\,(\widetilde{E}/F^q)$ and the corresponding subextension $F^q \subset \widetilde{K} \subset \widetilde{E}$, we see that $[\widetilde{K} : F^q] = 4$ or 8. Since $\widetilde{K} \subset \widetilde{E}$, we have

$$|\mathrm{Gal}\,(\widetilde{E}/F^q)| \vdots |\mathrm{Gal}\,(\widetilde{K}/F^q)|.$$

Moreover, $\mathrm{Gal}\,(\widetilde{K}/F^q)| \neq 8$, otherwise F^q would have a quadratic extension. Now it is evident that $|\mathrm{Gal}\,(\widetilde{K}/F^q)| = 12$ or 24, which contradicts our assumption.

2) $\implies$ 3). Consider any two quadratic forms f and g in 3 variables over F^q. By Lemma 1 it is sufficient to show that the form $(f + tg)$ has a nontrivial zero over $F^q(t)$. Let

$$\begin{pmatrix} L_{11} & L_{12} & L_{13} \\ L_{21} & L_{22} & L_{23} \\ L_{31} & L_{32} & L_{33} \end{pmatrix}$$

be a matrix of the form $(f + tg)$, where L_{ij} are some linear functions. Let

$$D_1 = L_{11}, \quad D_2 = \det \begin{pmatrix} L_{11} & L_{12} \\ L_{21} & L_{22} \end{pmatrix}, \quad D_3 = \det \begin{pmatrix} L_{11} & L_{12} & L_{13} \\ L_{21} & L_{22} & L_{23} \\ L_{31} & L_{32} & L_{33} \end{pmatrix}.$$

If $D_3 = 0$, then $(f + tg)$ is a degenerate form and, of course, is an isotropic one. If $D_2 = 0$ or $D_1 = 0$, then

$$\sum_{1 \leq i,j \leq 2} L_{ij} x_i x_j \quad \text{or} \quad L_{11} x_1^2$$

is a degenerate form and the original form $(f + tg)$ is isotropic too. So we may suppose that $D_1, D_2, D_3 \neq 0$. Since

$$(f + tg) \simeq \langle D_1, D_1 D_2, D_2 D_3 \rangle$$

(see [L]), the form $(f + tg)$ is a subform of $D_1 \langle\!\langle D_2, D_1 D_3 \rangle\!\rangle$. It is well known that

$$I^2(F^q(t)) \cong \coprod_f I(F_f^q) \oplus I^2(F^q),$$

where f runs over the set of all monic polynomials over F^q (see [L]). It is evident that $I^2(F^q) = 0$ and that all the irreducible divisors of D_1 and D_2 are linear. The polynomial D_3 is either irreducible of degree 3 or decomposes into a product of some linear polynomials. Therefore, all the invariants of $[\langle\!\langle D_2, D_1 D_3\rangle\!\rangle]$ in $\coprod_f I(F_f^q)$ are trivial, and so the form $\langle\!\langle D_2, D_1 D_3 \rangle\!\rangle$ is hyperbolic. Hence

$$(f + tg) \perp D_3 \simeq D_1 \langle\!\langle D_1 D_3 \rangle\!\rangle \simeq \langle D_3, -D_3, 1, -1 \rangle$$

and $(f + tg) \simeq \langle -D_3, 1, -1 \rangle$, and so the form $f + tg$ is isotropic over $F^q(t)$.

3) $\implies$ 2). Let there exist a field L ($F \subset L \subset F^q$) and a polynomial f over L such that $[L_f^n : L] = 12$ or 24 and $\deg f = 4$ or 8 (here we denote by L_f^n the normal closure of L_f over L). We claim that no root of f lies in F^q. Suppose this is not so. Since F^q is normal over F and so over L, all the roots of f lie in F^q. But in this case $[L_f^n : L] = 2^s$, which is impossible by our assumption. So the polynomial f over F^q decomposes into a product of irreducible polynomials of degree $3 \geq 5$ (for there are no irreducible quadratic polynomials over F^q). Now we show that f cannot be

decomposed into a product of polynomials of degree 3 and 5. Indeed, suppose this is not so. Then we consider the following diagram of extensions of fields

$$\begin{array}{c} L_f^n F^q \\ / \quad \backslash \\ L_f^n \quad F^q \\ \backslash \quad / \\ L \end{array}$$

Going along the left side of the diagram, we see that $|\operatorname{Gal}(L_f^n F^q/L)|$ is not divisible by 5, which is impossible by our assumption that $[L_f^n F^q : F^q]$ is divisible by 5.

So f decomposes over F^q into a product of two polynomials of degree 4. Take one of them. The order of its Galois group G over F^q is a divisor of 24 and is divided by 3 (otherwise its corresponding Galois extension would decompose into a composition of quadratic extensions, which F^q does not have). Now consider a Sylow 2-subgroup H of the group G and the corresponding field. Evidently, this field is an extension of degree 3 of F^q, and it has a quadratic extension.

Now we consider two arbitrary forms f and g in 3 variables over F^q. Since they have a common zero, the form $(f + tg)$ is isotropic over $F^q(t)$, the corresponding Pfister form $\langle\!\langle D_2, D_1 D_3 \rangle\!\rangle$ is all the more isotropic, and so hyperbolic, and has no nontrivial invariant in

$$\coprod_f I(F_f^q).$$

We shall come to a contradiction if we show that we may choose forms f and g so that D_3 is an irreducible polynomial of degree 3 over F^q and $D_2 \notin (F_{D_3}^q)^{*2}$.

LEMMA 4. *Suppose K is an arbitrary field, L/K is an extension of degree 3, $L = K_f$ is not quadratically closed, $\theta \in L$, $f(\theta) = 0$. Then there exists a polynomial g over K such that $\det g \leqslant 1$ and $g(\theta) \notin L^{*2}$.*

PROOF. Suppose $g(\theta) \notin L^{*2}$, $\deg g = 2$, and all the linear polynomials of θ are squares in L, $g(t) = t^2 + at + b$, $f(t) = t^3 + ct^2 + dt + e$. Then

$$(t^2 + at + b)(t + c - a) = (a(c - a) + b - d)t + b(c - a) - e$$

is a linear polynomial that is not a square in L, which contradicts our assumption.

Carrying out a linear substitution of the variable, we see that there exists a polynomial f over F^q such that $f(\theta) = 0$ and $\theta \notin (F_f^q)^{*2}$. Now consider the symmetric matrix

$$\begin{pmatrix} t & 0 & z \\ 0 & 1 & y \\ z & y & x \end{pmatrix}$$

and choose x, y, z properly. It is clear that

$$D_1 = D_2 = t, \qquad D_3 = xt - y^2 t - z^2 = (x - y^2)t - z^2.$$

When x and y run over all linear polynomials over F^q, the element $x - y^2$ runs over (in particular) all monic quadratic polynomials (we use the fact that $(-1) \in (F^q)^{*2}$) and the element z^2 runs over (in particular) all constants. So by means of the parameters x, y, and z, we can realize all monic polynomials of degree 3. The proof of the theorem is complete.

COROLLARY 5. *There exist two quadratic forms in 3 variables over $\mathbb{Q}^q$ that have no nontrivial common zero.*

PROOF. Indeed, there exists an irreducible polynomial of degree 4 over $\mathbb{Q}$ with Galois group isomorphic to S_4 (see [**La**]).

THEOREM 6. *Let F be a field such that the 2-subgroup of its Galois group is infinite and all its finite extensions are cyclic. Then any two quadratic forms in 5 variables over F have a common zero.*

PROOF. Let L/F be some maximal odd extension of F. Every finite extension of L is a cyclic 2-extension decomposed into a composition of quadratic extensions and so $\text{Gal}(\overline{L}/L) \cong \mathbb{Z}_2$. By Lemma 1, the following four statements are equivalent:
1) The forms f and g have a common zero over F.
2) The forms f and g have a common zero over L.
3) The form $(f + tg)$ is isotropic over $F(t)$.
4) The form $(f + tg)$ is isotropic over $L(t)$.

So from the very beginning, we may suppose that $\text{Gal}(\overline{F}/F) \cong \mathbb{Z}_2$. In particular, the field F has no odd extension. Let f and g be two quadratic forms in 5 variables over F.

If g is a nondegenerate form, then $\det(f + tg)$ is a polynomial of degree 5 in t. Since F has no odd extension, every polynomial of odd degree over F has a root which lies in F. Let μ be a root of $\det(f + tg)$. Now from the pair of forms, $\binom{f}{g}$ we go to the pair of forms $\binom{g}{f+\mu g}$. It is evident that it suffices to show that g and $(f + \mu g)$ have a common zero. The form $(f + \mu g)$ is degenerate, and so there is a basis $(e_1, e_2, e_3, e_4, e_5)$ of its vector space such that $(f + \mu g)(e_i, e_5) = 0$ for each i (we use the same notation for a quadratic form and for the associated bilinear symmetric form).

If $g(e_5, e_5) = 0$, then everything is proved. Otherwise, we make a linear substitution of the basis $e_i' = e_i - x_i e_5$ for each i from 1 to 4, where $x_i = g(e_i, e_5)/g(e_5, e_5)$. Since $g(e_i', e_5) = 0$ and $(f + \mu g)(e_i', e_5) = 0$ for each i from 1 to 4, from the very beginning we may suppose that

$$(f + tg) \simeq \widehat{(f + tg)} \perp \langle c_5 \rangle,$$

where $\widehat{f + tg}$ is some quadratic form whose coefficients are linear polynomials and $c_5 \in F$ (we have substituted the forms f and g for the forms g and $f + \mu g$, respectively). Let

$$d_k = \det \|f_{ij} + tg_{ij}\| \qquad k = 1, 2, 3, 4, 5, \quad 1 \leqslant i, j \leqslant k,$$

where $f_{ij} + tg_{ij}$ are the coefficients of the form $\widehat{(f + tg)}$. If $d_k = 0$ for some k, then we conclude that the form $(f + tg)$ is isotropic. So we may suppose that $d_k \neq 0$ for every k. In this case we have the following isomorphism $(f + tg) \simeq \langle d_1, d_1 d_2, d_2 d_3, d_3 d_4, d_4 d_5 \rangle$

Now we shall consider two cases.

1) The element d_5 belongs $D(f + tg)$. Then $(f + tg) \simeq \langle d_5 \rangle \perp \phi$, where ϕ is a form in $I^2(F(t))$ and $\dim \phi = 4$. It is well known that

$$I^3(F(t)) \cong \coprod_p I^2(F_p) \oplus I^3(F),$$

where p runs over all irreducible monic polynomials over F (see [**L**]). Since $\text{cd}_2 F = \text{cd}_2 F_p = 1$ for all p, $I^2(F_p) = I^3(F) = 0$ for all p, and so $I^3(F(t)) = 0$, $F(t)^* \subset D(\phi)$, and this implies that the form $(f + tg)$ is isotropic.

2) The element d_5 does not belong to $D(f+tg)$. Then we consider the anisotropic form

$$h = \langle d_1, d_1d_2, d_2d_3, d_3d_4, d_4d_5, -d_5 \rangle \in I^2(F(t)).$$

It is easy to see that

$$[h] = [\langle\!\langle d_2, d_3d_1 \rangle\!\rangle] + [\langle\!\langle -d_4, -c_5d_3 \rangle\!\rangle]$$
$$= [\langle\!\langle d_2d_4, d_3d_1 \rangle\!\rangle] + [\langle\!\langle -d_4, d_3d_1 \rangle\!\rangle] + [\langle\!\langle -d_4, -c_5d_3 \rangle\!\rangle]$$
$$= [\langle\!\langle d_2d_4, d_3d_1 \rangle\!\rangle] + [\langle\!\langle -d_4, c_5d_1 \rangle\!\rangle]$$

(we have used the equality $[\langle\!\langle x, z \rangle\!\rangle] + [\langle\!\langle y, z \rangle\!\rangle] = [\langle\!\langle -xy, z \rangle\!\rangle]$ (mod $I^3(K)$) for any field K and $x, y, z \in K$).

We claim that case 2) is impossible. To show that, it is sufficient to prove that the form h is equivalent in the Witt ring of $F(t)$ to some Pfister form of dimension 4.

Thus $[h] = [\langle\!\langle d_2d_4, d_3d_1 \rangle\!\rangle] + [\langle\!\langle -d_4, c_5d_1 \rangle\!\rangle]$. It is evident that $u(F) = 2$. Since

$$g\left(\sum_{i=1}^{4} x_i e_i\right)$$

is a quadratic form in 4 variables over F, it is isotropic over F. Therefore, carrying out a linear substitution of the vectors e_1, e_2, e_3, and e_4, we may suppose that $g_{11} = 0$ and so $d_1 = f_{11} + tg_{11} = f_{11} \in F$. Therefore, $c_5d_1 \in F$. Now we shall consider two cases again.

a) The polynomial d_4 has no linear divisor. Then the form $\langle\!\langle -d_4, c_5d_1 \rangle\!\rangle$ has no invariant in $\coprod_p F_p^*/F_p^{*2}$, for every element of F has a square root in every nontrivial extension of F (we have used the isomorphisms

$$I(F_p)/I^2(F_p) \cong I(F_p) \cong F_p^*/F_p^{*2}$$

for all p). So the form h is equivalent to the form $\langle\!\langle d_2d_4, d_3d_1 \rangle\!\rangle$ and the theorem is proved.

b) The polynomial d_4 has a linear divisor.

Let v be a root of d_4, $v \in F$. Then we shall substitute the forms f and g for the forms g and $f + vg$ respectively. Carrying out a linear transformation, we may suppose from the very beginning that

$$(f + tg) \simeq \widetilde{(f + tg)} \perp \langle c_4 \rangle \perp \langle c_5 t \rangle,$$

where $\widetilde{f + tg}$ is some form of dimension 3 with linear coefficients and $c_4 \in F$.

Let $\widetilde{f + tg} = \tilde{f} + t\tilde{g}$, where $\tilde{f}$ and $\tilde{g}$ are some forms over F. If $\tilde{g}$ is a nondegenerate form over F, then $\det(\widetilde{f + tg})$ is a polynomial of degree 3 in t. It has a root ρ in F. We shall substitute the forms f and g for the forms g and $f + \rho g$ respectively. Again, carrying out a linear transformation, we may suppose from the very beginning that

$$f + tg \simeq d_1 x^2 + 2exy + f_1 y^2 + c_3 z^2 + c_4 t u^2 + (c_5 + \rho c_5) v^2,$$

where d_1, e, and f are some linear polynomials, $c_3, c_4, c_5, \rho \in F$, and x, y, z, u, v are the variables of the form.

The polynomial $d_2 = d_1 f_1 - e^2$ is either quadratic or linear and if it decomposes to linear multipliers, then $[h]$ has in $\coprod_p F_p^*/F_p^{*2}$ only linear invariants. Let ε be an element of F^* such that $\varepsilon \notin F^{*2}$. Then

$$[h] = [\langle\!\langle -\varepsilon, \prod_i (t - \alpha_i) \rangle\!\rangle],$$

where $\alpha_i \in F$ for each i. Thus we may suppose that d_2 is an irreducible quadratic polynomial. It is clear that

$$[h] = [\langle\!\langle d_2, d_3 d_1 \rangle\!\rangle] + [\langle\!\langle -d_4, -c_5' d_3 \rangle\!\rangle],$$

where $c_5' = c_5 + pc_5 t$. From this equality it is easy to see that $[h]$ can have invariants only in $\coprod_p F_p^*/F_p^{*2}$, where p runs over the set $\{d_1, c_4 t, c_5 + pc_5 t, d_2\}$. Since $d_2 = d_1 f_1 - e^2$ and so $d_2 \equiv -e^2 \pmod{d_1}$, the invariant of $[h]$ in $F_{d_1}^*/F_{d_1}^{*2}$ is zero. Thus $[h]$ can have only two linear and one quadratic invariants (and no others). To complete the proof of Theorem 6, it suffices to prove the following

LEMMA 7. *Let $\alpha \in k_2(F(t))$ and suppose α has only two linear and one quadratic invariants in $\coprod_p k_1(F_p) = \coprod_p F_p^*/F_p^{*2}$. Then there exist $a, b \in F(t)$ such that $\alpha = \{a, b\}$.*

PROOF. Suppose r and s are monic polynomials of degree 2 and 1 over F respectively and r is irreducible. Then $r \in F_s^{*2}$ iff $s \in F_r^{*2}$. Indeed, we may use the reciprocity law in $K_2(F(t))$ for the symbol $\{r, s\}$: $N_{F_r/F}(s) = N_{F_s/F}(r)$. If $r \in F_s^{*2}$, then $N_{F_r/F}(s) \in F^{*2}$. The map $N_{F_r/F}: F_r^*/F_r^{*2} \to F^*/F^{*2}$ is bijective because $|F_r^*/F_r^{*2}| = |F^*/F^{*2}| = 2$ and $u(F) = 2$, and so $s \in F_r^{*2}$.

Now denote by r the quadratic polynomial of Lemma 7 and by $(t - a)$, $(t - b)$ the linear ones. Let $\varepsilon \in F^*$, $\varepsilon \notin F^{*2}$ and consider the three possible cases:

a) $r \in F_{(t-a)}^{*2}$, $r \in F_{(t-b)}^{*2}$. Since r is quadratic and $u(F) = 2$, there exists a $c \in F$ such that $r(c) \notin F^{*2}$, and so we conclude that the symbol $\{\varepsilon r, (t-a)(t-b)(t-c)\}$ has the same invariants as α and therefore is equal to α.

b) $r \notin F_{(t-a)}^{*2}$, $(t-b) \in F_{(t-a)}^{*2}$. Then the symbol $\{(t-b)r, (a-t)\varepsilon\}$ has the same invariants as α.

c) $r \notin F_{(t-a)}^{*2}$, $(t-b) \notin F_{(t-a)}^{*2}$. Then the symbol $\{\varepsilon(t-b)r, (a-t)\}$ has the same invariants as α. The proof of the lemma is completed.

Since $k_2(F(t)) \cong I^2(F(t))$, Theorem 6 is proved (the cases when some of the three invariants are trivial are considered similarly and are even easier).

In connection with the theorem above, let us consider the following problem. Let F be a field as in Theorem 6. What can one say about $u(F(t))$? Let us show that if $\alpha \in I^2(F(t))$ and has an only nontrivial invariant in $\coprod_p F_p^*/F_p^{*2}$, then α is a symbol. Indeed, if p is the only polynomial corresponding to the invariant, we consider the following finite sequence of polynomials: $p_1 = p$ and p_n is any polynomial of minimal degree such that $p_n \notin F_{p_{n-1}}^{*2}$ (the last polynomial of the sequence has degree 0). Then it is easy to see that

$$\alpha = [\langle\!\langle p_1 p_3 p_5 \cdots, p_2 p_4 p_6 \cdots \rangle\!\rangle],$$

because it has only a nontrivial invariant (in F_p^*/F_p^{*2}).

Now let us suppose that in the ring $F[t]$ the analog of the prime number theorem for arithmetic progressions takes place (i.e., for any $f, g \in F[t]$ having no common divisor, there exists a polynomial $h \in F[t]$ such that the polynomial $f + tg$ is irreducible).

Then $u(F(t)) \leqslant 6$. Indeed, let $\alpha \in I^2(F(t))$ have nontrivial invariants in $\coprod_i F_{f_i}^*/F_{f_i}^{*2}$ (i runs over a finite set). There exists a polynomial f such that $f \notin F_{f_i}^{*2}$ for every i. Let an irreducible polynomial p satisfy $p \equiv f \pmod{\prod_i f_i}$ and let β be the element of $I^2(F(t))$ that has the only nontrivial invariant precisely in F_p^*/F_p^{*2}. Then

$$\alpha = [\langle\!\langle f_1 f_2 \cdots f_n, f \rangle\!\rangle] + \beta$$

and so $\dim \alpha \leqslant 6$ (n is the order of the set of i).

Unfortunately, there exist fields that satisfy the above conditions, but do not satisfy the prime number theorem. For example, consider the field $\mathbb{Q}_p$ of p-adic numbers ($p \neq 2$) and add to it all roots of p. Denote the resulting field by K. Then consider the maximal odd extension of K, which we denote by F. It is easy to see that the field F satisfies all the necessary conditions. Let us show that the prime number theorem does not hold in this case. The counterexample is as follows. Let f and g be two irreducible polynomials over F such that $\alpha \in F_f^{*2}$, $\beta \in F_g^{*2}$, $\deg f, g \geqslant 2$, $v(\alpha) \neq (\beta)$, where α and β are roots of f and g, and v is the rational valuation induced by the p-adic valuation of $\mathbb{Q}_p$. Let h be an arbitrary irreducible polynomial of $F[t]$ and $h(\gamma) = 0$, $\gamma \in F_h$. Either $v(\gamma) \neq v(\alpha)$ or $v(\gamma) \neq v(\beta)$. Let $v(\gamma) \neq v(\alpha)$.

Evidently $h(\alpha) = \prod_i (\alpha - \gamma_i)$, where γ_i runs over all roots of H in $\overline{F}$, $v(\gamma_i) = v(\gamma_j)$ for all i, j. If $v(\gamma) < v(\alpha)$, then evidently

$$h(\alpha) = \prod_i (\alpha - \gamma_i) \equiv \prod_i \gamma_i = N_{F_h/F}(\gamma) \in F_f^{*2} \pmod{\wp},$$

where $\wp$ is the ideal of the valuation. If $v(\gamma) > v(\alpha)$, then

$$h(\alpha) = \prod_i (\alpha - \gamma_i) \equiv \alpha^{\deg h} \in F_f^{*2},$$

and so if s is a polynomial such that $s \notin F_f^{*2}$ and $s \notin F_g^{*2}$, then $s \not\equiv h \pmod{fg}$ for any irreducible polynomial h.

References

[B] A. Brumer, *Remarques sur les couples de formes quadratiques*, C. R. Acad. Sci. Paris Sér. A **286** (1978), no. 16, 679–681.
[P] R. S. Pierce, *Associative algebras*, Graduate Texts in Math., vol. 88, Springer-Verlag, Heidelberg, 1982.
[L] T. Y. Lam, *The algebraic theory of quadratic forms*, Addison-Wesley, Reading, MA, 1973.
[La] S. Lang, *Algebra*, Addison-Wesley, Reading, MA, 1965.

Translated by THE AUTHOR

St. Petersburg Electrotechnical University, St. Petersburg, Russia

Isogenies of Height One and Filtrations of Formal Groups

A. L. Smirnov

Dedicated to Boarding School #45 on the occasion of its 30th anniversary

ABSTRACT. We study the boundary homomorphism mapping the group of the points of a formal group, defined over the ring of integers of a local field, in the Galois cohomology with coefficients in the kernel of an isogeny of height one. The action of the homomorphism on the natural filtrations and its image are calculated.

Introduction

When calculating arithmetic invariants of an abelian variety A defined over a number field K, we come to the following division problem: for a given isogeny f and a point $P \in A(K)$, find a point $Q \in A(\bar{K})$ such that $f(Q) = P$. Of course, we are interested not in the coordinates of Q, but in invariant characteristics such as the field of definition $K(Q)$ or the class $\delta(P)$ of the cocycle $\sigma \mapsto (\sigma - 1)Q$ in the cohomology group $H^1(K, N)$ (here $N = \ker f$).

In the present paper we deal with this problem for a local field K. In this case the division problem can be made more precise. One can assign to the abelian variety A a formal group F defined over the ring of integers $\mathcal{O}_K$. The points of $F(\mathcal{O}_K)$ constitute a substantial part of all the points of $A(K)$. On the groups $F(\mathcal{O}_K)$ and $H^1(K, N)$ there are natural filtrations, and the problem is to find the location of the class $\delta(P)$ with respect to the filtration of $H^1(K, N)$, if the location of the point P with respect to the filtration of $F(\mathcal{O}_K)$ is known.

For isogenies of one-dimensional formal groups, the division problem is completely solved (see [**Se**] for the case $F = \mathbb{G}_m$, $f = [p]$, see [**Be**] for the case $\mathrm{ht}\, f = 1$, and [**BK**] for the general case).

In the present paper, we solve the division problem for isogenies of height one of formal groups of arbitrary dimension. In §1 we recall certain definitions and fix the notation. Section 2 contains a construction of a special coordinate system (with respect to a given isogeny) on a formal group defined over a field k with $\mathrm{char}\, k = p$. In §3 we use the results of §2 to construct a special coordinate system over a local field

---

1991 *Mathematics Subject Classification.* Primary 14L05.

This research was partially supported by Grant No. 94-01-00915 from the Russian Foundation for Fundamental Research and a grant from the Volkswagen Foundation.

©1996, American Mathematical Society

K. In §4 we describe the natural filtrations on the groups $F(\mathcal{O}_K)$ and $H^1(K,N)$ and prove Theorem 4.5.1, which is our main result. An example of the application of this theorem can be found in [**Sm**].

A part of these results was announced in 1984 in a preprint. The delay with the present publication is due to the desire to obtain the results in a final form.

I use this occasion to thank M. I. Bashmakov and A. N. Kirillov for useful discussions.

§1. Notation and preliminaries

We deal with a ring R, which is a field or a local Dedekind ring. This is why the following point of view is quite convenient for us: a formal variety over R is determined by a ring A, which is isomorphic to the ring $R[[X_1,\ldots,X_d]]$. In the general situation this condition should be true only locally. Let us recall some definitions (see [**L**]).

1.1. Formal series. By X we denote the collection $(X_1,\ldots,X_d)$, where X_i is a variable. By $R[[X]]$ we denote the ring of the formal power series $R[[X_1,\ldots,X_d]]$. If $\alpha = (\alpha_1,\ldots,\alpha_d)$, then $|\alpha| = \sum \alpha_i$ and $X^\alpha = X_1^{\alpha_1}\cdots X_\alpha^{\alpha_d} \in R[[X]]$.

Each $f \in R[[X]]$ can be uniquely written in the form $f = \sum a_\alpha X^\alpha$, where $a_\alpha \in R$. The natural projection $R[[X]] \to R$, $\sum a_\alpha X^\alpha \mapsto a_0$, is denoted as $f \mapsto f(0)$.

1.2. Formal varieties. Let a couple (A,ε) be given, where A is an R-algebra and $\varepsilon: A \to R$ is a homomorphism of R-algebras. We say that the couple (A,ε) defines a *formal variety of dimension d*, if this couple is isomorphic to the couple $(R[[X]], f \mapsto f(0))$.

For a formal variety F, we denote the corresponding algebra A by $\mathcal{O}_F$, the corresponding homomorphism ε by ε_F, and the ideal $\ker \varepsilon_F$ by I_F. With each formal variety F we associate its cotangent space $\Omega_F = I_F/I_F^2$ regarded as an R-module.

Let us also recall the notions of morphisms and products of formal varieties.

(a) A *morphism* $f: F_1 \to F_2$ is defined by a homomorphism of R-algebras $f^*: \mathcal{O}_{F_2} \to \mathcal{O}_{F_1}$ such that $f^* \varepsilon_{F_2} = \varepsilon_{F_1} f^*$ (or, in other words, $f^*(I_{F_2}) \subset I_{F_1}$).

(b) The formal variety $F_1 \times F_2$ corresponds to the R-algebra $\widehat{\mathcal{O}_{F_1} \otimes \mathcal{O}_{F_2}}$, i.e., to the tensor product of the R-algebras $\mathcal{O}_{F_1}$ and $\mathcal{O}_{F_2}$ completed with respect to the ideal $I_{F_1} \otimes \mathcal{O}_{F_2} + \mathcal{O}_{F_1} \otimes I_{F_2}$.

Besides, each homomorphism $f: F_1 \to F_2$ defines a R-linear map $f^*: \Omega_{F_1} \to \Omega_{F_2}$.

1.3. Coordinate systems. Fix a d-dimensional formal variety F and set $A = \mathcal{O}_F$, $I = I_F$.

DEFINITION 1.3.1. A *coordinate system* on F is a collection
$$X = (X_1,\ldots,X_d)$$
of elements of I such that the canonical map, from the ring of formal series over R in d variables to A that sends the ith variable to X_i, is an isomorphism.

Each coordinate system X on F allows us to identify the R-algebra A with $R[[X]]$ and gives a certain basis of the space Ω_F: $dX_i - X_i \bmod I^2$.

If G is another formal variety and $Y = (Y_1,\ldots,Y_e)$ is a coordinate system on G, then we identify the set of morphisms from F to G with the set of collections $(f_1,\ldots,f_e)$ of elements of I. Namely, for a morphism $f: F \to G$, we put $f_i = f^*(Y_i)$. The morphism $f^*: \Omega_F \to \Omega_G$ will be denoted by $f \bmod \deg 2$.

Below we need the following easy lemma.

LEMMA 1.3.2. *Let $X = (X_1, \ldots, X_d)$, $X_i \in I$. The system X is a coordinate system on F iff the set $(dX_1, \ldots, dX_d)$ is a basis of Ω_F.*

1.4. Formal groups. Let F be a formal variety over R. To define the notion of formal group, we need the morphisms $\mathrm{id}_F \colon F \to F$, $0 \colon F \to F$, and $\tau \colon F \times F \to F \times F$. They correspond to the homomorphisms of the R-algebras $\mathcal{O}_F$, $\mathcal{O}_F$, and $\widehat{\mathcal{O}_F \otimes \mathcal{O}_F}$ given by $(\mathrm{id}_F)^*(f) = f$, $0^*(f) = f(0)$, and $\tau^*(f \otimes g) = g \otimes f$.

DEFINITION 1.4.1. A (commutative) *formal group* over R is a (commutative) group object in the category of formal varieties over R. In other words, a formal group is a formal variety F provided with morphisms $\mu \colon F \times F \to F$ (addition) and $i \colon F \to F$ (inverse element) satisfying the following conditions:
(a) $i^2 = \mathrm{id}_F$;
(b) $\mu(\mu \times \mathrm{id}_F) = \mu(\mathrm{id}_F \times \mu)$ (these two morphisms map $F \times F \times F$ to F);
(c) $\mu(\mathrm{id}_F \times 0) = \mathrm{id}_F$ (these morphisms map F to F);
(d) $\mu(\mathrm{id}_F \times i) = 0$ (these morphisms map F to F);
(e) $\mu = \mu\tau$ (these morphisms map $F \times F$ to F).

Let F be a formal group and $X = (X_1, \ldots, X_d)$ a coordinate system on F. The system X gives the coordinate system $(Y, Z) = (Y_1, \ldots, Y_d, Z_1, \ldots, Z_d)$ on $F \times F$, where $Y_i = X_i \otimes 1$, $Z_i = 1 \otimes X_i$. The morphisms i and μ give the systems of formal series $(I_1, \ldots, I_d)$ and $(F_1, \ldots, F_d)$, where $I_k(X) = i^*(X_k) \in R[[X]]$ and $F_k(Y, Z) = \mu^*(X_k) \in R[[Y, Z]]$. The conditions (a)–(e) of Definition 1.4.1 imply the following identities:
(a) $I_k(I_1(X), \ldots, I_d(X)) = X_k$, $F_k(X, I_1(X), \ldots, I_d(X)) = 0$, and $F_k(X, 0) = X_k$;
(b) $F_k(F_1(X, Y), \ldots, F_d(X, Y), Z) = F_k(X, F_1(Y, Z), \ldots, F_d(Y, Z))$;
(c) $F_k(Y, Z) = F_k(Z, Y)$;
(d) $I_k(X) \equiv -X_k \bmod \deg 2$, $F_k(Y, Z) \equiv Y_k + Z_k \bmod \deg 2$.

1.5. Isogenies.

DEFINITION 1.5.1. Let F and G be formal groups. A morphism $f \colon F \to G$ of formal varieties is a *homomorphism of formal groups* if the following diagrams are commutative:

$$\begin{array}{ccc} F \times F & \xrightarrow{f \times f} & G \times G \\ \mu \downarrow & & \downarrow \mu \\ F & \xrightarrow{f} & G \end{array} \qquad \begin{array}{ccc} F & \xrightarrow{f} & G \\ i \downarrow & & \downarrow i \\ F & \xrightarrow{f} & G \end{array}$$

DEFINITION 1.5.2. A homomorphism $f \colon F \to G$ of formal groups is an *isogeny* if $\mathcal{O}_F/(f^*I_G)$ is a finite flat R-algebra.

§2. Isogenies in characteristic p

Let k be a field, $\mathrm{char}\, k = p > 0$.

2.1. Structure of homomorphisms. Let us recall a result of Lazard concerning homomorphisms of formal groups in characteristic p.

PROPOSITION 2.1.1 [**L**, 4.3, p. 174]. *Let $f \colon F \to G$ be a homomorphism of (commutative) formal groups defined over k. Let $\dim F = m$ and $\dim G = n$. There exists* a uniquely defined integer u such that $0 \leqslant u \leqslant \min(m,n)$, and coordinate systems $X = (X_1, \ldots, X_m)$ on F and $Y = (Y_1, \ldots, Y_n)$ on G such that

$$Y_1(X) = X_1^{p^{h_1}}, \ldots, Y_u(X) = X_u^{p^{h_u}}, Y_{u+1}(X) = 0, \ldots, Y_n(X) = 0,$$

where $Y_i(X) = f^*(Y_i) \in R[[X]]$ and the h_i's are nonnegative integers.

It is clear that $u = m = n$ for an isogeny f.

DEFINITION 2.1. The *height* of an isogeny f is the number

$$\text{ht } f = h_1 + \cdots + h_n,$$

where the numbers h_i are defined in Proposition 2.1.1.

2.2. Coordinate system X. Let F be a commutative formal group over k, $d = \dim F$. Let $f: F \to F$ be an isogeny of height one. We denote by r the multiplicity of 0 as of an eigenvalue of the operator f mod deg 2. Let v be the following $(r \times r)$-matrix:

$$v = \begin{pmatrix} 0 & 1 & 0 & \cdots & 0 \\ 0 & 0 & 1 & \cdots & 0 \\ \vdots & \vdots & \vdots & \ddots & \vdots \\ 0 & 0 & 0 & \cdots & 1 \\ 0 & 0 & 0 & \cdots & 0 \end{pmatrix}.$$

PROPOSITION 2.2.1. *There exists a coordinate system* $X = (X_1, \ldots, X_d)$ *on* F, *such that*

(i) f mod deg 2 $= \begin{pmatrix} v & 0 \\ 0 & \lambda \end{pmatrix}$, *where λ is an invertible* $(d - r) \times (d - r)$-*matrix*;

(ii) *the series* $f_1(X) = f^*(X_1)$ *depends on* $X^p = (X_1^p, \ldots, X_d^p)$ *only.*

PROOF. (i) According to Lazard's result (see Proposition 2.1.1), there exist coordinate systems $Y = (Y_1, \ldots, Y_d)$ and $Z = (Z_1, \ldots, Z_d)$ on F such that:

$$f^*: Y_1 \mapsto Z_1^{p^{h_1}}, \ldots, Y_d \mapsto Z_d^{p^{h_d}}.$$

The condition $\text{ht } f = 1$ implies that all the h_i's are equal to 0, except one, which is equal to 1: let $h_1 = 1, h_2 = \cdots = h_d = 0$. We denote by D the operator f mod deg 2. Its form with respect to Y and Z shows that $\text{rk } D = d - 1$ and $\dim \ker D = 1$. Let $\Omega = \Omega_0 \oplus \Omega_1$ be the decomposition of Ω into the root subspaces corresponding to 0 and to all the other eigenvalues of D. Since the operator D is nilpotent on Ω_0 and $\dim \ker D = 1$, there is a basis $v_1, \ldots, v_r$ of Ω_0 with respect to which the matrix of D on Ω_0 is v. Let $v_{r+1}, \ldots, v_d$ be a basis of Ω_1. The matrix of D on Ω in the basis $v_1, \ldots, v_d$ has the form indicated in (i). Take X_i so that $v_i = X_i$ mod deg 2. The coordinate system $X = (X_1, \ldots, X_d)$ satisfies (i). The relation $f^*(Y_1) = Z_1^p$ implies that $D(dY_1(0)) = 0$ and, consequently, $\lambda dY_1(0) = v_1$ for an invertible λ, because $\dim \ker D = 1$. The system $(\lambda Y_1, X_2, \ldots, X_d)$ is a coordinate system on F and satisfies (i) and (ii). □

2.3. Coordinate system Z. Fix the following system $Z = (Z_1, \ldots, Z_d)$ of elements of A: $Z_1 = X_r, Z_2 = f_2, \ldots, Z_d = f_d$. Proposition 2.2.1 shows that Z is a coordinate system on F, and there are series $\xi_1, \ldots, \xi_d \in R[[Z]]$ such that $X_i = \xi_i(Z)$. Put

$$\Phi_1(Z) = f_1(\xi_1(Z), \ldots, \xi_d(Z)) \quad \text{and} \quad \Phi(T) = \Phi_1(T, 0, \ldots, 0).$$

LEMMA 2.3.1. *Let* $\Phi(T) = b_1 T + b_2 T^2 + \cdots$. *Then* $b_p \neq 0$.

PROOF. We have the isomorphisms

$$k[[X]]/(f_1,\ldots,f_d) = k[[Z]]/(\Phi_1, Z_2, \ldots, Z_d) = k[[Z_1]]/(\Phi_1(Z_1, 0, \ldots, 0)).$$

Since ht $f = 1$, we have $\dim k[[Z_1]]/(\Phi(Z_1, 0, \ldots, 0)) = p$ and, thus, $\Phi_1(Z_1, 0, \ldots, 0)$ starts with the term Z_1^p.

§3. Isogenies over the ring of integers of a local field

Let K be a finite extension of $\mathbb{Q}_p$. Let $R = \mathcal{O}_K$ be the ring of integers in K, $\mathfrak{M}$ the maximal ideal in R, k the residue field $R/\mathfrak{M}$. Let $v \colon K^* \to \mathbb{Z}$ be the discrete valuation of K compatible with the standard valuation of $\mathbb{Q}_p$ and normalized so that $v(K^*) = \mathbb{Z}$. We assume that the valuation v is extended to a $\mathbb{Q}$-valued valuation of $\bar{K}$.

3.1. Coordinate system X. Below F is a commutative formal group over R, $\dim F = d$, and $f \colon F \to F$ is an isogeny. We assume that f is an isogeny of height one, i.e., $f \bmod \mathfrak{M}$ is an isogeny of height one over k (see 2.1). By Ω we denote the cotangent space Ω_F and by D the operator $f \bmod \deg 2$ on Ω. Consider the operator $\bar{D} = D \otimes k$ on the space $\bar{\Omega} = \Omega \otimes k$. We denote by r the dimension of the maximal subspace in $\bar{\Omega}$, where $\bar{D}$ acts as a nilpotent operator. It is clear that there are exactly r eigenvalues of D on Ω that are not invertible in the ring of integers of $\bar{K}$. Denote them by $v_1, \ldots, v_r$. We assume that $v_1, \ldots, v_r \in K$.

PROPOSITION 3.1.1. *There exists a coordinate system X on F such that*

(i) *the matrix of D with respect to X has the form* $D = \begin{pmatrix} V & * \\ 0 & U \end{pmatrix}$, *where*

$$U \in \mathrm{GL}_{d-r}(R), \quad * \equiv 0 \bmod \mathfrak{M}, \quad \text{and} \quad V = \begin{pmatrix} v_1 & 1 & 0 & \cdots & 0 \\ 0 & v_2 & 1 & \cdots & 0 \\ \vdots & \vdots & \vdots & \ddots & \vdots \\ 0 & 0 & 0 & \cdots & 1 \\ 0 & 0 & 0 & \cdots & v_r \end{pmatrix};$$

(ii) *the series $f_1 \bmod \mathfrak{M}$ depends on $(X_1^p, \ldots, X_d^p)$ only.*

PROOF. We construct the required system X in several steps. First, we consider a coordinate system Y on F such that the coordinate system $Y \bmod \mathfrak{M}$ satisfies the conclusion of Proposition 2.2.1 (obviously, each coordinate system on $F \bmod \mathfrak{M}$ can be lifted to a coordinate system on F).

The system Y gives the basis $e_i = dY_i(0)$ of the space Ω and the basis $\bar{e}_i$ of $\bar{\Omega}$. Consider the subspace $\bar{\Omega}_k$ of $\bar{\Omega}$ generated by $\bar{e}_1, \ldots, \bar{e}_k$. The form of the matrix of $\bar{D}$ with respect to the coordinate system $Y \bmod \mathfrak{M}$ shows (see 2.2.1) that

(3.1.2) $$\bar{\Omega}_k = \ker \bar{D}^k \quad \text{for } 1 \leqslant k \leqslant r.$$

Let us choose elements $v_1, \ldots, v_r$ of Ω as follows: v_1 is a primitive v_1-eigenvector of D on Ω (an element v of a free R-module M is called *primitive* if M/Rv is also a free R-module). If vectors $v_1, \ldots, v_k$ are given, we choose v_{k+1} so that its image in $\Omega/\langle v_1, \ldots, v_k \rangle$ is a primitive v_{k+1}-eigenvector of the natural action of D on this space (here $\langle v_1, \ldots, v_k \rangle = Rv_1 \oplus \cdots \oplus Rv_k$). Since $v_{k+1} \equiv 0 \bmod \mathfrak{M}$, the construction shows that $\overline{Dv_{k+1}} \in \langle \overline{v}_1, \ldots, \overline{v}_k \rangle$, where $\overline{v}_i = v_i \bmod \mathfrak{M}$. Besides, $\overline{Dv_1} = 0$, and the vectors $\overline{v}_1, \ldots, \overline{v}_r$ are linearly independent. Thus, (3.1.2) yields

(3.1.3) $\qquad \langle \overline{e}_1, \ldots, \overline{e}_k \rangle = \langle \overline{v}_1, \ldots, \overline{v}_k \rangle, \quad \text{for} \quad 1 \leqslant k \leqslant r.$

Let us write the vectors v_i in the basis $e_1, \ldots, e_d$:

$$(v_1, \ldots, v_r)^t = \begin{pmatrix} \lambda \\ \mu \end{pmatrix} (e_1, \ldots, e_d)^t,$$

where λ is an $r \times r$-matrix, μ a $(d-r) \times r$-matrix, t the transposition. The equalities (3.1.3) imply that $\mu \equiv 0 \bmod \mathfrak{M}$ and that $\overline{\lambda} = \lambda \bmod \mathfrak{M}$ is an invertible matrix with zeros under the main diagonal.

Consider the coordinate system $Z = (Z_1, \ldots, Z_d)$ on F given by

$$Z^t = \begin{pmatrix} \lambda & 0 \\ 0 & 1 \end{pmatrix} Y^t,$$

where 1 is the unit $(d-r) \times (d-r)$-matrix. Write the matrix of the operator D with respect to Z in the following form:

$$D = \begin{pmatrix} \alpha & \beta \\ \gamma & \delta \end{pmatrix},$$

where the matrices α, β, γ, and δ are of size $r \times r$, $r \times (d-r)$, $(d-r) \times r$, and $(d-r) \times (d-r)$ respectively. It is easy to see that:
(a) $\gamma = 0$; $\beta \equiv 0 \bmod \mathfrak{M}$; δ is an invertible matrix;
(b) the matrix α has the form

$$\alpha = \begin{pmatrix} v_1 & \alpha_1 & & & * \\ 0 & v_2 & \alpha_2 & & \\ \vdots & \vdots & \vdots & \ddots & \\ 0 & 0 & 0 & \ldots & \alpha_{r-1} \\ 0 & 0 & 0 & \ldots & v_r \end{pmatrix},$$

where $* \equiv 0 \bmod \mathfrak{M}$ and $\alpha_1, \ldots, \alpha_{r-1}$ are invertible;
(c) the series $\overline{f}_1(Z) = f_1^*(Z) \bmod \mathfrak{M}$ depends on Z^p only.

LEMMA 3.1.4. *Let $a \colon R^r \to R^r$ be the linear operator given by the matrix α. There exists a basis $c_1, \ldots, c_r$ of R^r such that*
(a) $ac_1 = v_1 c_1$; $ac_i = v_i c_i + c_{i-1}$ for $i \geqslant 2$;
(b) $c_i \in \langle e_1, \ldots, e_i \rangle$ for $1 \leqslant i \leqslant r$, where $e_1, \ldots, e_r$ is the standard basis of R^r and $\langle e_1, \ldots, e_i \rangle$ is the subspace of R^r generated by $e_1, \ldots, e_i$.

PROOF OF THE LEMMA. Using induction on r, we easily construct a basis $c'_1, \ldots, c'_r$ satisfying condition (b) and the following condition (a'): $ac'_i = v_i c'_i + \lambda_{i-1} c'_{i-1}$ for $i \geqslant 2$, where $\lambda_i \in R^*$.

To construct the basis $c_1, \ldots, c_r$ put: $c_{r-k} = \lambda_{r-1}, \ldots, \lambda_{r-k} c'_{r-k}$. $\square$

We continue with the proof of Proposition 3.1.1. Let ε be an invertible $(r \times r)$-matrix such that $(e_1, \ldots, e_r) \cdot \varepsilon = (c_1, \ldots, c_r)$, where $(e_1, \ldots, e_r)$ and $(c_1, \ldots, c_r)$ are the bases from Lemma 3.1.4.

Put $(X_1,\ldots,X_d) = (Z_1,\ldots,Z_d)\begin{pmatrix}\varepsilon & 0 \\ 0 & 1\end{pmatrix}$, where, as before, 1 denotes the unit $(d-r)\times(d-r)$-matrix. It is clear that the coordinate system $X = (X_1,\ldots,X_d)$ meets the requirements of the Proposition.

3.2. Coordinate system Z. Let us fix the following system $Z = (Z_1,\ldots,Z_d)$ of elements of $\mathcal{O}_F$:
$$Z_1 = X_r, \ Z_2 = f_2, \ \ldots, \ Z_d = f_d.$$

Proposition 3.1.1 shows that Z is a coordinate system on F and, consequently, there are series $\xi_1,\ldots,\xi_d \in R[[Z]]$ such that $X_i = \xi_i(Z)$. In §4 we use Z (resp., X) as the coordinate system on the left (resp., right) hand side of the arrow $F \xrightarrow{f} F$. Then f has the following simple form:

(3.2.1) $$f_1(Z) = \Phi_1(Z), \ f_2(Z) = Z_2, \ \ldots, \ f_d(Z) = Z_d,$$

where $\Phi_1(Z) = f_1(\xi_1(Z),\ldots,\xi_d(Z))$.

3.3. Newton diagrams. We recall some well-known properties of Newton diagrams (see, e.g., [K, BK]). Let $s(T) = a_1T + a_2T^2 + \cdots \in R[[T]]$. Consider the subset S in $\mathbb{R}^2$,
$$S = \{(x,y) : \text{there exists } n \text{ such that } x \geq n, \ y \geq v(a_n)\}.$$
The closed convex hull of S is called the *Newton diagram* of $s(T)$.

Let $(q_0,v_0),\ldots,(q_m,v_m)$ be all the vertices of the Newton diagram ($q_0 < \cdots < q_m$). Set $\alpha_i = -(v_i - v_{i-1})/(q_i - q_{i-1})$ ($1 \leq i \leq m$) and $t_i = \alpha_{m+1-i}$. Set also $t_0 = 0$ and $t_{m+1} = \infty$.

Consider the function $\varphi_s(x)$ given by
$$\varphi_s(x) = q_m t_1 + (t_2 - t_1)q_{m-1} + \cdots + (x - t_k)q_{m-k} \quad \text{for } t_k \leq x \leq t_{k+1}.$$

PROPOSITION 3.3.1. (i) *If $x \in \mathfrak{M}$, then $v(s(x)) \geq \varphi_s(v(x))$. This inequality turns into an equality for $v(x) \notin \{t_1,\ldots,t_m\}$.*

(ii) *If $x \in \mathfrak{M}_{\overline{K}}$ and $s(x) = 0$, then $v(x) \in \{\alpha_1,\ldots,\alpha_m\}$. In this case, there are exactly $(q_i - q_{i-1})$ elements $x \in \mathfrak{M}_{\overline{K}}$ such that $s(x) = 0$ and $v(x) = \alpha_i$.*

3.4. The series $\Phi(T)$. Set $\Phi(T) = \Phi_1(T,0,\ldots,0)$ and let $b_1,b_2,\ldots$ be such that $\Phi(T) = b_1T + b_2T^2 + \cdots$.

PROPOSITION 3.4.1. $v(b_1) = v(\det D)$; $b_i \equiv 0 \bmod (b_1)$ for $i \leq p-1$; $v(b_p) = 0$.

PROOF. Lemma 2.3.1 implies $v(b_p) = 0$. Let us verify that $v(b_1) = v(\det D)$. Consider the ideal J generated by $X_r^2, f_2,\ldots,f_d$ in the ring $R[[X]]$. By the definition of $\Phi(T)$, we have $b_1X_r \equiv \Phi_1(X_r, f_2,\ldots,f_d) \bmod J$. Using $f \bmod \deg 2$ (see Proposition 3.1.1), we obtain the following congruences mod J:

$$X_{r-1} \equiv f_r - v_r X_r \equiv -v_r X_r,$$
$$X_{r-2} \equiv -v_{r-1}X_{r-1} \equiv (-1)^2 v_r v_{r-1} X_r,$$
$$X_{r-3} \equiv -v_{r-2}X_{r-2} \equiv (-1)^3 v_r v_{r-1} v_{r-2} X_r,$$
$$\cdots\cdots\cdots\cdots\cdots\cdots\cdots\cdots$$
$$X_1 \equiv -v_2 X_2 \equiv (-1)^r v_2 \cdots v_r X_r,$$
$$f_1 \equiv v_1 X_1 \equiv (-1)^r v_1 \cdots v_r X_r.$$

Consequently, $v(b_1) = v(v_1 \cdots v_r) = v(\det D)$.

Here we use the notation from 4.1 and 4.3. Let $N = \{0, P_1, \ldots, P_{p-1}\}$. Since multiplication by n is an automorphism for $(n, p) = 1$, we have $v_F(P_i) = v_F(P_j)$. Since $Z_2(P_i) = \cdots = Z_d(P_i) = 0$, we have $v_F(P_i) = v(y_i)$, where $y_i = X_r(P_i)$. Consequently, the equation $\Phi(y) = 0$ has $p - 1$ solutions $y_1, \ldots, y_{p-1}$, and $v(y_i) = v(y_j)$.

The assertion $b_i \equiv 0 \bmod b_1$ follows from the connection between the zeros of the series $\Phi(T)$ and its Newton diagram (Proposition 3.3.1). $\square$

Proposition 4.3.1 means that the Newton diagram for Φ (see 3.3) has the form shown in Figure 1.

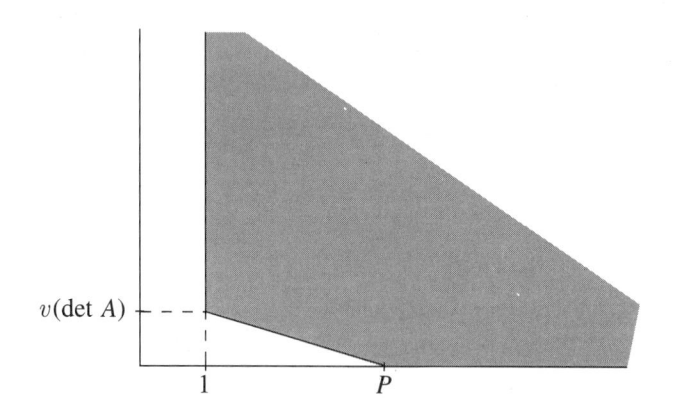

FIGURE 1

COROLLARY 3.4.2. (1) *For any* $P \in N \setminus \{0\}$, *we have* $v(P) = t$, *where* $t = v(\det D)/(p - 1)$.

(2) *The function* φ_Φ *(see 3.2) has the form*

$$\varphi_\Phi(i) = \begin{cases} pi, & \text{if } 0 \leqslant i \leqslant t, \\ pt + i - t, & \text{if } i \geqslant t. \end{cases}$$

§4. Isogenies and filtrations

We keep the notation of §3.

4.1. Filtration on formal varieties. Let F be a formal variety over R.

DEFINITION 4.1.1. An *R-point* of F is a homomorphism of R-algebras $P: \mathcal{O}_F \to R$ taking I_F into $\mathfrak{M}$ continuously with respect to the I_F-adic topology of $\mathcal{O}_F$ and the $\mathfrak{M}$-adic topology of R.

The set of all the R-points of F is denoted by $F(R)$.

DEFINITION 4.1.2. Let P be a R-point of F. We set

$$v_F(P) = \min\{v(x) \mid x \in P(I_F)\} \quad \text{and} \quad F(R)^n = \{P \in F(R) \mid v_F(P) \geqslant n\}.$$

Since the group F is fixed, we often write $v(P)$ instead of $v_F(P)$, and we write $v_L(P)$ if F is considered over the ring $\mathcal{O}_L$, where L is an finite extension of K.

If $X = (X_1, \ldots, X_d)$ is a coordinate system on F, then we identify $F(R)$ with the set $\mathfrak{M} \times \cdots \times \mathfrak{M}$ (d times): $P \mapsto (P(X_1), \ldots, P(X_d))$. In this case we write $P = (x_1, \ldots, x_d)$ and $x_i = X_i(P)$. It is clear that $v_F(P) = \min\{v(x_i) \mid 1 \leqslant i \leqslant d\}$ and $F(R)^n \approx \mathfrak{M}^n \times \cdots \times \mathfrak{M}^n$.

For a morphism $f: F \to G$ of formal varieties over R, we have a natural map $F(R) \to G(R)$, $P \mapsto P \circ f^*$, which is also denoted by f.

If $Y = (Y_1, \ldots, Y_n)$ is a coordinate system on G and the morphism f is given by series $f_1, \ldots, f_n \in R[[X]]$ (see 1.3.1), then $f(P) = (y_1, \ldots, y_n)$, where $y_i = f_i(x_1, \ldots, x_d)$ (it is clear that these series are convergent).

4.2. Filtration on formal groups. Let F be a formal group over R, $\dim F = d$.

The morphisms μ and i (see 1.4) introduce a group structure on $F(R)$ and $F^n(R)$ is a subgroup in $F(R)$ since $(F \times F)^n(R) = F^n(R) \times F^n(R)$. If $X = (X_1, \ldots, X_d)$ is a coordinate system on F and $x = (x_1, \ldots, x_d)$, $y = (y_1, \ldots, y_d)$ are two points of $F(R)$, then their sum (difference) in the group $F(R)$ is denoted by $x \underset{F}{\pm} y$.

LEMMA 4.2.1. *Let $x, y \in F^n(R)$. Then $x \underset{F}{\pm} y \equiv x \pm y \mod F^{n+1}(R)$, where $F^k(R)$ is identified with $\mathfrak{M}^k \times \cdots \times \mathfrak{M}^k$ and the addition (subtraction) in the right-hand side is the usual addition (subtraction) in $\mathfrak{M} \times \cdots \times \mathfrak{M}$.*

PROOF. The lemma follows immediately from the property (d) of 1.4. $\square$

PROPOSITION 4.2.2. *Let L be a finite Galois extension of K, let $G = \mathrm{Gal}(L/K)$, and $\sigma \in G_i \setminus G_{i+1}$, where G_j is the lower filtration (see [Se, Chapter IV]), $i \geqslant 1$. Then for $P \in F(\mathcal{O}_L)$ we have*

$$v_L(\sigma P \underset{F}{-} P) \geqslant v_L(P) + i.$$

Here the equality holds if $(v_L(P), p) = 1$.

PROOF. We need the following assertions (1) and (2).

(1) Let $x = (x_1, \ldots, x_d)$, $y = (y_1, \ldots, y_d) \in \mathfrak{M}_L \times \cdots \times \mathfrak{M}_L$. If $|\alpha| \geqslant 2$, then $v_L(x^\alpha - y^\alpha) > v_L(x - y)$.

To verify this inequality, consider the identity

$$x^\alpha - y^\alpha = (x_1^{\alpha_1} - y_1^{\alpha_1})x_2^{\alpha_2} \cdots x_d^{\alpha_d} + y_1^{\alpha_1}(x_2^{\alpha_2} - y_2^{\alpha_2})x_3^{\alpha_3} \cdots x_d^{\alpha_d} + \cdots$$
$$+ y_1^{\alpha_1} \cdots y_{d-2}^{\alpha_{d-2}}(x_{d-1}^{\alpha_{d-1}} - y_{d-1}^{\alpha_{d-1}})x_d^{\alpha_d} + y_1^{\alpha_1} \cdots y_{d-1}^{\alpha_{d-1}}(x_d^{\alpha_d} - y_d^{\alpha_d}),$$

and note that the value of v_L for each summand is greater than $v_L(x - y)$.

(2) Let $x \in \mathfrak{M}_L$. Then $v_L(x^\sigma - x) \geqslant i + v_L(x)$ and the equality holds if $(v_L(x), p) = 1$.

To verify this, write $x = \Pi^a \cdot \xi$, where $a = v_L(x)$, $\xi \in \mathcal{O}_L^*$, $v_L(\Pi) = 1$. Since $\sigma \in G_i$ and $i \geqslant 0$, we have $\Pi^\sigma - \Pi = \eta \Pi^{i+1}$, where $\eta \in \mathcal{O}_L^*$ (see [Se, Chapter IV, Prop. 5]). Besides, by the definition of the groups G_k (loc. cit.) $\xi^\sigma = \xi + y$, where $v_L(y) \geqslant i + 1$. Thus,

$$x^\sigma - x = (\Pi + \eta \Pi^{i+1})^a \cdot (\xi + y) - \Pi^a \cdot \xi = a \cdot \eta \Pi^{a+i} \xi + \cdots.$$

This proves (2).

Consider the series $F_k(X, Y)$ (see 1.4):

$$F_k(X, Y) = X_k + Y_k + \sum_{|\alpha+\beta| \geqslant 2} a_k(\alpha, \beta) X^\alpha Y^\beta.$$

Substituting $I(Y) = (I_1(Y), \ldots, I_d(Y))$ (see 1.4) for Y we get:

(4.2.3) $$F_k(X, I(Y)) = X_k + I_k(Y) + \sum_{|\alpha+\beta|\geq 2} a_k(\alpha, \beta) X^\alpha I(Y)^\beta.$$

Substituting Y for X in (4.2.3) and taking relation (d) of 1.4 into account, we get

(4.2.4) $$0 = Y_k + I_k(Y) + \sum_{|\alpha+\beta|\geq 2} a_k(\alpha, \beta) Y^\alpha I(Y)^\beta.$$

Subtracting (4.2.4) from (4.2.3) and substituting x^σ for X and x for Y we get:

(4.2.5) $$(x^\sigma \underset{F}{-} x)_k = x_k^\sigma - x_k + \sum_{|\alpha+\beta|\geq 2} a_k(\alpha, \beta)((x^\sigma)^\alpha - x^\alpha) I(x)^\beta.$$

Let us show that the v_L of each term in the right-hand side of (4.2.5) is greater than $v_L(x) + i$. This is true for $x_k^\sigma - x_k$ by assertion (2) (see above). If $|\alpha| = 1$, then $v_L(x^\sigma)^\alpha - x^\alpha = v_L(x_j^\sigma - x_j)$ for some j. Therefore,

$$v_L(a_k(\alpha, \beta)((x^\sigma)^\alpha - x^\alpha) I(x)^\beta) \geq v_L(x_j^\sigma - x_j) \geq v_L(x) + i.$$

If $|\alpha| \geq 2$, our inequality follows by assertion (1) (see above). Thus, assertion (i) is proved. Assertion (ii) can be proved by the same arguments, taking into account the second part of (2) and the fact that $v_L(I(x)^\beta) > 0$ for $|\beta| > 0$. □

4.3. Filtrations on the groups C and H. We set $C = F(R)/f(F(R))$. The natural filtration $F^i(R)$ (see 4.2) gives a natural filtration on the group C:

$$C^i = \text{image of } F^i(R) \text{ in } C = F^i(R)/[F^i(R) \cap f(F(R))].$$

We set $w(P) = i$ if $P \in C^i \setminus C^{i+1}$ and $w(0) = \infty$.

Consider the group $N = \ker[f \colon F(\mathcal{O}_{\bar{K}}) \to F(\mathcal{O}_{\bar{K}})]$. Since ht $f = 1$, we have $N \cong \mathbb{Z}/p\mathbb{Z}$.

We set $H = H^1(G, N)$, where $G = \text{Gal}(\bar{K}/K)$. The upper filtration G^u (see [**Se**, Chapter IV]) gives the following filtration on the group H:

$$H_u = \ker[\text{res} \colon H^1(G, N) \to H^1(G^u, N)].$$

We define $u(\delta) = u$, if $\delta \in H_u \setminus H_{u-\varepsilon}$ for each $\varepsilon > 0$.

4.4. Description of H. In the case $\mu_p \subset K$ and $N \subset F(R)$, the filtration H_u can be explicitly described. Namely, in this case we have

$$H \approx H^1(G, \mu_p) \approx K/K^p \quad \text{(Hilbert 90)}.$$

We have the following filtration on the group $K/K^p = V$: $V^i = $ image of U_i, where $U_i = \{u \in K^* \mid u \equiv 1 \bmod \mathfrak{M}^i\}$. The following description of the filtration H_u can be easily deduced from results of [**Se**, Chapter XV, §2].

THEOREM 4.4.1. *Let $e_0 = v(p)/(p-1)$. Then*

(i) $V^i/V^{i+1} = \begin{cases} 0, & \text{if } i \geq pe_0 + 1 \text{ or } (i \equiv 0 \bmod p, i \neq 0, i \neq pe_0), \\ \mathbb{Z}/p\mathbb{Z}, & \text{if } i = 0 \text{ or } i = pe_0, \\ R/\mathfrak{M}, & \text{otherwise;} \end{cases}$

(ii) $V^i = H_u$, *where* $u = pe_0 - i$.

4.5. Main theorem. The exact sequence of G-modules $0 \to N \to F(\mathcal{O}_{\overline{K}}) \xrightarrow{f} F(\mathcal{O}_{\overline{K}}) \to 0$ gives the boundary homomorphism $\delta\colon F(R)/\operatorname{Im}(f) \to H^1(G, N)$, i.e., the homomorphism $\delta\colon C \to H$.

THEOREM 4.5.1. *Assume that $N \subset F(R)$ and all the eigenvalues of f mod deg 2, which are not invertible in $\mathcal{O}_{\overline{K}}$, lie in K. Then*

(i) $C^i/C^{i+1} = \begin{cases} 0, & \text{if } i \geqslant pt+1 \text{ or } (i \equiv 0 \bmod p,\ i \neq pt), \\ \mathbb{Z}/p\mathbb{Z}, & \text{if } i = pt, \\ R/\mathfrak{M}, & \text{otherwise;} \end{cases}$

(ii) $u(\delta(P)) = pt - w(P)$;
(iii) *if $\mu_p \in K$, then $\delta(C) = H_u$, where $u = 1 + p(e_0 - t)$.*

PROOF. Take the coordinate systems $X = (X_1, \ldots, X_d)$ as in 3.1 and $Z = (Z_1, \ldots, Z_d)$ as in 3.2. The isogeny f is given by $f\colon (z_1, \ldots, z_d) \mapsto (x_1, \ldots, x_d)$, where $x_1 = \Phi_1(z_1, \ldots, z_d)$ and $x_i = z_i$ for $i \geqslant 2$.

LEMMA 4.5.2. *Let $P \in F(R)$. There exists a point $P' \in F(R)$ such that:*
(a) $P' \equiv P(\operatorname{Im}(f))$;
(b) *if $x = (x_1, \ldots, x_d) = X(P')$, then $v(x_i) > v(x_1)$ for $i \geqslant 2$;*
(c) $v(x_1) = w(P)$.

PROOF OF THE LEMMA. Consider the following set $S \subset F(R)$:

$$Q \in S \text{ iff } Q \equiv P \bmod \operatorname{Im}(f) \text{ and } v(Q) = w(P).$$

It is clear that S is a closed subset of $F(R)$. Consider the following function m on S: $m(Q) = \min\{v(X_i(Q)) \mid 2 \leqslant i \leqslant d\}$. Let the function m be maximal at the point $P' \in S$. By construction, P' satisfies property (a). Let us show that P' also satisfies property (b). If $m(P') \leqslant v(X_1(P'))$, then consider the point

$$P'' = P' \underset{F}{-} (\Phi_1(0, x_2, \ldots, x_d), x_2, \ldots, x_d).$$

It is clear that $v(\Phi_1(0, x_2, \ldots, x_d)) \geqslant m$ and, consequently, we have the following congruence:

$$P'' \equiv (y_1, 0, \ldots, 0) \bmod F^{m+1}(R).$$

We have $P'' \equiv P' (\bmod \operatorname{Im}(f))$, because the point

$$(\Phi_1(0, x_2, \ldots, x_d), x_2, \ldots, x_d)$$

lies in $\operatorname{Im}(f)$ (namely, it is $f(Q)$, where $Z_1(Q) = 0$, $Z_i(Q) = x_i$). Besides, $v(P'') \geqslant \min(v(P'), v((\Phi_1(0, x_1, \ldots, x_d), x_2, \ldots, x_d))) = \min(w, m) = w$ and, consequently, $v(P'') = w$ (let us recall that w is the maximum v on $P \underset{F}{+} \operatorname{Im}(f)$). Thus $P'' \in S'$.

But we have $m(P'') > m$. This contradicts the maximality $m(P')$. Consequently, the assumption that $m(P') \leqslant v(x_1(P'))$ is false and P' satisfies (b). However, (b) implies that $v(x_1) = v(P') = w(P)$. □

LEMMA 4.5.3. *Let $P \in F(\mathcal{O}_L)$, $v_L(X_i(P)) > v_L(X_1(P))$ for $i \geqslant 2$. If $P = \varphi(Q)$, then $v_L(\Phi(z_1)) = v_L(x_1)$, $v_L(z_i) > v_L(z_1)$ for $i \geqslant 2$. Here $z_i = Z_i(Q)$ and $x_i = X_i(P)$.*

PROOF OF THE LEMMA. We have $z_i = x_i$ for $i \geq 2$ and, consequently, $x_1 \equiv \Phi(Z_1)$ mod Ω, where Ω is the ideal generated by $x_2, \ldots, x_d$. Since $x_1 \not\equiv 0 \mod \Omega$, we have $\Phi(z_1) \not\equiv 0 \mod \Omega$ and $z_1 \not\equiv 0 \mod \Omega$. □

Let us continue the proof of the theorem.

(i) Let $P = (x_1, \ldots, x_d)$, $v(P) > pt + 1$. Let us show that $P \in \text{Im}(f)$. We can assume that P satisfies Lemma 4.5.2. By Proposition 3.3.1, there exists a $z \in \mathfrak{M}$ such that $\Phi(z) = x_1$. Consider the point $P_1 = P - \underset{F}{-}(\Phi(z), 0, \ldots, 0)$. We have $v(P_1) > v(P)$. But $v(P) = w(P)$ is the maximal value of v on $P + \text{Im}(f)$. Therefore, $w(P) = \infty$ and $P \in \text{Im}(f)$. The same argument shows that $C^{kp} = C^{kp+1}$ for $k \neq t$.

Let us show that $C^i/C^{i+1} \cong R/\mathfrak{M}$ for $i < pt$, $(i, p) = 1$. Consider the projection $M/F^{i+1}(R) \to C^i/C^{i+1}$, where

$$M = \{x \mid v(x_1) \geq i, v(x_j) \geq i + 1 \text{ for } j \geq 2\}.$$

Lemma 4.5.2 shows that it is a surjection. Let us show that it is also on injection. Let $x = (x_1, \ldots, x_d) \in M$ and $x \equiv \varphi(Q) \mod F^{i+1}(R)$. By Lemma 4.5.3, we have $x_1 \equiv \Phi(z_1) \mod \mathfrak{M}^{i+1}$ (here $z_i = Z_i(Q)$) and Proposition 3.3.1 shows that $x_1 \in \mathfrak{M}^{i+1}$. Since $M/F^{i+1}(R) \cong R/\mathfrak{M}$, we get the isomorphism $C^i/C^{i+1} \cong R/\mathfrak{M}$. A similar argument shows that $C^{pt}/C^{pt+1} \cong \mathbb{Z}/p\mathbb{Z}$.

(ii) Let $w(P) = w$, $K_P = K(z_1, \ldots, z_d)$, where $z_i = Z_i(Q)$, $f(Q) = P$. We assume that P satisfies Lemma 4.5.2. Then by Lemma 4.5.3 we have $v_L(Q) = v_L(z_1) = w$ (take into account Proposition 3.3.1).

We have $v_L(Q^\sigma - Q) = v_L(R) = pt$ (here σ is a nontrivial element of $\text{Gal}(k_p/k) \cong \mathbb{Z}/p\mathbb{Z}$, R is a generator of N). To see that, use Proposition 3.3.1. According to Proposition 4.2.2 we have $\sigma \in (G(p))_j \setminus (G(p))_{j+1}$, where $j = pt_1 - w$, $G(P) = \text{Gal}(K_p/K)$. Taking into account Theorem 4.4.1, we complete the proof of (ii).

(iii) This follows by (i) and (ii). □

4.6. Natural filtration. In the definition of the filtration H_u, we took into account only the upper filtration on the Galois group G. However, the G-module N arises from a certain group scheme over R and, consequently, N can also be equipped by a filtration. Thus, it is natural to set $H_{(v)} = H_u$, where $v = u - pt$. It is easy to see that t depends on N itself, but not on F and f.

Besides, Theorem 4.5.1 compares the decreasing filtration C^i with the increasing filtration H_u. It is natural to convert the latter into a decreasing filtration and set

$$H^v = H_u, \quad \text{where } v = pt - u \text{ and } v(\delta) = pt - u(\delta).$$

Now, the assertions (ii) and (iii) of Theorem 4.5.1 take the following natural form:

(ii') $v(\delta(P)) = w(P)$,

(iii') $\delta(C) = H^{>0}$, where $H^{>0}$ is the union of all H^ε's with $\varepsilon > 0$.

4.7. Behavior of $w(P)$ with respect to extensions. Let L be a finite Galois p-extension of K, $\psi_{L/K}$ is the Herbrand function of L/K. The function c related to L is denoted by c_L.

PROPOSITION 4.7.1. *Let $P \in F(R)$. If $c(P)$ is not a break point of $\psi_{L/K}$, then $c_L(P) = \psi_{L/K}(c(P))$.*

PROOF. Since $\psi_{L_2/K} = \psi_{L_2/L_1} \cdot \psi_{L_1/K}$ (for $L_2 \supset L_1 \supset K$), we can assume that $\mathrm{Gal}(L/K) \cong \mathbb{Z}/p\mathbb{Z}$. Let $Q \in F(\mathcal{O}_{\overline{K}})$, $f(Q) = P$. Denote by $K(Q)$ the definition field of Q. Let $G_1 = \mathrm{Gal}(K(Q)/K)$ and $G_2 = \mathrm{Gal}(L/K)$. Let u and v be the jumps of the upper filtration of G_1 and G_2. Theorem 4.5.1 shows that $u = c(P)$. Since $c(P) \neq v$, $\mathrm{Gal}(L(Q)/K) \cong G_1 \times G_2$ and we know its upper filtration. Consequently, we can learn the Herbrand function $\psi_{L(Q)/L}$ and its jump, which gives $c_L(P)$ (Theorem 4.5.1). We omit the details of the calculation. □

References

[Se] J.-P. Serre, *Local fields*, Springer-Verlag, Berlin–Heidelberg–New York, 1979.

[L] M. Lazard, *Commutative formal groups*, Lecture Notes in Math., vol. 443, Springer-Verlag, Berlin–Heidelberg–New York, 1975.

[Be] V. Berkovich, *Division by an isogeny of points on an elliptic curve*, Mat. Sb. **93** (1974), 465–486; English transl. in Math. USSR-Sb. **22** (1974).

[BK] M. I. Bashmakov and A. N. Kirillov, *The Lutz filtration in formal groups*, Izv. Akad. Nauk SSSR Ser. Mat. **39** (1975), 1227–1239; English transl., Math. USSR Izv. **9** (1975), no. 6, 1155–1167.

[K] N. Koblitz, *p-adic numbers, p-adic analysis, and zeta-functions*, Springer-Verlag, Berlin–Heidelberg–New York, 1977.

[Sm] A. Smirnov, *Fontaine's inequality for torsors*, Algebra i Analiz **4** (1992), no. 5, 219–226; English transl., St. Petersburg Math. J. **4** (1993), no. 5, 1021–1027.

Translated by THE AUTHOR

POMI, FONTANKA 27, ST. PETERSBURG 191011, RUSSIA

Homology Stability for H-Unital $\mathbb{Q}$-Algebras

A. A. Suslin

Dedicated to Boarding School #45 on the occasion of its 30th anniversary

Introduction

Homology stability for rings with a unit is a well-known and well-studied problem. It concerns the sequence of canonical homomorphisms

$$H_i(\mathrm{GL}_1(R)) \to \cdots \to H_i(\mathrm{GL}_n(R)) \to \cdots \to H_i(\mathrm{GL}(R))$$

(all the unspecified homology groups are always taken with integer coefficients). The main results show that for rings of finite stable rank (e.g., for commutative rings of finite Krull dimension) this sequence stabilizes from a certain point depending on R (see [4, 6, 9]). On the other hand, it has been known for a while (see [2]) that homology stability for the congruence subgroup $\mathrm{GL}(A) = \mathrm{GL}(R, A)$ (here A is a two-sided ideal in R) implies that a nonunital ring A satisfies excision in algebraic K-theory. Thus in general there is no hope to prove results concerning homology stability for $\mathrm{GL}(A)$ unless A satisfies excision in algebraic K-theory. The excision problem in the class of $\mathbb{Q}$-algebras was settled in [11]. It turned out that a $\mathbb{Q}$-algebra A satisfies excision if and only if it is H-unital. Here we show that for H-unital $\mathbb{Q}$-algebras homology stability is also valid. The range of stability obtained in this paper is hardly the best possible (presumably the range of stability for H-unital $\mathbb{Q}$-algebras should be the same as for rings with unit, see [9]), however it is not very far from the best possible one and anyway is sufficient for the applications. The main application of our results is the following nonstable version of the Karoubi conjecture [5]: if A is a stable C^*-algebra, then the natural map $\mathrm{BGL}_n(A) \to \mathrm{BGL}_n(A)^{\mathrm{top}}$ induces isomorphism on homology (and hence $\mathrm{BGL}_n(A)^+ \to \mathrm{BGL}_n(A)^{\mathrm{top}}$ is a homotopy equivalence) for any n.

In the proof of our stability theorem, we combine the approaches to this problem developed in [4] and [9]. The main ingredients of the proof are the acyclicity theorem for the simplicial complex of special unimodular functions (proved in §2) and the acyclicity theorem for the sum of triangular complexes (proved in §4). In [9] the first theorem was used to prove stability for homology of the Volodin space, while the second was used to relate homology of GL_n to the homology of the corresponding Volodin space. One new difficulty appearing in the present situation is that we are not able to separate these problems. The only way out is to prove (using a rather complicated induction) all stability results simultaneously. We do it in §6. In the course of the proof, we use certain comparsion results for spectral sequences. Though these results are certainly well known to specialists, I found it easier to give a detailed proof than to give a reference.

The excision problem in the class of all rings was recently settled in [10]. It is tempting to prove a more general stability theorem, which would apply to all rings satisfying excision (this amounts to the proof of the acyclicity theorem for the sum of triangular complexes). So far I was not able to do it (probably being too lazy).

§1. Simplicial complexes

Recall that a *simplicial complex* is a pair $(V, \mathcal{D})$ consisting of a set V and a set $\mathcal{D}$ of finite nonempty subsets of V (called *simplices*) subject to the only condition that a nonempty subset of a simplex is again a simplex (see [6, Chapter 3]). Usually one demands additionally that every one-element subset of V be a simplex. However it will be more convenient for our purposes to waive this requirement. An *ordered* simplicial complex is a simplicial complex $(V, \mathcal{D})$ together with a partial ordering on the set V such that every simplex is totally ordered.

For any simplicial complex $(V, \mathcal{D})$, we denote by $C_*(V, \mathcal{D})$ the oriented chain complex of $(V, \mathcal{D})$ (see [8, Chapter 4, §1]) and by $\widetilde{C}_*(V, \mathcal{D})$ we denote the corresponding augmented complex (i.e., $\widetilde{C}_{-1} = \mathbb{Z}$ and $\widetilde{C}_n = C_n$ for $n \geq 0$). For an ordered simplicial complex $(V, \mathcal{D})$ the group $C_n(V, \mathcal{D})$ may be identified with a free abelian group generated by n-simplices of $(V, \mathcal{D})$. After this identification the differential takes the form

$$d(D) = \sum_{i=0}^{n}(-1)^i d_i(D),$$

where d_i is the ith face operator given by

$$d_i(D) = \{v_0, \ldots, \widehat{v_i}, \ldots, v_n\} \quad \text{if } D = \{v_0 < v_1 < \cdots < v_n\}.$$

We shall say that a simplicial complex $(V, \mathcal{D})$ is t-*acyclic* if $\widetilde{H}_i(V, \mathcal{D}) = H_i(\widetilde{C}_*(V, \mathcal{D})) = 0$ for $i \leq t$. Thus the condition of t-acyclicity is void for $t \leq -2$, while (-1)-acyclicity is equivalent to nonemptiness.

Let $(V', \mathcal{D}')$ and $(V'', \mathcal{D}'')$ be two simplicial complexes. Set $V = V' \coprod V''$ and let $\mathcal{D}$ be the set of those nonempty subsets $D \subset V$ for which $D \cap V'$ and $D \cap V''$ are either empty or belong to $\mathcal{D}'$ and $\mathcal{D}''$ respectively. The simplicial complex $(V, \mathcal{D})$ is called the *join* of $(V', \mathcal{D}')$ and $(V'', \mathcal{D}'')$ and is denoted $(V', \mathcal{D}') * (V'', \mathcal{D}'')$. Note that we have a canonical isomorphism of chain complexes

$$\widetilde{C}_*((V', \mathcal{D}') * (V'', \mathcal{D}''))[-1] = \widetilde{C}_*(V', \mathcal{D}')[-1] \otimes \widetilde{C}_*(V'', \mathcal{D}'')[-1].$$

This isomorphism together with the Künneth formula immediately implies the following lemma.

LEMMA 1.1. *Assume that $(V', \mathcal{D}')$ is s-acyclic and $(V'', \mathcal{D}'')$ is t-acyclic. Then $(V', \mathcal{D}') * (V'', \mathcal{D}'')$ is $(s + t + 2)$-acyclic.*

Let X be a set and let $m \geq 1$ be an integer. Denote by $F_m(X)$ the set of X-valued functions defined on nonempty subsets of $\{1, \ldots, m\}$. For any function $f \in F$, we shall denote by $\mathrm{dom}(f)$ the domain of definition of f. We shall say that g is a subfunction of f if and only if $\mathrm{dom}(g) \subset \mathrm{dom}(f)$ and $g = f|_{\mathrm{dom}(g)}$. If $f_1, \ldots, f_k \in F_m(X)$ are such that $\mathrm{dom}(f_i) \cap \mathrm{dom}(f_j) = \emptyset$ for $i \neq j$, then $(f_1, \ldots, f_k)$ will denote the only function f such that $\mathrm{dom}(f) = \bigcup_{i=1}^{k} \mathrm{dom}(f_i)$ and $f|_{\mathrm{dom}(f_i)} = f_i$.

We shall say (following van der Kallen [4]) that $F \subset F_m(X)$ *satisfies the chain condition* if F contains all subfunctions of every function $f \in F$.

Assume that $F \subset F_m(X)$ satisfies the chain condition. Set $V = \{1, \ldots, m\} \times X$ and let $\mathcal{D}$ be the set of graphs of all functions $f \in F$. The resulting simplicial complex will be denoted by F_* in the sequel. Identifying functions and their graphs, one may say that n-simplices of F_* are functions $f \in F$ with $|\mathrm{dom}(f)| = n+1$. If $f \in F$ is any function, then we shall use the symbol f_* (resp. $\dot{f}_*$) to denote the subcomplex of F_* consisting of all subfunctions of f (resp. of all proper subfunctions of f). Note that $f_* \cong \Delta^n$ and $\dot{f}_* = \dot{\Delta}^n = S^{n-1}$, where $n = |\mathrm{dim}(f)| - 1$, in particular, f_* is ∞-acyclic and $\dot{f}$ is $(|\mathrm{dom}(f)| - 3)$-acyclic. Finally, set $F_f = \{g \in F : \mathrm{dom}(g) \cap \mathrm{dom}(f) = \emptyset, (f, g) \in F\}$.

LEMMA 1.2. *Let $F \subset G$ be two subsets of $F_m(X)$ satisfying the chain condition and let d be an integer. Assume further that the following conditions hold*:
 a) *if $f \in F$ and $f|_{\{i\}} \in G$ for all $i \in \mathrm{dom}(f)$, then $f \in G$ (in other words every simplex of F_* all vertices of which are in G_* belongs to G_*)*;
 b) *G_* is d-acyclic*;
 c) *If $f \in F$ is a function such that $f|_{\{i\}} \notin G$ for every $i \in \mathrm{dom}(f)$, then $(G \cap F_f)_*$ is $(d - |\mathrm{dom}(f)|)$-acyclic*.

Then F_* is also d-acyclic.

PROOF. For any function $f \in F$ set
$$l(f) = |\{i \in \mathrm{dom}(f) : f|_{\{i\}} \notin G\}|, \qquad F^{(k)} = \{f \in F : l(f) \leqslant k\}.$$

It is clear that all $F^{(k)}$ satisfy the chain condition and we have the following chain of inclusions: $G = F^{(0)} \subset F^{(1)} \subset \cdots \subset F^{(m)} = F$. Using induction on k, we shall show that $(F^{(k)})_*$ is d-acyclic for all k. The statement is true for $k = 0$ in view of condition b). Further, we have the following formula:
$$(F^{(k)})_* = (F^{(k-1)})_* \cup \bigcup_{k = l(f) = \mathrm{dom}(f)} (G \cap F_f)_* * (f_*).$$

Moreover, the intersection of $(G \cap F_f)_* * (f_*)$ with the union of all other members of the above formula coincides with its intersection with $(F^{(k-1)})_*$ and is equal to $(G \cap F_f)_* * (\dot{f}_*)$. Thus $(F^{(k)})_*$ may be obtained from $(F^{(k-1)})_*$ attaching consecutively $(G \cap F_f)_* * (f_*)$ along $(G \cap F_f)_* * (\dot{f}_*)$. Since $(G \cap F_f)_* * (f_*)$ is ∞-acyclic (actually contractible) and $(G \cap F_f)_* * (\dot{f}_*)$ is $(d - 1)$-acyclic (according to condition c) and Lemma 1.1) the exact sequence of Mayer–Vietoris shows that such attaching does not affect the d-acyclicity.

REMARK 1.3. A typical situation when the condition a) of the above lemma is satisfied is the following one: we are given the subsets $X_1, \ldots, X_m$ of X and G consists of those functions $f \in F$ for which $f(i) \in X_i$ for all $i \in \mathrm{dom}(f)$. Each time when we use Lemma 1.2 in the sequel, the relationship between F and G will be of this kind.

§2. The acyclicity theorem

For a (unital) ring R denote by R^∞ the free left R-module with basis $e_1, \ldots, e_n, \ldots$ and by R^n denote the submodule of R with basis $e_1, \ldots, e_n$. A vector $v \in R^n$

will be identified with the corresponding column of coordinates of height n. We say that vectors $v_1, \ldots, v_m \in R^\infty$ are *jointly unimodular* (or form an *unimodular frame*) if they form a basis of a free direct summand of R^∞. Denote by $U_m(R^n)$ the subset of $F_m(R^n)$ consisting of functions f such that $f(i_1), \ldots, f(i_k)$ is a unimodular frame (here $\text{dom}(f) = \{i_1, \ldots, i_k\}$). If A is a two-sided ideal in R, then we set

$$U_{m,A}(R^n) = \{f \in U_m(R^n) : f(i) \equiv e_i \pmod{A} \text{ for all } i \in \text{dom}(f)\}.$$

Note that if A is a proper ideal, then conditions $f(i) \in R^n$ and $f(i) \equiv e_i \pmod{A}$ are incompatible provided that $i > n$. This shows that the set $U_{m,A}(R^n)$ is of interest only when $m \leqslant n$ (this inequality will always be satisfied in the sequel). Furthermore, one easily sees that the set $U_{m,A}(R^n)$ is independent of the choice of the ambient unital ring R and depends only on the structure of A considered as a ring without unit (provided that $m \leqslant n$).

The group of elementary matrices $E_n(R, A)$ acts on the set $U_{m,A}(R^n)$ by left multiplications, there is also a natural action of the permutation group Σ_m on $U_{m,A}(R^n)$ arising from the diagonal action of Σ_m on $\{1, \ldots, m\} \times R^n$. Thus $\text{dom}(\sigma f) = \sigma(\text{dom}(f))$ and $(\sigma f)(\sigma i) = \sigma f(i)$.

Recall that the *stable rank* of an ideal A (denoted $\text{sr}(A)$) is the least integer r for which the following condition holds:

SR$_r$: If $v = (x_1, \ldots, x_{r+1})^T \in R^{r+1}$ is a unimodular column congruent to e_1 modulo A, then there exist $y_1, \ldots, y_r \in A$ such that the column $(x_1 + y_1 x_{r+1}, \ldots, x_r + y_r x_{r+1})^T$ is still unimodular (see [12]).

Below we shall need the following elementary property of the stable rank (see [12]).

LEMMA 2.1. *Assume that $n \geqslant \text{sr}(A) + 1$. Then for any $i = 1, \ldots, n$ the group $E_n(R, A)$ acts transitively on the set of unimodular columns $v \in R^n$ congruent to e_i modulo A.*

THEOREM 2.2 (The acyclicity theorem). *The simplicial complex $U_{n,A}(R^n)_*$ is $(n - \text{sr}(A) - 1)$-acyclic.*

PROOF. We proceed by induction on n. To simplify the notations, set $r = \text{sr}(A)$. The theorem is trivial for $n = 0$ and $n = 1$. Now assume that the theorem is already proved for all $n < N$. Consider the following auxiliary sets (satisfying the chain condition)

$$U'_{n,A}(R^n) = \{f \in U_{n,A}(R^n) : f(n) \in e_n + A^{n-1} \text{ provided that } n \in \text{dom}(f)\},$$
$$U''_{n,A}(R^n) = \{f \in U'_{n,A}(R^n) : f(i) \in R^{n-1} \text{ for } i \in \text{dom}(f) \cap \{1, \ldots, n-1\}\}.$$

LEMMA 2.3. *For any $n \leqslant N$ the simplicial complex $U''_{n,A}(R^n)_*$ is $(n-r-1)$-acyclic.*

PROOF. The complex $U''_{n,A}(R^n)_*$ coincides with the joining of $U_{n-1,A}(R^{n-1})_*$ and the nonempty discrete simplicial complex $(n, e_n + A^{n-1})$. Thus our statement follows from the induction hypothesis and Lemma 1.1.

LEMMA 2.4. *Assume that $n \leqslant N$ and $f \in U_{n,A}(R^\infty)$. Then the simplicial complex $(U_{n,A}(R^n) \cap U_{n,A}(R^\infty)_f)_*$ is $(n - r - 1 - |\text{dom}(f)|)$-acyclic.*

PROOF. We proceed by induction on n. We may suppose that
$$n - r - 1 - |\text{dom}(f)| \geq -1$$
and hence $n \geq r + |\text{dom}(f)| \geq r + 1$. Acting by an appropriate permutation, we may reduce the general case to the following special one: $\text{dom}(f) = n - k + 1, \ldots, n$ (where $k = |\text{dom}(f)|$). Adding coordinates with numbers $n + 1, \ldots$ to the first n coordinates (this does not change the set under consideration) and using the definition of the stable rank, we may assume further that $f(n) = (x_1, \ldots, x_n, \ldots)^T$ where the column $(x_1, \ldots, x_n)^T$ is unimodular. Now using Lemma 2.1 we may assume finally that $f(n) = (0, \ldots, 0, 1, \ldots)^T$ where 1 stands at the nth position. For any vector $v \in R^\infty$ set $\widetilde{v} = v - v_n \cdot f(n)$ (here v_n denotes the nth coordinate of v), for any function $g \in F_{n-1}(R^\infty)$ let $\widetilde{g}$ be the function with $\text{dom}(\widetilde{g}) = \text{dom}(g)$, given by $\widetilde{g}(i) = \widetilde{g(i)}$. Set
$$U_{n,A}(R^n) \cap U_{n,A}(R^\infty)_f = F, \qquad U_{n-1,A}(R^{n-1}) \cap U_{n,A}(R^\infty)_f = G,$$
$$d = n - r - k - 1.$$

The set G coincides with the set $U_{n-1,A}(R^{n-1}) \cap U_{n-1,A}(R^\infty)_{f'}$, where $f' = \widetilde{f}|_{\{n-k+1,\ldots,n-1\}}$, if $k > 1$ and coincides with $U_{n-1,A}(R^{n-1})$ if $k = 1$. In any case G is d-acyclic according to our hypotheses. If $g \in F$ is any function, then
$$G \cap F_g = U_{n-1,A}(R^{n-1}) \cap U_{n,A}(R^\infty)_{(g,f)} = U_{n-1,A}(R^{n-1}) \cap U_{n-1,A}(R^\infty)_{(\widetilde{g},f')}.$$

Thus $(G \cap F_g)$ is $(d - |\text{dom}(g)|)$-acyclic according to the induction hypothesis and hence we may apply Lemma 1.2. This lemma shows that F is d-acyclic.

LEMMA 2.5. *Assume that $n \leq N$ and $f \in U_{n-1,A}(R^\infty)$. Then the simplicial complex $(U''_{n,A}(R^n) \cap U_{n,A}(R^\infty)_f)_*$ is $(n - r - 1 - |\text{dom}(f)|)$-acyclic.*

The proof repeats that of Lemma 2.4 (with minor changes) and we omit it.

LEMMA 2.6. *The simplicial complex $U'_{N,A}(R^N)_*$ is $(N - r - 1)$-acyclic.*

PROOF. Set $F = U'_{N,A}(R^N)$, $G = U''_{N,A}(R^N)$, $d = N - r - 1$. Then G_* is d-acyclic according to Lemma 2.3. Furthermore, if $f \in F$ is a function such that $f|_{\{i\}} \notin G$ for any $i \in \text{dom}(f)$, then $N \notin \text{dom}(f)$ and hence $f \in U_{N-1,A}(R^\infty)$. Thus
$$(G \cap F_f)_* = (U''_{N,A}(R^N) \cap U_{N,A}(R^\infty)_f)_*$$
is $(d - |\text{dom}(f)|)$-acyclic according to Lemma 2.5. Applying Lemma 1.2, we conclude that F_* is d-acyclic.

LEMMA 2.7. *The simplicial complex $U_{N,A}(R^N)_*$ is $(N - r - 1)$-acyclic.*

PROOF. Set $F = U_{N,A}(R^N)$, $G = U'_{N,A}(R^N)$, $d = N - r - 1$. The simplicial complex G_* is d-acyclic according to Lemma 2.6. Let $f \in F$ be a function such that $f|_{\{i\}} \notin G$ for any $i \in \text{dom}(f)$. This implies immediately that
$$\text{dom}(f) = \{N\} \quad \text{and} \quad G \cap F_f = F_f = U_{N,A}(R^N) \cap U_{N,A}(R^\infty)_f.$$

Thus $(G \cap F_f)_*$ is $(d - |\text{dom}(f)|)$-acyclic according to Lemma 2.4. Using Lemma 1.2 once again, we conclude that F_* is d-acyclic.

Denote by $SU_{n,A}(R^n)$ the subset of $U_{n,A}(R^n)$ consisting of those functions f for which there exists a matrix $\alpha \in E_n(R, A)$ such that $f(i) = \alpha e_i$ for all $i \in \text{dom}(f)$. It is clear that $SU_{n,A}(R^n)$ satisfies the chain condition. Furthermore, $SU_{n,A}(R^n)$ is obviously invariant with respect to actions of $E_n(R, A)$ and Σ_n.

COROLLARY 2.8. *The simplicial complex $SU_{n,A}(R^n)_*$ is $(n - \max(r, 2) - 1)$-acyclic.*

PROOF. Note first of all that p-simplices of the simplicial complexes $U_{n,A}(R^n)_*$ and $SU_{n,A}(R^n)_*$ are the same when $p \leq n - r - 1$ (this follows easily from Lemma 2.1). Thus Theorem 2.2 implies that $SU_{n,A}(R^n)$ is $(n - r - 2)$-acyclic. This ends the argument for $r = 1$. Now assume that $r \geq 2$. Every $(n - r - 1)$-dimensional cycle of $SU_{n,A}(R^n)_*$ is a boundary in $U_{n,A}(R^n)_*$. Thus to prove our statement, we must show that for every function $f \in U_{n,A}(R^n)$ with $|\text{dom}(f)| = n - r + 1$, $d(f)$ is a boundary in $SU_{n,A}(R^n)_*$. Using the actions of Σ_n and $E_n(R, A)$ we may assume further that

$$\text{dom}(f) = \{1, \ldots, n - r + 1\}, \qquad f(i) = e_i \quad \text{for } i = 1, \ldots, n - r,$$
$$f(n - r + 1) = (-a, -a, \ldots, -a, 1 + a, \ldots)^T;$$

here $1 + a$ stands in the $(n - r + 1)$th position, $a \in A$. Consider a new function $g \in F_n(R^n)$:

$$\text{dom}(g) = \{1, \ldots, n - r + 2\}, \qquad g(i) = f(i) \quad \text{if } i < n - r + 2,$$
$$g(n - r + 2) = e_{n-r+2}.$$

Then $f = d_{n-r+1}(g)$ and, moreover, one easily verifies that $d_i(g) \in SU_{n,A}(R^n)$ for $i \leq n - r$. The relation

$$0 = d^2(g) = d\left(\sum_{i=0}^{n-r+1} (-1)^i d_i(g)\right)$$

implies that $d(f) = \sum_{i=0}^{n-r} (-1)^{n-r+i} d(d_i(g))$ and hence $d(f)$ is a boundary in $SU_{n,A}(R^n)_*$.

§3. H-unital $\mathbb{Q}$-algebras

All rings considered in this section are $\mathbb{Q}$-algebras (possibly without unit), all the unspecified tensor products are taken over $\mathbb{Q}$.

Let A be a $\mathbb{Q}$-algebra. We denote by $B_*(A)$ the following chain complex:

$$B_q(A) = A^{\otimes q} \qquad (q \geq 1),$$
$$d(a_1 \otimes \cdots \otimes a_q) = \sum_{i=1}^{q-1} (-1)^{i-1} (a_1 \otimes \cdots \otimes a_i a_{i+1} \otimes \cdots \otimes a_q).$$

For a left A-module M we denote by $B'_*(A; M)$ the following chain complex:

$$B'_q(A; M) = A^{\otimes q} \otimes M \qquad (q \geq 0),$$
$$d(a_1 \otimes \cdots \otimes a_q \otimes m) = \sum_{i=1}^{q-1} (-1)^{i-1} (a_1 \otimes \cdots \otimes a_i a_{i+1} \otimes \cdots \otimes m)$$
$$+ (-1)^{q-1} (a_1 \otimes \cdots \otimes a_{q-1} \otimes a_q m).$$

Following M. Wodzicki ([13], see also [11, §7]) we shall say that the algebra A is H-*unital* if the complex $B_*(A)$ is acyclic. A left A-module M is called H-*unitary* if the complex $B'_*(A; M)$ is acyclic. Until the end of this section, we assume that A is an H-unital Q-algebra. Below we shall need the following obvious properties of H-unitary A-modules.

LEMMA 3.1. a) *For any vector space $V/\mathbb{Q}$, the left A-module $A \otimes V$ is H-unitary.*
b) *If in an exact sequence of left A-modules*

$$0 \to M_0 \to M_1 \to \cdots \to M_k \to 0$$

all modules but one are H-unitary, then so is the last one.
c) *For an H-unitary left A-module M, the structure map $A \otimes M \to M$ is surjective.*

Let X be a finite set. Denote by C_q the $\mathbb{Q}$-vector space with the basis consisting of all $(p-1)$-tuples of different points of X (there is exactly one 0-tuple, so that $C_1 = \mathbb{Q}$). Let $C_X(A)$ denote the complex $C_X(A)_p = C_p \otimes A^{\otimes p}$ $(p \geq 1)$ with the boundary operator

$$d((x_1, \ldots, x_{p-1}) \otimes (a_1 \otimes \cdots \otimes a_p))$$
$$= \sum_{i=1}^{p-1} (-1)^{i-1} (x_1, \ldots, \widehat{x_1}, \ldots, x_{p-1}) \otimes (a_1 \otimes \cdots \otimes a_i a_{i+1} \otimes \cdots \otimes a_p).$$

PROPOSITION 3.2. *Assume that $|X| = n$. Then the complex $C_X(A)$ is n-acyclic.*

PROOF. We proceed by induction on n. There is nothing to prove for $n = 0$, so assume that $n > 0$ and the statement is already proved for smaller values of $|X|$. Let $z \in X$ be any point and set $Y = X - \{z\}$. The complex $C_Y(A)$ is $(n-1)$-acyclic according to the induction hypothesis and the group $H_n(C_Y(A)) = Z_n(C_Y(A))$ fits into the exact sequence

$$0 \to H_n(C_Y(A)) \to C_Y(A)_n \xrightarrow{d} \cdots \xrightarrow{d} C_Y(A)_1 \to 0.$$

Left multiplication of the first position in $A^{\otimes p}$ by elements of A defines a left A-module structure on $C_Y(A)_p$ and the differential of $C_Y(A)$ is clearly A-linear. Thus $H_n(C_Y(A))$ has a natural A-module structure and the above sequence in conjunction with Lemma 3.1 shows that this module is H-unitary and, in particular, the structure morphism $A \otimes H_n(C_Y(A)) \to H_n(C_Y(A))$ is surjective. Now consider the filtration of $C_X(A)$ according to the position of the point z:

$$F_p(C_X(A)_m) = \coprod_{\substack{x_i \neq z \\ \text{for } i > p}} (x_1, \ldots, x_{m-1}) \otimes A^{\otimes m}.$$

Thus $F_0(C_X(A)) = C_Y(A)$, $F_{1/0}(C_X(A)) = A \otimes C_Y(A)[-1]$

$$F_{p/p-1}(C_X(A)) = \coprod_{\substack{y_1, \ldots, y_{p-1} \\ \text{different points of } Y}} A^{\otimes p} \otimes C_{Y - \{y_1, \ldots, y_{p-1}\}}(A)[-p].$$

The induction hypothesis implies that the E^1-term of the corresponding spectral sequence looks as follows:

$$H_n(C_Y(A)) \xleftarrow{d_1} A \otimes H_n(C_Y(A)) \qquad 0 \qquad \ldots$$
$$0 \qquad\qquad 0 \qquad\qquad \coprod_{y \in Y} A^{\otimes 2} \otimes H_{n-1}(C_{Y-\{y\}}(A)) \ldots$$
$$0 \qquad\qquad 0 \qquad\qquad 0 \qquad \ldots$$

Furthermore, one easily sees that the differential $d^1 \colon A \otimes H_n(C_Y(A)) \to H_n(C_Y(A))$ coincides with the structure morphism of the left A-module $H_n(C_Y(A))$ and hence is surjective. Thus $E^2_{pq} = 0$ for $p + q \leqslant n$ and hence $C_X(A)$ is n-acyclic.

§4. Acyclicity of the sum of triangular complexes

For any Lie algebra $\mathfrak{g}/\mathbb{Q}$, denote by $C_*(\mathfrak{g})$ the standard complex computing the reduced homology of $\mathfrak{g}$ with coefficients $\mathbb{Q}$. Thus $C_n(\mathfrak{g}) = \Lambda^n(\mathfrak{g})$ ($n \geqslant 1$) and the differential is given by

$$d(g_1 \wedge \cdots \wedge g_n) = \sum_{1 \leqslant i < j \leqslant n} (-1)^{i+j-1}([g_i, g_j] \wedge g_1 \wedge \cdots \wedge \widehat{g_i} \wedge \cdots \wedge \widehat{g_j} \wedge \cdots \wedge g_n).$$

If $\mathfrak{g}'$ is a subalgebra of $\mathfrak{g}$, then the complex $C_*(\mathfrak{g}')$ may be identified with a subcomplex of $C_*(\mathfrak{g})$.

Let A be an algebra over $\mathbb{Q}$ (nonunital). Consider the Lie algebra $\mathfrak{gl}_n(A)$ of $n \times n$ matrices with entries from A. If σ is a partial ordering of the set $\{1, \ldots, n\}$, then by $t_n^\sigma(A)$ we shall denote the subalgebra of $\mathfrak{gl}_n(A)$ consisting of those matrices $(a_{ij})_{1 \leqslant i,j \leqslant n}$ for which $a_{ij} = 0$ if $i \not\stackrel{\sigma}{<} j$. There is an obvious action of the algebra $\mathfrak{gl}_n(A)$ on the A-module A^n of columns of height n. Along with algebra $\mathfrak{gl}_n(A)$ and its subalgebras $t_n^\sigma(A)$, we shall be considering the corresponding affine algebras $\mathfrak{gl}_n(A) \times_{\text{s.d.}} (A^n)^k$, $t_n^\sigma(A) \times_{\text{s.d.}} (A^n)^k$. The main objective of this section is the proof of the following theorem.

THEOREM 4.1. *Let A be an H-unital $\mathbb{Q}$-algebra; then*
a) *The canonical homomorphism*

$$H_m\left(\sum_\sigma C_*(t_n^\sigma(A))\right) \to H_m\left(\sum_\sigma C_*(t_n^\sigma(A) \times_{\text{s.d.}} (A^n)^k)\right)$$

is an isomorphism provided that $n \geqslant 2m$.
b) $H_m(\sum_\sigma C_*(t_n^\sigma(A))) = 0$ *provided that $n \geqslant 2m + 1$.*

We start with a series of preliminary remarks.

For any vector space $V/\mathbb{Q}$, the group Σ_p acts on $V^{\otimes p}$ by permuting the tensor positions. Apart from this canonical action, we shall also consider the signed action given by

$$s(v_1 \otimes \cdots \otimes v_p) = \varepsilon(s)(v_{s^{-1}(1)} \otimes \cdots \otimes v_{s^{-1}(p)}),$$

the pth exterior power $\Lambda^p(V)$ coinciding with the coinvariants of the signed action.

The vector space $C_p(\mathfrak{gl}_n(A))$ coincides with

$$\Lambda^p(\mathfrak{gl}_n(A)) = (\mathfrak{gl}_n(\mathbb{Q})^{\otimes p} \otimes A^{\otimes p})_{\Sigma_p}.$$

In what follows, we shall always consider the canonical action of Σ_p on $\mathfrak{gl}_n(\mathbb{Q})^{\otimes p}$, while $A^{\otimes p}$ will be endowed with the signed action of Σ_p (so that the diagonal action of Σ_p on $\mathfrak{gl}_n(\mathbb{Q})^{\otimes p} \otimes A^{\otimes p}$ will be the signed one).

The basis of the vector space $\mathfrak{gl}_n(\mathbb{Q})^{\otimes p}$ consists of pure tensors of the form $e_{i_1,j_1} \otimes \cdots \otimes e_{i_p,j_p}$. To each such tensor one can associate a labelled oriented graph with vertices $1, \ldots, n$ in which an arrow labelled with k goes from i_k to j_k. In this way we get a one-to-one correspondence between the basis elements of $\mathfrak{gl}_n(\mathbb{Q})^{\otimes p}$ and labelled oriented graphs with vertices $1, \ldots, n$ and labels ranging from 1 to p (each label should appear exactly once). The canonical action of Σ_p on $\mathfrak{gl}_n(\mathbb{Q})^{\otimes p}$ corresponds to the label permutation action of Σ_p on graphs.

The $\mathbb{Q}$-vector space $C_p(t_n^\sigma(A))$ coincides with

$$\Lambda^p(t_n^\sigma(A)) = (t_n^\sigma(\mathbb{Q})^{\otimes p} \otimes A^{\otimes p})_{\Sigma_p}.$$

The basis of $t_n^\sigma(\mathbb{Q})^{\otimes p}$ consists of those oriented labelled graphs for which the end of each arrow is σ-bigger than its origin. The $\mathbb{Q}$-vector space $\sum_\sigma C_p(t_n^\sigma(A))$ coincides with

$$\left(\left(\sum_\sigma t_n^\sigma(\mathbb{Q})^{\otimes p}\right) \otimes A^{\otimes p}\right)_{\Sigma_p}.$$

We shall denote the subspace $\sum_\sigma t_n^\sigma(\mathbb{Q})^{\otimes p}$ of $\mathfrak{gl}_n(\mathbb{Q})^{\otimes p}$ by $V(p)$. From what was said above, it is clear that the basis of $V(p)$ consists of labelled oriented graphs without cycles.

All graphs considered below will have $1, \ldots, n$ as the set of vertices. Let us introduce some notation and terminology pertaining to graphs. Let u be an oriented graph and $i \in \{1, \ldots, n\}$. Set

$$\text{indeg}_u(i) = \text{the number of arrows of } u \text{ ending at } i,$$
$$\text{outdeg}_u(i) = \text{the number of arrows of } u \text{ starting at } i.$$

We shall say that i *does not appear* in u if $\text{indeg}_u(i) = \text{outdeg}_u(i) = 0$ and we shall say that i is an *inessential vertex* of u if $\text{indeg}_u(i) = \text{outdeg}_u(i) = 1$ (here our terminology differs from that of Hanlon [3]).

A graph obtained from u by replacing a pair of composable arrows by their composition will be called a *face* of u. Usually there is no obvious way to label the faces of u, so we leave them unlabelled.

A *simple path* in a graph u is a sequence of arrows such that the endpoint of each arrow coincides with the origin of the next one and all these intermediate vertices are inessential. Each arrow of a graph without cycles appears in a unique maximal simple path. We denote the number of maximal simple paths by $\text{sp}(u)$.

We shall say that u is *nicely labelled* if labels increase by one along each maximal simple path. To give a nice labelling of u is essentially the same as to give a total ordering on the set of maximal simple paths of u. This shows that the group Σ_q ($q = \text{sp}(u)$) acts naturally on the set of nice labellings of u. To be more precise, let u be a nicely labelled graph with p arrows. Set $q = \text{sp}(u)$ and let $\tau \in \Sigma_q$. The given labelling defines a total ordering on the set of maximal simple paths of u and we denote by $l_1, \ldots, l_q$ the lengths of these maximal simple paths (thus $l_i > 0$, $\sum_{i=1}^q l_i = p$). Define $s_{u,\tau} \in \Sigma_p$ by

$$s_{u,\tau}(i) = \left(\sum_{\tau(k)<\tau(j)} l_k\right) + i - l_1 - \cdots - l_{j-1}$$

provided that $l_1 + \cdots + l_{j-1} < i \leq l_1 + \cdots + l_j$.

The following lemma is obvious from the definitions.

LEMMA 4.2. a) $s_{u,\tau} \cdot u$ *is a nicely labelled graph.*
b) *If the graph* $s \cdot u$ *is nicely labelled, then* $s = s_{u,\tau}$ *for a unique* $\tau \in \Sigma_q$.
c) $s_{u,\tau_1\tau_2} = s_{v,\tau_1} \cdot s_{u,\tau_2}$, *where* $v = s_{u,\tau_2} \cdot u$.

Denote by $V(p,q)$ (resp. $W(p,q)$) the subspace of $V(p)$ spanned by labelled graphs (resp. nicely labelled graphs) u with $\mathrm{sp}(u) = q$. Let M be any $\mathbb{Q}[\Sigma_p]$-module. Define an action of Σ_q on $W(p,q) \otimes M$ by $^\tau(u \otimes m) = s_{u,\tau} \cdot (u \otimes m)$.

LEMMA 4.3. *For any* $\mathbb{Q}[\Sigma_p]$-*module* M, *we have a canonical isomorphism*
$$(W(p,q) \otimes M)_{\Sigma_q} = (V(p,q) \otimes M)_{\Sigma_p}.$$

PROOF. The canonical map $W(p,q) \otimes M \hookrightarrow V(p,q) \otimes M \twoheadrightarrow (V(p,q) \otimes M)_{\Sigma_p}$ is clearly Σ_q-invariant and hence defines a homomorphism $(W(p,q) \otimes M)_{\Sigma_q} \to (V(p,q) \otimes M)_{\Sigma_p}$. To provide a map in the opposite sense, start with any labelled graph u and choose $s \in \Sigma_p$ such that su is nicely labelled. Now set $\phi(u \otimes m) = (su \otimes sm)$ mod Σ_p. Lemma 4.2 shows that ϕ is well defined and Σ_p-invariant. Thus we get a homomorphism
$$(V(p,q) \otimes M)_{\Sigma_p} \to (W(p,q) \otimes M)_{\Sigma_q}$$
inverse to the one constructed above.

PROOF OF ITEM b) OF THEOREM 4.1.

As was shown above, we have
$$\sum_\sigma C_p(t_n^\sigma(A)) = (V(p) \otimes A^{\otimes p})_{\Sigma_p}$$
$$= \coprod_{q=1}^p (V(p,q) \otimes A^{\otimes p})_{\Sigma_p} = \coprod_{q=1}^p (W(p,q) \otimes A^{\otimes p})_{\Sigma_q}.$$

We will filter the complex $C_* = \sum_\sigma C_*(t_n^\sigma(A))$ according to the number of maximal simple paths in the involved graphs. Thus we set
$$F_q(C_p) = \coprod_{j \leq q} (V(p,j) \otimes A^{\otimes p})_{\Sigma_p} = \coprod_{j \leq q} (W(p,j) \otimes A^{\otimes p})_{\Sigma_q}.$$

Note that all the terms appearing in $d((u \otimes A^{\otimes p})_{\Sigma_p})$ correspond to the faces of u, and if v is any face of u, then $\mathrm{sp}(v) \leq \mathrm{sp}(u)$. Moreover $\mathrm{sp}(v) < \mathrm{sp}(u)$ unless v is obtained from u by composing two subsequent arrows in a maximal simple path. Thus $F_q(C_*)$ are subcomplexes of C_* and $F_{q/q-1}(C_p) = (W(p,q) \otimes A^{\otimes p})_{\Sigma_q}$. It turns out that
$$W(q) = \{W(p,q) \otimes A^{\otimes p}\}_{p=1}^\infty$$
may be endowed with a Σ_q-equivariant differential in such a way that $F_{q/q-1}(C_*) = W(q)_{\Sigma_q}$. Let u be a nicely labelled graph with p arrows and $\mathrm{sp}(u) = q$. Denote by $l_1, \ldots, l_q$ the lengths of the maximal simple paths of u. There are $l_k - 1$ faces of u obtained by taking compositions of subsequent arrows in the kth maximal simple path.

Each of these faces has a canonical labelling that we denote by $\partial_1^k(u), \ldots, \partial_{l_k-1}^k(u)$. The differential of $W(q)$ is given by

$$d(u \otimes (a_1 \otimes \cdots \otimes a_p)) = \sum_{k=1}^{q} (-1)^{l_1+\cdots+l_k} \sum_{i=1}^{l_k-1} (-1)^{i-1} (\partial_i^k(u) \otimes (a_1 \otimes$$
$$\cdots \otimes a_{l_1+\cdots+l_{k-1}+i} \cdot a_{l_1+\cdots+l_{k-1}+i+1} \otimes \cdots \otimes a_p)).$$

To prove the acyclicity of $W(q)$, we filter it further according to the number $(l_2 - 1) + \cdots + (l_q - 1)$. Since the faces ∂_i^k with $k > 1$ diminish this number, they do not appear in the differentials of the factor complexes. The factor complex $F_{s/s-1}(W(q))$ is a direct sum over all nicely labelled graphs v with $\text{sp}(v) = q, l_1(v) = 1$, $(l_2(v) - 1) + \cdots + (l_q(v) - 1) = s$. Here the summand corresponding to such a v coincides in degree p with $\coprod u \otimes A^{\otimes p}$, the sum being taken over all nicely labelled graphs u with p arrows such that v coincides with the graph obtained from u by contracting the first maximal simple path to a single arrow (such u will be called a subdivision of the first arrow of v). To give a subdivision with p arrows of v is the same as to give a sequence of $p - (q+s)$ different vertices not appearing in v. Thus denoting by X the set of vertices not appearing in v, we see immediately that the complex under consideration coincides with $C_X(A)[-(s+q-1)] \otimes A^{\otimes(s+q-1)}$. The number $s+q$ coincides with the number of arrows in v and hence $|X| \geq n - 2(s+q)$. Proposition 3.2 shows that the above complex is $(|X| + (s+q-1))$-acyclic. Furthermore,

$$|X| + (s+q-1) \geq \max(s+q-1, n-(s+q)-1)$$
$$\geq \frac{(s+q-1)+(n-(s+q)-1)}{2} = \frac{n-2}{2}.$$

Since the minimal integer which is $\geq (n-2)/2$ coincides with $[(n-1)/2]$, we conclude that $F_{s/s-1}(W(q))$ and hence also $W(q)$ are $[(n-1)/2]$-acyclic. Since $H_*(W(q)_{\Sigma_q}) = H_*(W(q))_{\Sigma_q}$, we conclude that $F_{q/q-1}(C_*)$ and hence also C_* are $[(n-1)/2]$-acyclic.

PROOF OF ITEM a) OF THEOREM 4.1. Note first of all that $t_n^\sigma(A) \times_{\text{s.d.}} (A^n)^k = t_{n+k}^{\sigma(k)}(A)$, where $\sigma(k)$ is the partial ordering of the set $\{1, \ldots, n+k\}$ characterized by the following properties:
a) $\sigma(k)|_{\{1,\ldots,n\}} = \sigma$,
b) $i \stackrel{\sigma(k)}{<} j$ provided that $i \leq n < j$,
c) $n+i$ is incomparable with $n+j$ if $1 \leq i \neq j \leq k$.
This shows that

$$\sum_\sigma C_p(t_n^\sigma(A) \times_{\text{s.d.}} (A^n)^k) = (V^k(p) \otimes A^p)_{\Sigma_p},$$

where $V^k(p)$ is a $\mathbb{Q}$-vector space spanned by labelled oriented graphs (with p arrows) u which have the set $\{1, \ldots, n+k\}$ as the set of vertices, have no cycles and satisfy the following additional property: $\text{outdeg}_u(j) = 0$ for $j > n$. There is a canonical retraction

$$\sum_\sigma C_*(t_n^\sigma(A) \times_{\text{s.d.}} (A^n)^k) \twoheadrightarrow \sum_\sigma C_*(t_n^\sigma(A)).$$

Denoting the kernel of this retraction by D^k we get a direct sum decomposition

$$\sum_\sigma C_*(t_n^\sigma(A) \times_{\text{s.d.}} (A^n)^k) = \sum_\sigma C_*(t_n^\sigma(A)) \oplus D^k.$$

Thus our statement is equivalent to $[n/2]$-acyclicity of the complex D^k. The group D_p^k corresponds under the above isomorphism to a vector subspace of $V^k(p)$ spanned by those graphs u in which at least one of the vertices $n+1, \ldots, n+k$ appears. Applying the same procedure as in the proof of part b) to the complex D^k, we come to complexes corresponding to nicely labelled graphs v with vertices $1, \ldots, n+k$, $\text{outdeg}_v(j) = 0$ for $j > n$, $\text{indeg}_u(j) > 0$ for at least one $j > n$, $\text{sp}(v) = q$, $l_1(v) = 1$, $(l_2(v) - 1) + \cdots + (l_q(v) - 1) = s$. The complex corresponding to such v is isomorphic to $C_X(A)[-(s+q-1)] \otimes A^{\otimes(s+q-1)}$, where X is a subset of $\{1, \ldots, n\}$ consisting of vertices not appearing in v. The range of acyclicity of this complex is equal to $|X| + (s + q - 1)$. Once again $s + q$ is the number of arrows in v and hence the number of vertices appearing in v is $\leqslant 2(s+q)$. Since at least one of the vertices $n+1, \ldots, n+k$ must appear in v, we conclude that $|X| \geqslant n - 2(s+q) + 1$. This easily implies (see the end of the proof of item b)) that the corresponding complex is $[n/2]$-acyclic.

§5. Volodin spaces

For any set X we denote by $P(X)$ the (contractible) simplicial set whose p-simplices are $(p+1)$-tuples of elements of X and the ith face (resp. the ith degeneracy) omits (resp. repeats) the ith entry. If G is a group, then there is a canonical free left action of G on $P(G)$ and the quotient simplicial set $G \setminus P(G)$ (denoted BG) is the standard simplicial model for a classifying space of G.

For any family $\{G_i\}_{i \in I}$ of subgroups of G, we denote by $V(G, \{G_i\}_{i \in I})$ the simplicial subset of $P(G)$ formed by simplices $(g_0, \ldots, g_p)$ satisfying the following condition:

There exists an $i \in I$ such that for all $0 \leqslant j, k \leqslant p$ one has $g_j^{-1} g_k \in G_i$.

The subset $V(G, \{G_i\}_{i \in I})$ is invariant under the action of G and the corresponding quotient simplicial set coincides with $\bigcup_{i \in I} BG_i \subset BG$. Since the action of G on $V(G, \{G_i\}_{i \in I})$ is free, we get the following lemma.

LEMMA 5.1. *There is a natural spectral sequence*

$$E_{pq}^2 = H_p(G, H_q(V(G, \{G_i\}_{i \in I}))) \Rightarrow H_{p+q}\left(\bigcup_{i \in I} BG_i\right).$$

We refer the reader to [11, §2] for further information about the general properties of Volodin spaces.

Let A be a ring without unit. Embed A as a two-sided ideal into a unital ring R and set

$$\text{GL}_n(A) = \text{GL}_n(R, A) = \{\alpha \in \text{GL}_n(R) : \alpha \equiv 1_n \mod A\},$$
$E_n(A) = $ subgroup of $\text{GL}_n(A)$ generated by elementary matrices $E_{ij}(a)$
$$(1 \leqslant i \neq j \leqslant n, \, a \in A).$$

Note that $E_n(A)$ is always contained in the relative elementary group $E_n(R, A)$ of Bass (see [1]). Furthermore, one has the following result due to Vasserstein.

LEMMA 5.2. *Assume that $A = A \cdot A$ and $n \geq 3$. Then $E_n(A) = E_n(R, A)$ (for any ambient ring R).*

PROOF. The group $E_n(R, A)$ is the normal closure of $E_n(A)$ in $E_n(R)$. Thus, we must show that $E_{ij}(r) \cdot \alpha \cdot E_{ij}(-r) \in E_n(A)$ for any $\alpha \in E_n(A)$ and any $1 \leq i \neq j \leq n, r \in R$. Writing α as a product of $E_{kl}(a)$ and using the standard commutator relations for elementary matrices, one sees immediately that the only nontrivial case is $\alpha = E_{ji}(a)$. Since $A = A \cdot A$, we may assume further that $a = b \cdot c$ for some $b, c \in A$. Now it suffices to choose $k \neq i, j$ and to use the following relation

$$\begin{aligned} E_{ij}(r) E_{ji}(a) E_{ij}(-r) &= E_{ij}[E_{jk}(b), E_{ki}(c)] E_{ij}(-r) \\ &= [E_{ij}(r) E_{jk}(b) E_{ij}(-r), E_{ij}(r) E_{ki}(c) E_{ij}(-r)] \\ &= [E_{ik}(rb) E_{jk}(b), E_{ki}(c) E_{kj}(-cr)]. \end{aligned}$$

If $n \leq m$, then we shall identify $\mathrm{GL}_n(A)$ with a subgroup of $\mathrm{GL}_m(A)$ by means of the following standard embedding:

$$\alpha \mapsto \begin{pmatrix} \alpha & 0 \\ 0 & 1_{m-n} \end{pmatrix}.$$

Finally we set $\mathrm{GL}(A) = \varinjlim \mathrm{GL}_n(A)$, $E(A) = \varinjlim E_n(A)$.

The stability theorem of Vasserstein [12] and Lemma 5.2 imply the following corollary.

COROLLARY 5.3. *Assume that $A = A \cdot A$ and $n \geq \max(\mathrm{sr}(A), 2) + 1$. Then $E_n(A) = \mathrm{GL}_n(A) \cap E(A)$.*

For any partial ordering σ of the set $\{1, \ldots, n\}$, we denote by $T_n^\sigma(A)$ the subgroup of σ-triangular matrices in $\mathrm{GL}_n(A)$:

$$T_n^\sigma(A) = \{(a_{ij}) \in \mathrm{GL}_n(A) : a_{ij} = 0 \text{ if } i \not\leq^\sigma j, a_{ii} = 1\}.$$

Set $V_n(A) = V(E_n(A), \{T_n^\sigma(A)\}_{\sigma \in \Pi_n})$, where Π_n is the set of all partial orderings of the set $\{1, \ldots, n\}$. The following fact is proved in [11, Lemma 2.5].

LEMMA 5.4. *Let G be any subgroup of $\mathrm{GL}_n(A)$ containing $E_n(A)$ and let k be any nonnegative integer. Then the canonical embedding and projection*

$$G \hookrightarrow \begin{pmatrix} G & 0 \\ * & 1_k \end{pmatrix}, \quad \begin{pmatrix} G & 0 \\ * & 1_k \end{pmatrix} \twoheadrightarrow G$$

induce mutually inverse homotopy equivalences (for any $\Pi \subset \Pi_n$)

$$V(G, \{T_n^\sigma(A)\}_{\sigma \in \Pi}) \leftrightarrows V\left(\begin{pmatrix} G & 0 \\ * & 1_k \end{pmatrix}, \left\{\begin{pmatrix} T_n^\sigma(A) & 0 \\ * & 1_k \end{pmatrix}\right\}_{\sigma \in \Pi}\right).$$

Every embedding $\phi: \{1, \ldots, n\} \hookrightarrow \{1, \ldots, m\}$ defines a homomorphism $\mathrm{GL}_n(A) \to \mathrm{GL}_m(A)$: $\alpha \mapsto {}^\phi\alpha$, where ${}^\phi\alpha$ is the matrice given by

$$({}^\phi\alpha)_{ij} = \begin{cases} \alpha_{kl} & \text{if } i = \phi(k), j = \phi(l), \\ 1 & \text{if } i = j \notin \mathrm{im}(\phi), \\ 0 & \text{if } i \neq j \text{ and either } i \notin \mathrm{im}(\phi) \text{ or } j \notin \mathrm{im}(\phi). \end{cases}$$

This homomorphism takes $E_n(A)$ to $E_m(A)$ and takes triangular subgroups in $\mathrm{GL}_n(A)$ to triangular subgroups in $\mathrm{GL}_m(A)$, thus defining a map of simplicial sets $V_n(A) \to V_m(A)$, which we denote again exponentially: $x \mapsto {}^\phi x$. This construction defines in particular a natural action of the permutation group Σ_n on $V_n(A)$.

LEMMA 5.5. *Assume that $m > n$ and let $\phi\colon \{1,\ldots,n\} \hookrightarrow \{1,\ldots,m\}$ be any embedding. If $A = A \cdot A$, then the action of $E_m(A)$ on the image of the homomorphism $H_*(V_n(A)) \to H_*(V_m(A))$ induced by ϕ is trivial.*

PROOF. It suffices to treat the case $m = n+1$. If ϕ is the standard embedding, then our statement is proved in [**11**, Lemma 2.8]. In the general case, denote the standard embedding by ϕ_0 and find $s \in \Sigma_n$ such that $\phi = s\phi_0$. Now our statement follows from the following obvious formula

$$\alpha \cdot {}^\phi u = {}^s({}^{s^{-1}}\alpha \cdot {}^{\phi_0}u) = {}^s({}^{\phi_0}u) = {}^\phi u \qquad (\alpha \in E_m(A),\ u \in H_*(V_n(A))).$$

A classical and well-known theory uniquely relates divisible nilpotent groups with nilpotent Lie algebras over $\mathbb{Q}$. The following fact is proved in [**11**, Theorem 5.13].

THEOREM 5.6. *Let A be any $\mathbb{Q}$-algebra and let Π be any set of partial orderings of $\{1,\ldots,n\}$. Then there are canonical isomorphisms*

$$\widetilde{H}_*\left(\bigcup_{\sigma \in \Pi} BT_n^\sigma(A)\right) = H_*\left(\sum_{\sigma \in \Pi} C_*(t_n^\sigma(A))\right).$$

Now using Theorem 4.1 we come to the following result.

COROLLARY 5.7. *Assume that A is an H-unital $\mathbb{Q}$-algebra. Then*
a) *The canonical homomorphism*

$$H_m\left(\bigcup_{\sigma \in \Pi_n} BT_n^\sigma(A)\right) \to H_m\left(\bigcup_{\sigma \in \Pi_n} B\begin{pmatrix} T_n^\sigma(A) & 0 \\ * & 1_k \end{pmatrix}\right)$$

is an isomorphism for any k provided that $n \geq 2m$.
b) $\widetilde{H}_m(\bigcup_{\sigma \in \Pi_n} BT_n^\sigma(A)) = 0$ *provided that $n \geq 2m + 1$.*

The Volodin space gives us a link between the homology of $E_n(A)$ and the homology of $\bigcup_{\sigma \in \Pi_n} BT_n^\sigma(A)$ via Lemma 5.1. Following this approach, we obtain, in particular, the following result.

COROLLARY 5.8. *Assume that A is an H-unital $\mathbb{Q}$-algebra. Assume further that n and L are positive integers such that $n \geq 2L + 2$ and $E_n(A)$ acts trivially on $H_q(V_n(A))$ for $q \leq L$. In this case the canonical homomorphism*

$$H_p(E_n(A)) \to H_p\left(\begin{pmatrix} E_n(A) & 0 \\ * & 1_k \end{pmatrix}\right)$$

is an isomorphism for all k and all $p \leq L + 1$.

PROOF. We set

$$\widetilde{E}_n(A) = \begin{pmatrix} E_n(A) & 0 \\ * & 1_k \end{pmatrix} \quad \text{and} \quad \widetilde{T}_n^\sigma(A) = \begin{pmatrix} T_n^\sigma(A) & 0 \\ * & 1_k \end{pmatrix}.$$

Further we shall denote by $\widetilde{V}_n(A)$ the simplicial set $V(\widetilde{E}_n(A), \{\widetilde{T}_n^\sigma(A)\}_{\sigma \in \Pi_n})$. According to Lemma 5.4, the canonical embedding and projection $V \rightleftarrows \widetilde{V}_n(A)$ are mutually inverse homotopy equivalences. These maps of simplicial sets are compatible with the canonical embedding and projection of groups $E_n(A) \rightleftarrows \widetilde{E}_n(A)$. This implies in particular that the action of $\begin{pmatrix} 1_n & 0 \\ * & 1_k \end{pmatrix}$ on $H_*(\widetilde{V}_n(A))$ is trivial and hence the action of $\widetilde{E}_n(A)$ on $H_q(\widetilde{V}(A))$ is trivial for $q \leqslant L$. Consider the spectral sequences associated with the actions of $E_n(A)$ on $V_n(A)$ and of $\widetilde{E}_n(A)$ on $\widetilde{V}_n(A)$:

$$E^2_{pq} = H_p(E_n(A), H_q(V_n(A))) \Rightarrow H_{p+q}\left(\bigcup_{\sigma \in \Pi_n} BT^\sigma_n(A)\right),$$

$$E^2_{pq} = H_p(\widetilde{E}_n(A), H_q(\widetilde{V}_n(A))) \Rightarrow H_{p+q}\left(\bigcup_{\sigma \in \Pi_n} B\widetilde{T}^\sigma_n(A)\right).$$

The embedding $V_n(A) \hookrightarrow \widetilde{V}_n(A)$ induces a homomorphism $E \to \widetilde{E}$. The homomorphism $E^\infty_i \to \widetilde{E}^\infty_i$ is an isomorphism for $i \leqslant L+1$ in view of Corollary 5.7. The homomorphism $E_{0q} \to \widetilde{E}^2_{0q}$ is an isomorphism for all q (since the action of $\begin{pmatrix} 1_n & 0 \\ * & 1_k \end{pmatrix}$ on $H_q(\widetilde{V}_n(A)) = H_q(V_n(A))$ is trivial). Using induction on p, it suffices to show that $H_{L+1}(E_n(A)) = H_{L+1}(\widetilde{E}_n(A))$ assuming that $H_p(E_n(A)) = H_p(\widetilde{E}_n(A))$ for $p \leqslant L$. Our assumptions imply that $E^2_{pq} \xrightarrow{\sim} \widetilde{E}^2_{pq}$ for all $p \leqslant L, q \leqslant L$. Now Theorem A.4 shows that

$$H_{L+1}(E_n(A)) = E^2_{L+1,0} \xrightarrow{\sim} \widetilde{E}^2_{L+1,0} = H_{L+1}(E_n(A)).$$

§6. The main theorem

Throughout this section A is an H-unital $\mathbb{Q}$-algebra, $r = \operatorname{sr}(A), \widetilde{r} = \max(r, 2)$.

THEOREM 6.1. *Assume that* $n \geqslant 2l + \widetilde{r}$. *Then*
1) $H_l(V_n(A)) \twoheadrightarrow H_l(V_{n+1}(A)) \xrightarrow{\sim} H_l(V_{n+2}(A))$;
2) *the action of* $E_{n+1}(A)$ *on* $H_l(V_{n+1}(A))$ *is trivial*;
3) *the action of the permutation group* Σ_{n+1} *on* $H_l(V_{n+1}(A))$ *is trivial*;
4) *for any* $k \geqslant 0$, $H_{l+1}\left(\begin{pmatrix} E_n(A) & 0 \\ * & 1_k \end{pmatrix}\right) = H_{l+1}(E_n(A))$;
5) *the action (by conjugation) of* Σ_{n+1} *on* $H_{l+1}(E_{n+1}(A))$ *is trivial*;
6) $H_{l+1}(E_n(A)) \twoheadrightarrow H_{l+1}(E_{n+1}(A)) \xrightarrow{\sim} H_{l+1}(E_{n+2}(A))$.

PROOF. We proceed by induction on l. If $l = 0$ the statement is trivial. Assume that $L > 0$ and the theorem is true for all $l < L$.

(6.2) For any partial ordering σ, denote by $\operatorname{Max}(\sigma)$ the set of elements maximal with respect to σ. Denote further by $V_{n+1}(A)^i$ the simplicial subset of $V_{n+1}(A)$ whose p-simplices are those $(p+1)$-tuples $(\alpha_0, \ldots, \alpha_p)$ $(\alpha_i \in E_{n+1}(A))$ for which there exists a partial ordering σ such that $\alpha_k^{-1} \cdot \alpha_j \in T^\sigma_{n+1}$ for all j and k and such that $i \in \operatorname{Max}(\sigma)$. Since $V_{n+1}(A) = \bigcup_{i=1}^{n+1} V_{n+1}(A)^i$, we get a spectral sequence

$$E^1_{pq} = \coprod_{i_0 < \cdots < i_q} H_p(V_{n+1}(A)^{i_0, \ldots, i_q}) \Rightarrow H_{p+q}(V_{n+1}(A)).$$

If $(\alpha_0, \ldots, \alpha_p)$ is a p-simplex of a simplicial set $V_{n+1}(A)^{i_0, \ldots, i_q}$, then $(\alpha_k^{-1}\alpha_j) \cdot e_{i_s} = e_{i_s}$ for all $s = 0, \ldots, q$ and hence $\alpha_j \cdot e_{i_s} = \alpha_k \cdot e_{i_s}$. Thus associating with $(\alpha_0, \ldots, \alpha_q)$

a special unimodular function f with $\mathrm{dom}(f) = \{i_0,\ldots,i_q\}$, $f(i_s) = \alpha_0 \cdot e_{i_s}$ we get a map from the simplicial set $V_{n+1}(A)^{i_0,\ldots,i_q}$ to the discrete set of special unimodular functions $f \in SU_{n+1,A}(R^{n+1})$ such that $\mathrm{dom}(f) = \{i_0,\ldots,i_q\}$ (here R is any unital ring containing A as a two-sided ideal). The fiber of this map over a function f_0 given by $f_0(i_s) = e_{i_s}$ coincides with the simplicial set

$$V(E_{n+1}(A)^{i_0,\ldots,i_q}, \{T^\sigma_{n+1}(A)\}_{\sigma:i_s \in \mathrm{Max}(\sigma) \text{ for all } s=0,\ldots,q}),$$

where $E_{n+1}(A)^{i_0,\ldots,i_q} = \{\alpha \in E_{n+1}(A) : \alpha \cdot e_{i_s} = e_{i_s} \text{ for all } s = 0,\ldots,q\}$.

The map under consideration is clearly $E_{n+1}(A)$-equivariant and since the action of $E_{n+1}(A)$ on the above functions is transitive, we conclude that all fibers of this map are isomorphic to the above Volodin space. The isomorphism of the fiber over a function f to the one over f_0 depends only on the choice of a matrix $\alpha \in E_{n+1}(A)$ such that $\alpha \cdot f(i_s) = e_{i_s}$ and hence is well defined up to the action of the group $E_{n+1}(A)^{i_0,\ldots,i_q}$. For any sequence $i_0 < \cdots < i_q$ there exists a unique $(p-q, q+1)$-shuffle taking the set $\{n-q+1,\ldots,n+1\}$ to $\{i_0,\ldots,i_q\}$. Conjugation by the corresponding permutation matrix defines isomorphisms $V_{n+1}(A)^{n-q+1,\ldots,n+1} \xrightarrow{\sim} V_{n+1}(A)^{i_0,\ldots,i_q}$ and

$$V\left(\begin{pmatrix} \overline{E}_{n-q}(A) & 0 \\ * & 1 \end{pmatrix}, \left\{\begin{pmatrix} T^\tau_{n-q}(A) & 0 \\ * & 1 \end{pmatrix}\right\}_{\tau \in \Pi_{n-q}}\right)$$
$$\xrightarrow{\sim} V(E_{n+1}(A)^{i_0,\ldots,i_q}, \{T^\sigma_{n+1}(A)\}),$$

where $\overline{E}_{n-q}(A) = \mathrm{GL}_{n-q}(A) \cap E(A)$. Denote the simplicial set

$$V(\overline{E}_{n-q}(A), \{T^\tau_{n-q}(A)\}_{\tau \in \Pi_{n-q}})$$

by $\overline{V}_{n-q}(A)$. The embedding

$$\overline{V}_{n-q}(A) \hookrightarrow V\left(\begin{pmatrix} \overline{E}_{n-q}(A) & 0 \\ * & 1 \end{pmatrix}, \left\{\begin{pmatrix} T^\tau_{n-q}(A) & 0 \\ * & 1 \end{pmatrix}\right\}_{\tau \in \Pi_{n-q}}\right)$$

is a homotopy equivalence according to Lemma 5.4. Finally, $\overline{V}_{n-q}(A) \supset V_{n-q}(A)$ and this inclusion becomes an equality if $n - q \geq r + 1$ (see Corollary 5.3). Gathering all the above remarks together, we get

(6.2.1) $$E^1_{pq} = C_q(SU_{n+1,A}(R^{n+1})_*) \otimes H_p(\overline{V}_{n-q}(A)).$$

Denote by $\delta_i : \{1,\ldots,n-q-1\} \hookrightarrow \{1,\ldots,n-q\}$ the nondecreasing embedding missing i ($i = 1,\ldots,n-q$). One easily verifies that the differential $d^1 : E^1_{p,q+1} \to E^1_{pq}$ is given by

$$d^1(f \otimes u) = \sum_{k=0}^{q}(-1)^k d_k(f) \otimes (\alpha_k \cdot {}^{\delta_{i_k - k}} u).$$

Here $f \in SU_{n+1,A}(R^{n+1})$, $\mathrm{dom}(f) = \{i_0,\ldots,i_q\}$, $u \in H_p(\overline{V}_{n-q-1}(A))$ and $\alpha_k \in \overline{E}_{n-q}(A)$ are certain matrices (depending on the function f). The above formula becomes simpler provided that $p < L$, $n - q - 1 \geq 2p + \widetilde{r}$. In this case the action of $\overline{E}_{n-q}(A) = E_{n-q}(A)$ and Σ_{n-q} on the group $H_p(\overline{V}_{n-q}(A)) = H_p(V_{n-q}(A))$ is trivial according to the induction hypothesis, so that the formula for d^1 takes the form $d^1(f \otimes u) = d(f) \otimes {}^\delta u$, where $\delta = \delta_{n-q}$ is the standard embedding. The inductive hypothesis implies further that

$$H_p(\overline{V}_{n-q-1}(A)) \twoheadrightarrow H_p(V_{n-q}(A)) \xrightarrow{\sim} H_p(V_{n-q+1}(A)).$$

Thus
$$\ker d^1_{pq} = Z_q(SU_{n+1,A}(R^{n+1})) \otimes H_p(V_{n-q}(A)),$$
$$\operatorname{im} d^1_{p,q+1} = B_q(SU_{n+1,A}(R^{n+1})) \otimes H_p(V_{n-q}(A)).$$

Since $q \leq n - 1 - \tilde{r}$, Corollary 2.8 shows that $H_q(SU_{n+1,A}(R^{n+1})_*) = 0$ and we get

LEMMA 6.3. $E^2_{pq} = 0$ provided that $p < L$, $q > 0$ and $n - q - 1 \geq 2p + \tilde{r}$.

Lemma 6.3 shows that all E^2_{pq}-terms with $p + q = L$ vanish except for E^2_{0L} and hence the edge homomorphism $\varepsilon \colon E^1_{0L} \to H_L(V_{n+1}(A))$ is surjective. The homomorphism ε is given by $\varepsilon(f \otimes u) = \alpha \cdot \delta_i(u)$, where $f \in SU_{n+1,A}(R^{n+1})$, $\operatorname{dom}(f) = \{i\}$, $u \in H_L(V_n(A))$, and $\alpha \in E_{n+1}(A)$. The action of $E_{n+1}(A)$ on the image of δ_i is trivial according to Lemma 5.5, and we conclude that $H_L(V_{n+1}(A))$ coincides with the sum of the images of the homomorphisms $\delta_i \colon H_L(V_n(A)) \to H_L(V_{n+1}(A))$. Using Lemma 5.5 once more, we get

LEMMA 6.4. *The action of $E_{n+1}(A)$ on $H_L(V_{n+1}(A))$ is trivial.*

Thus we proved the validity of $2)_L$. Now Corollary 5.8 implies the validity of $4)_L$.

COROLLARY 6.5. *The natural action by conjugation of the group $\operatorname{GL}_{n+1}(\mathbb{Z})$ on*
$$\operatorname{im}(H_{L+1}(E_n(A)) \to H_{L+1}(E_{n+1}(A)))$$
is trivial.

PROOF. The group $\operatorname{GL}_{n+1}(\mathbb{Z})$ is generated by matrices
$$E_{i,n+1}(1), \quad E_{n+1,i}(1), \quad \operatorname{diag}(1,1,\ldots,-1).$$
The action of the last matrix on $\operatorname{im}(H_L(E_n(A)) \to H_L(E_{n+1}(A)))$ is obviously trivial. Let us check the triviality of the action of $E_{n+1,i}(1)$. Set
$$G = \begin{pmatrix} E_n(A) & 0 \\ * & 1 \end{pmatrix}.$$
Then $E_n(A) \subset G \subset E_{n+1}(A)$ and G is normalized by $E_{n+1,i}(1)$. It suffices to check that $E_{n+1,i}(1)$ acts trivially on $\operatorname{im}(H_L(E_n(A)) \to H_L(G))$. According to $4)_L$ the canonical embedding and projection $E_n(A) \hookrightarrow G$ induce mutually inverse isomorphisms on H_L. Now our statement follows from the observation that the composition of the group homomorphisms
$$E_n(A) \hookrightarrow G \xrightarrow{E_{n+1,i}(1)} G \twoheadrightarrow E_n(A)$$
is the identity map.

PROPOSITION 6.6. $H_{L+1}(E_n(A)) \twoheadrightarrow H_{L+1}(E_{n+1}(A)) \xrightarrow{\sim} H_{L+1}(E_{n+2}(A))$.

PROOF. Consider the hyperhomology spectral sequences associated with the action of the matrix $E_{n+1}(A)$ on the complex $C_*(SU_{n+1,A}(R^{n+1})_*)$. Since
$$\widetilde{H}_p(SU_{n+1,A}(R^{n+1})_*) = 0 \quad \text{for } p \leq n - \tilde{r},$$
we conclude that $E^\infty_p = H_p(E_{n+1}(A))$ for $p \leq n - \tilde{r}$. On the other hand
$$E^1_{pq} = H_p(E_{n+1}(A), C_q(SU_{n+1,A}(R^{n+1})_*)).$$

Denote by $C_q^{i_0,\ldots,i_q}$ the $E_{n+1}(A)$-submodule of $C_q = C_q(SU_{n+1,A}(R^{n+1})_*)$ generated by functions f with $\text{dom}(f) = \{i_0,\ldots,i_q\}$. Thus

$$C_q = \coprod_{i_0,\ldots,i_q} C_q^{i_0,\ldots,i_q},$$

the action of $E_{n+1}(A)$ on the basis of $C_q^{i_0,\ldots,i_q}$ is transitive and the stabilizer of the function f_0 ($f_0(i_s) = e_{i_s}$) coincides with $E_{n+1}(A)^{i_0,\ldots,i_q}$. According to the Shapiro lemma, we get

(6.6.1) $$E_{pq}^1 = \coprod_{i_0\ldots i_q} H_p(E_{n+1}(A)^{i_0,\ldots,i_q}).$$

For any set of indices $\{i_0,\ldots,i_q\}$ there exists a unique $(n-q, q+1)$-shuffle taking $n-q+1,\ldots,n+1$ to $i_0,\ldots,i_q$. Conjugation by the corresponding permutation matrix defines an isomorphism

$$\begin{pmatrix} \overline{E}_{n-q}(A) & 0 \\ * & 1 \end{pmatrix} = E_{n+1}(A)^{n-q+1,\ldots,n+1} \xrightarrow{\sim} E_{n+1}(A)^{i_0,\ldots,i_q}.$$

Using the induction hypothesis, Corollary 5.3 and the (already proved) item 4)$_L$, we conclude that

(6.6.2) $$E_{pq}^1 = \coprod_{i_0\ldots i_q} H_p(E_{n-q}(A)) \quad \text{if } 1 \leq p \leq L+1 \text{ and } n-q \geq 2(p-1) + \tilde{r}.$$

The same formula holds for $p = 0$ and any q.

Assume that $p \leq L+1$, $n-q-1 \geq 2(p-1) + \tilde{r}$, so that formula (6.6.2) applies to E_{pq}^1 and $E_{p,q+1}^1$. In this case one verifies immediately that the differential $d^1: E_{p,q+1}^1 \to E_{pq}^1$ is given by

$$d^1((i_0,\ldots,i_q) \otimes u) = \sum_{k=0}^{q} (-1)^k (i_0,\ldots,\widehat{i_k},\ldots,i_q) \otimes \delta_{i_k - k} u.$$

According to (6.5) and the induction hypothesis, the action of the permutation group Σ_{n+1} on $\text{im}(H_p(E_{n-q-1}(A)) \to H_p(E_{n-q}(A)))$ is trivial. Thus, the above formula becomes simpler and we get

(6.6.4) $$d^1((i_0,\ldots,i_q) \otimes u) = \sum_{k=0}^{q} (-1)^k (i_0,\ldots,\widehat{i_k},\ldots,i_q) \otimes \delta u,$$

where $\delta = \delta_{n-q}$ is the canonical embedding of $E_{n-q-1}(A)$ to $E_{n-q}(A)$ (provided that $p \leq L+1, n-q-1 \geq 2(p-1) + \tilde{r}$).

Assume that $p \leq L$. Then the formula (6.6.4) and the induction hypothesis show that the pth column of our spectral sequence coincides up to dimension $n+1-2p-r$ with the complex computing homology of the partially ordered set $1 < 2 < \cdots < n+1$ with coefficients in $H_p(E(A))$, and furthermore the group $E_{p,n+2-2p-\tilde{r}}^1$ maps surjectively onto the corresponding term of the latter complex. Since the geometric realization of the partially ordered set $1 < 2 < \cdots < n+1$ is contractible we conclude that

(6.6.5) $$E_{pq}^2 = 0 \text{ provided that } p \leq L, 1 \leq q \leq n+1-2p-\tilde{r}.$$

The last formula implies that the only nonzero E^2 term on the diagonal $p + q = L + 1$ is $E^2_{L+1,0}$ and hence the edge homomorphism

$$E^1_{l+1,0} = \coprod_i H_{L+1}(E_n(A)) \to H_{L+1}(E_{n+1}(A))$$

is surjective. Finally, applying (6.5), we get

(6.6.6) $$H_{L+1}(E_n(A)) \twoheadrightarrow H_{L+1}(E_{n+1}(A)).$$

To conclude, consider a similar spectral sequence for the action of the group $E_{n+2}(A)$ on the complex $C_*(SU_{n+2,A}(R^{n+2})_*)$. In this case, using the same reasoning as above, we see that $E^2_{pq} = 0$ provided that $p + q = L + 2$, $q \geqslant 2$. Thus there are no differentials coming to $E^2_{L+1,0}$ and hence $H_{L+1}(E_{n+2}(A)) = E^\infty_{L+1,0} = E^2_{L+1,0}$. Finally, combining (6.6.6) and (6.6.4) (with n replaced by $n + 1$) we easily see that $E^2_{L+1,0} = H_{L+1}(E_{n+1}(A))$.

(6.7) The action of the permutation group Σ_{n+1} on $H_{L+1}(E_{n+1}(A))$ is trivial. This follows immediately from (6.6) and (6.5).

LEMMA 6.8. $H_L(V_{n+1}(A)) \xrightarrow{\sim} H_L(V_{n+2}(A))$.

PROOF. Consider the homomorphism of spectral sequences

$$E^2_{pq} = H_p(E_{n+1}(A)), H_q(V_{n+1}(A)) \Rightarrow H_{p+q}\left(\bigcup_{\sigma \in \Pi_{n+1}} BT^\sigma_{n+1}(A)\right)$$

$$\downarrow \qquad\qquad \downarrow \qquad\qquad \downarrow$$

$$\widetilde{E}^2_{pq} = H_p(E_{n+2}(A)), H_q(V_{n+2}(A)) \Rightarrow H_{p+q}\left(\bigcup_{\sigma \in \Pi_{n+2}} BT^\sigma_{n+2}(A)\right)$$

The induction hypothesis implies that the homomorphism $E^2_{pq} \to \widetilde{E}^2_{pq}$ is an isomorphism for $q \leqslant L - 1$, $p \leqslant L + 1$. The homomorphism $E^\infty_k \to \widetilde{E}^\infty_k$ is an isomorphism for $k \leqslant n/2$ (according to Corollary 5.7.b) and in particular for all $k \leqslant L + 1$. Theorem A.6 shows that $E^2_{0L} \xrightarrow{\sim} \widetilde{E}^2_{0L}$, i.e. (since the action of $E_{n+1}(A)$ on $H_L(V_{n+1}(A))$ is trivial), $H_L(V_{n+1}(A)) \xrightarrow{\sim} H_L(V_{n+2}(A))$.

LEMMA 6.9. *The action of the permutation group* Σ_{n+1} *on* $H_L(V_{n+1}(A))$ *is trivial.*

PROOF. Since the isomorphism $H_L(V_{n+1}(A)) \xrightarrow{\sim} H_L(V_{n+3}(A))$ is obviously Σ_{n+1}-equivariant, it suffices to show that the action of Σ_{n+3} on $H_L(V_{n+3}(A))$ is trivial. The permutation $(n + 2, n + 3)$ acts trivially on $H_L(V_{n+3}(A))$ because the homomorphism $H_L(V_{n+1}(A)) \to H_L(V_{n+3}(A))$ is surjective. Since the normal closure of $(n + 1, n + 2)$ coincides with the whole group Σ_{n+3}, this ends the proof.

COROLLARY 6.10. $H_L(V_n(A)) \twoheadrightarrow H_L(V_{n+1}(A))$.

PROOF. We have already seen (see Lemma 6.3 and the text immediately after it) that $H_L(V_{n+1}(A))$ is the sum of the homomorphisms $\delta_i \colon H_L(V_n(A)) \to H_L(V_{n+1}(A))$. Lemma 6.9 shows that all these homomorphisms coincide with the homomorphism induced by the standard embedding $V_n(A) \hookrightarrow V_{n+1}(A)$ and hence the latter is surjective.

This ends the inductive step and the proof of Theorem 6.1.

COROLLARY 6.11. *Assume that $n \geqslant 2l + \tilde{r}$. Then*

$$H_{l+1}(\mathrm{GL}_{n+1}(A)) \xrightarrow{\sim} H_{l+1}(\mathrm{GL}_{n+2}(A)).$$

PROOF. Embed A as a two-sided ideal into a ring with unit R. According to the stability theorem of Vasserstein [12], $E_{n+1}(A) = E_{n+1}(R, A)$ is a normal subgroup of $\mathrm{GL}_{n+1}(A)$ and $\mathrm{GL}_{n+1}(A)/E_{n+1}(A) \xrightarrow{\sim} \mathrm{GL}_{n+2}(A)/GE_{n+2}(A)$. Now our statement follows from Proposition A.7 applied to the homomorphism of the Hochschild–Serre spectral sequences

$$E^2_{pq} = H_p(\mathrm{GL}_{n+1}(A))/E_{n+1}(A), H_q(E_{n+1}(A)) \Rightarrow H_{p+q}(\mathrm{GL}_{n+1}(A))$$
$$\downarrow \qquad\qquad \downarrow \quad \downarrow$$
$$\widetilde{E}^2_{pq} = H_p(\mathrm{GL}_{n+2}(A))/E_{n+2}(A), H_q(E_{n+2}(A)) \Rightarrow H_{p+q}(\mathrm{GL}_{n+2}(A))$$

COROLLARY 6.12. *Let A be a stable C^*-algebra. Then $H_i(\mathrm{GL}_n(A)) \to H_i(\mathrm{GL}(A))$ is an isomorphism for all i and n.*

PROOF. It is known that C^*-algebras are H-unital (see [13]) and furthermore the stable rank of any stable C^*-algebra is equal to 2 ([7]). Corollary 6.11 shows that $H_i(\mathrm{GL}_n(A)) \xrightarrow{\sim} H_i(\mathrm{GL}(A))$ for sufficiently large n. Finally, the algebra A is isomorphic to $M_2(A)$ (see [5]). This isomorphism of rings defines isomorphisms of groups $\mathrm{GL}_n(A) \xrightarrow{\sim} \mathrm{GL}_{2n}(A)$, $\mathrm{GL}(A) \xrightarrow{\sim} \mathrm{GL}(A)$ such that the following diagram commutes

$$\begin{array}{ccc} \mathrm{GL}_n(A) & \hookrightarrow & \mathrm{GL}(A) \\ \downarrow \wr & & \downarrow \wr \\ \mathrm{GL}_{2n}(A) & \hookrightarrow & \mathrm{GL}(A) \end{array}$$

The inverse induction on n ends the proof.

COROLLARY 6.13. $B\mathrm{GL}_n(A)^+ \to B\mathrm{GL}_n(A)^{\mathrm{top}}$ *is a homotopy equivalence for all n, where plus construction on the left is taken with respect to the perfect normal subgroup* $[\mathrm{GL}_n(A), \mathrm{GL}_n(A)] \triangleleft \mathrm{GL}_n(A)$.

Appendix. Comparison theorems for spectral sequences

LEMMA A.1. *Let $f: C \to \widetilde{C}$ be a homomorphism of complexes (of degree -1). Assume that $C_{n-1} \hookrightarrow \widetilde{C}_{n-1}$ and $C_n \twoheadrightarrow \widetilde{C}_n$. Then $H_{n-1}(C) \hookrightarrow H_{n-1}(\widetilde{C})$ and $H_n(C) \twoheadrightarrow H_n(\widetilde{C})$.*

COROLLARY A.2. *Let $f: E \to \widetilde{E}$ be a homomorphism of (homological) spectral sequences and let $r \geqslant 2$ be an integer.*
  a) *If $E^{r-1}_{pq} \hookrightarrow \widetilde{E}^{r-1}_{pq}$ and $E^{r-1}_{p+r-1, q-r+2} \twoheadrightarrow \widetilde{E}^{r-1}_{p+r-1, q-r+2}$, then*

$$E^r_{pq} \hookrightarrow \widetilde{E}^r_{pq} \quad \text{and} \quad E^r_{p+r-1, q-r+2} \twoheadrightarrow \widetilde{E}^r_{p+r-1, q-r+2}.$$

b) If $E^{r-1}_{p-r+1,q+r-2} \hookrightarrow \widetilde{E}^{r-1}_{p-r+1,q+r-2}$, $E^{r-1}_{p+r-1,q-r+2} \twoheadrightarrow \widetilde{E}^{r-1}_{p+r-1,q-r+2}$, and $E^{r-1}_{pq} \xrightarrow{\sim} \widetilde{E}^{r-1}_{pq}$, then

$$E^r_{p-r+1,q+r-2} \hookrightarrow \widetilde{E}^r_{p-r+1,q+r-2}, \quad E^r_{p+r-1,q-r+2} \twoheadrightarrow \widetilde{E}^r_{p+r-1,q-r+2},$$
$$E^r_{pq} \xrightarrow{\sim} \widetilde{E}^r_{pq}.$$

LEMMA A.3. *Let B and $\widetilde{B}$ be abelian groups equipped with increasing filtrations*

$$0 = F_{-1}B \subset F_0B \subset \cdots \subset F_nB = B, \quad 0 = F_{-1}\widetilde{B} \subset F_0\widetilde{B} \subset \cdots \subset F_n\widetilde{B} = \widetilde{B}$$

and let $f : B \to \widetilde{B}$ be a homomorphism of filtered groups. The following conditions are equivalent:
  a) $F_{k/k-1}B \xrightarrow{\sim} F_{k/k-1}\widetilde{B}$ *for* $k = 0, 1, \ldots, n$;
  b) $F_{k/k-1}B \twoheadrightarrow F_{k/k-1}\widetilde{B}$ *for* $k = 0, 1, \ldots, n-1$ *and* $B \xrightarrow{\sim} \widetilde{B}$;
  c) $F_{k/k-1}B \hookrightarrow F_{k/k-1}\widetilde{B}$ *for* $k = 1, \ldots, n$ *and* $B \xrightarrow{\sim} \widetilde{B}$.

THEOREM A.4. *Let $f : E \to \widetilde{E}$ be a homomorphism of first quadrant homological spectral sequences and let L be a nonnegative integer. Assume that*
  1) $E^2_{pq} \hookrightarrow \widetilde{E}^2_{pq}$ *for* $p + q = L - 1$,
  2) $E^2_{pq} \xrightarrow{\sim} \widetilde{E}^2_{pq}$ *for* $p + q = L$,
  3) $E^2_{pq} \twoheadrightarrow \widetilde{E}^2_{pq}$ *for* $p + q = L + 1, q \geqslant 1$,
  4) $E^\infty_L \xrightarrow{\sim} \widetilde{E}^\infty_L$, $E^\infty_{L+1} \xrightarrow{\sim} \widetilde{E}^\infty_{L+1}$.

Then $E^2_{L+1,0} \xrightarrow{\sim} \widetilde{E}^2_{L+1,0}$.

PROOF. First of all, using induction on r and Lemma A.2, we prove that the following statements hold:
  a) $E^r_{pq} \hookrightarrow \widetilde{E}^r_{pq}$ for $p + q = L - 1$;
  b) $E^r_{pq} \twoheadrightarrow \widetilde{E}^r_{pq}$ for $p + q = L$;
  c) $E^r_{pq} \xrightarrow{\sim} \widetilde{E}^r_{pq}$ for $p + q = L, r \leqslant q + 1$;
  d) $E^r_{pq} \twoheadrightarrow \widetilde{E}^r_{pq}$ for $p + q = L + 1, q \geqslant 1$.

Item b) shows that $E^\infty_{pq} \twoheadrightarrow \widetilde{E}^\infty_{pq}$ for $p + q = L$. Since $E^\infty_L \xrightarrow{\sim} \widetilde{E}^\infty_L$, Lemma A.3 shows that $E^\infty_{pq} \xrightarrow{\sim} \widetilde{E}^\infty_{pq}$ for $p + q = L$. Next, using inverse induction on r, we prove that $E^r_{pq} \xrightarrow{\sim} \widetilde{E}^r_{pq}$ for $p + q = L, r \geqslant q + 2$. This follows by an easy diagram chase in the commutative diagram with exact rows

$$\begin{array}{ccccccc}
0 & \longrightarrow & E^{r+1}_{pq} & \longrightarrow & E^r_{pq} & \longrightarrow & E^r_{p-r,q+r-1} \\
& & \downarrow{\wr} & & \downarrow & & \downarrow \\
0 & \longrightarrow & \widetilde{E}^{r+1}_{pq} & \longrightarrow & \widetilde{E}^r_{pq} & \longrightarrow & \widetilde{E}^r_{p-r,q+r-1}
\end{array}$$

Thus $E^r_{pq} \xrightarrow{\sim} \widetilde{E}^r_{pq}$ for any r provided that $p + q = L$. Lemma A.3 in conjunction with item d) above imply that $E^\infty_{L+1,0} \xrightarrow{\sim} \widetilde{E}^\infty_{L+1,0}$. Now using inverse induction on r we prove that $E^r_{L+1,0} \xrightarrow{\sim} \widetilde{E}^r_{L+1,0}$ for any r. Consider the following commutative diagram

$$\begin{array}{ccccc} E^r_{L+1,0} & \longrightarrow & E^r_{L+1-r,r-1} & \longrightarrow & E^r_{L+1-2r,2r-2} \\ \downarrow & & \downarrow \wr & & \uparrow \\ \widetilde{E}^r_{L+1,0} & \longrightarrow & \widetilde{E}^r_{L+1-r,r-1} & \longrightarrow & \widetilde{E}^r_{L+1-2r,2r-2} \end{array}$$

We shall use the symbols Z, B, H (resp. $\widetilde{Z}, \widetilde{B}, \widetilde{H}$) to denote the cycles, boundaries, and homology of the top row (resp. bottom row). Thus $H = E^{r+1}_{L+1-r,r-1}$, $\widetilde{H} = \widetilde{E}^{r+1}_{L+1-r,r-1}$ and hence $H \xrightarrow{\sim} \widetilde{H}$. Since $Z \xrightarrow{\sim} \widetilde{Z}$, we conclude that $B \xrightarrow{\sim} \widetilde{B}$. Now applying the five lemma to the commutative diagram with exact rows

$$\begin{array}{ccccccccc} 0 & \longrightarrow & E^{r+1}_{L+1,0} & \longrightarrow & E^r_{L+1,0} & \longrightarrow & B & \longrightarrow & 0 \\ & & \downarrow \wr & & \downarrow & & \downarrow \wr & & \\ 0 & \longrightarrow & \widetilde{E}^{r+1}_{L+1,0} & \longrightarrow & \widetilde{E}^r_{L+1,0} & \longrightarrow & \widetilde{B} & \longrightarrow & 0 \end{array}$$

we conclude that $E^r_{L+1,0} \xrightarrow{\sim} \widetilde{E}^r_{L+1,0}$.

REMARK A.5. If we want to prove only the surjectivity of the homomorphism $E^2_{L+1,0} \to \widetilde{E}^2_{L+1,0}$, then we may slightly weaken the requirements of Theorem A.4: in item 3) one should demand the surjectivity of $E^2_{pq} \to \widetilde{E}^2_{pq}$ ($p+q = L+1$) only for $q \geq 1$, $p \geq 2$ and in item 4) the requirement $E_{L+1} \xrightarrow{\sim} \widetilde{E}_{L+1}$ should be replaced by $E_{L+1} \twoheadrightarrow \widetilde{E}_{L+1}$.

THEOREM A.6. *Let $f: E \to \widetilde{E}$ be a homomorphism of first quadrant homological spectral sequences and let L be a nonnegative integer. Assume that*
1) $E^2_{pq} \hookrightarrow \widetilde{E}^2_{pq}$ for $p+q = L$, $p \geq 1$,
2) $E^2_{pq} \xrightarrow{\sim} \widetilde{E}^2_{pq}$ for $p+q = L+1$, $p \geq 2$,
3) $E^2_{pq} \twoheadrightarrow \widetilde{E}^2_{pq}$ for $p+q = L+2$, $p \geq 3$,
4) $E^\infty_L \xrightarrow{\sim} \widetilde{E}^\infty_L$, $E^\infty_{L+1} \xrightarrow{\sim} \widetilde{E}^\infty_{L+1}$.

Then $E^2_{0L} \xrightarrow{\sim} \widetilde{E}^2_{0L}$.

PROOF. First of all we use induction on r and Lemma A.2 to show that
a) $E^r_{pq} \hookrightarrow \widetilde{E}^r_{pq}$ for $p+q = L$, $p \geq 1$,
b) $E^r_{pq} \hookrightarrow \widetilde{E}^r_{pq}$ for $p+q = L+1$, $p \geq 2$,
c) $E^r_{pq} \xrightarrow{\sim} \widetilde{E}^r_{pq}$ for $p+q = L+1$, $p \geq r$,
d) $E^r_{pq} \twoheadrightarrow \widetilde{E}^r_{pq}$ for $p+q = L+2$, $p \geq r+1$.

Item b) implies that $E^\infty_{pq} \hookrightarrow \widetilde{E}^\infty_{pq}$ for $p+q = L+1$, $p \geq 2$. Applying Lemma A.3 to the homomorphism $E^\infty_{L+1} \to \widetilde{E}^\infty_{L+1}$ and filtrations $0 = F_{-1} \subset F_1 \subset \cdots \subset F_{L+1}$ and $0 = \widetilde{F}_{-1} \subset \widetilde{F}_1 \subset \cdots \subset \widetilde{F}_{L+1}$ of the corresponding groups, we conclude that $E^\infty_{pq} \xrightarrow{\sim} \widetilde{E}^\infty_{pq}$ for $p+q = L+1$, $p \geq 2$. Using inverse induction on r and a diagram chase in the commutative diagram with exact rows

$$\begin{array}{ccccccc}
E^r_{p+r,q-r+1} & \longrightarrow & E^r_{pq} & \longrightarrow & E^{r+1}_{pq} & \longrightarrow & 0 \\
\downarrow & & \uparrow & & \downarrow \wr & & \\
\widetilde{E}^r_{p+r,q-r+1} & \longrightarrow & \widetilde{E}^r_{pq} & \longrightarrow & \widetilde{E}^{r+1}_{pq} & \longrightarrow & 0
\end{array}$$

we conclude further that $E^r_{pq} \xrightarrow{\sim} \widetilde{E}^r_{pq}$ for $p + q = L + 1$, $p \geqslant 2$. Thus $E^r_{pq} \xrightarrow{\sim} \widetilde{E}^r_{pq}$ for all r provided that $p + q = L + 1$ and $p \geqslant 2$. Finally, applying the same procedure as in the end of the proof of Theorem A.4, we come to the desired conclusion.

Using Lemmas A.2 and A.3, one can also prove the following standard result.

PROPOSITION A.7. *Let $f : E \to \widetilde{E}$ be a homomorphism of first quadrant homological spectral sequences and let L be a nonnegative integer. Assume that*
1) $E^2_{pq} \hookrightarrow \widetilde{E}^2_{pq}$ *for* $p + q = L - 1$,
2) $E^2_{pq} \xrightarrow{\sim} \widetilde{E}^2_{pq}$ *for* $p + q = L$,
3) $E^2_{pq} \twoheadrightarrow \widetilde{E}^2_{pq}$ *for* $p + q = L + 1$, $p > 0$.

Then $E^\infty_L \xrightarrow{\sim} \widetilde{E}^\infty_L$.

References

1. H. Bass, *Algebraic K-theory*, Benjamin, New York–Amsterdam, 1968.
2. R. Charney, *On the problem of homology stability for congruence subgroups*, Comm. Algebra **12** (**17**) (1984), 2081–2123.
3. P. Hanlon, *On the decomposition of the tensor algebra of the classical Lie algebras*, Adv. Math. **56** (1985), 238–282.
4. W. van der Kallen, *Homology stability for general linear groups*, Invent. Math. **60** (1980), 269–295.
5. M. Karoubi, *Homologie de groupes discrets associés à des algebras d'opérateurs*, J. Operator Theory **15** (1986), 109–161.
6. Yu. Nesterenko and A. Suslin, *Homology of the full linear group over a local ring and Milnor K-theory*, Izv. Akad. Nauk SSSR Ser. Mat. **53** (1989), 121–146; English transl., Math. USSR-Izv. **34** (1990), 121–145.
7. M. Riefel, *Dimension and stable rank in the K-theory of C^*-algebras*, Proc. London Math. Soc. **46** (1983), 301–333.
8. E. Spanier, *Algebraic topology*, McGraw-Hill, New York–London, 1966.
9. A. Suslin, *Stability in algebraic K-theory*, Lecture Notes in Math., vol. 966, 1982, pp. 334–357.
10. _____, *Excision in integral algebraic K-theory*, Proc. Steklov Inst. Math. (to appear).
11. A. Suslin and M. Wodzicki, *Excision in algebraic K-theory*, Ann. Math. **136** (1992), 51–122.
12. L. Vasserstein, *On the stabilization of the general linear group over a ring*, Mat. Sb. **79** (1969), 405–424; English transl., Math. USSR-Sb. **8** (1969), 383–400.
13. M. Wodzicki, *Excision in cyclic homology and in rational algebraic K-theory*, Ann. Math. **129** (1989), 591–639.

Translated by THE AUTHOR

POMI, FONTANKA 27, ST. PETERSBURG 191011, RUSSIA

PART II

Geometry and Analysis

Willmore Surfaces, 4-Particle Toda Lattice and Double Coverings of Hyperelliptic Surfaces

M. V. Babich

Dedicated to Boarding School #45 on the occasion of its 30th anniversary

ABSTRACT. In this article we write the Gauss–Weingarten (GW) equations for the Conformal Gaussian Image (CGI) for an arbitrary surface in $\mathbb{S}^3$ in 4×4 matrices. In this representation the (GW) equations for the (CGI) of the so-called Willmore tori coincide with a certain reduction of the Toda lattice equations. This relationship considered is in detail. The formulas for the Willmore tori in terms of the Ψ-function for the Lax pair for the Toda lattice are constructed. In the finite-gap case, sufficient conditions for a Riemann surface to correspond to a Willmore surface are found. The classification of such surfaces is presented. Examples, for several classes of parameters of theta-functional solutions, are calculated.

§0. Introduction

This paper is devoted to the construction of the so-called Willmore surfaces $\mathcal{F}$, which are the extremals of the functional W:

$$W(\mathcal{F}) = \int H^2 \, dS,$$

where H is the mean curvature and dS is the area differential. We present the formulas for the immersion of a general Willmore surface in terms of solution of the linear representation (in 4×4 matrices) with additional ("spectral") parameter λ. In a special case (when $\mathcal{F}$ is a torus), the linear system becomes the well-known Lax pair for the two-dimensional Toda lattice with two additional reductions.

We investigate this case in detail. A rich class of solutions of the Toda lattice with the necessary properties in terms of the theta-functions of the double-coverings of hyperelliptic surfaces with some additional symmetries is found.

In §1 we shall write the Gauss–Weingarten (GW) equations for the conformal Gaussian image for an arbitrary surface in $\mathbb{S}^3$ (or for a surface in $\mathbb{R}^3$, since the conformal Gaussian images of the surface in $\mathbb{R}^3$ and the images of its stereographic projection to $\mathbb{S}^3$ coincide). We shall see that it is possible to introduce an additional parameter into

---

1991 *Mathematics Subject Classification.* Primary 53C42; Secondary 58F07.

the (GW)-system. In §2, we consider the case of W-surfaces, and present the explicit formulas for the Willmore tori in terms of the Ψ-function for some special solutions of the Toda lattice. We consider the so-called finite-gap solutions for the Toda lattice; they are parametrized by Riemann surfaces. The main problem is to find parameters, i.e., the Riemann surface and the divisor $\mathcal{D}$ such that the solutions of the Toda lattice have the necessary symmetries. In §3, we present sufficient conditions for these parameters in terms of the existence of certain objects, namely a function v and a differential $d\tilde{\Omega}_0$ with special properties. These conditions are not constructive, their direct verification is not convenient, and we present a more constructive version of such conditions. In §4 we develop some techniques for working with multi-sheeted coverings of $\mathbb{C}$. In §5, we give the classification of real double-sheeted coverings of hyperelliptic curves; some of them we consider as examples. We construct two examples, namely two families of surfaces for which all the conditions hold; so we present two families of Willmore surfaces. We note that generally these surfaces are not tori. For obtaining tori, some additional restrictions must be satisfied the periodicity of the Ψ-function (see [1]). We shall not discuss it here.

Acknowledgements. I am deeply grateful to Professor A. I. Bobenko, who drew my attention to this topic, and to Professor A. Fokas, whose assistance made it possible to complete this work. I am thankful to T. Lakoba for help with the proofreading of the text.

§1. Gauss–Weingarten equations for $\mathcal{F}_\gamma$

The aim of this section is to write down the Gauss–Weingarten system for the image of an arbitrarily smooth surface $\mathcal{F}$ in the spherical space $\mathbb{S}^3$, $\mathcal{F} \subset \mathbb{S}^3 \subset \mathbb{R}^4$, under the conformal Gaussian mapping[1] γ.

Let a surface $\mathcal{F}$ be given by the function f:

$$f: \mathcal{D} \to \mathbb{S}^3, \qquad \mathcal{D} \subset \mathbb{C}_z,$$

which gives a conformal parametrization of $\mathcal{F}$ by the points of $\mathbb{R}^2 \equiv \mathbb{C}_z$ with complex coordinates $z, \bar{z}$.

Let us consider the space $\mathbb{R}^{4,1}$ and a standard basis $\{\check{e}^k\}_{k=1}^5$ on it:

$$\check{e}^k = (\underbrace{0, \ldots, 0, 1}_{k}, 0, \ldots, 0).$$

We denote the scalar product on $\mathbb{R}^{4,1}$ by $\langle \cdot; \cdot \rangle$:

$$\langle a; b \rangle = a_1 b_1 + a_2 b_2 + a_3 b_3 + a_4 b_4 - a_5 b_5,$$
$$a_i = \varepsilon_i \langle a; \check{e}^i \rangle, \quad \varepsilon_1 = \cdots = \varepsilon_4 = 1, \quad \varepsilon_5 = -1.$$

We define the conformal Gaussian map γ of $\mathcal{F}$, $\gamma: \mathcal{F} \to \mathbb{R}^{4,1}$ by the formula

(1.1) $$Y = XH + N,$$

where

$X = (f; 1)$, $f \in \mathbb{S}^3$ is a point of the surface $\mathcal{F}$;
$N = (n; 0)$, $n \in \mathbb{S}^3$ is the unit normal vector to the $\mathcal{F}$ at the point f;
$H \in \mathbb{R}$ is the mean curvature of $\mathcal{F}$ at the point f.

---

[1] About the conformal Gaussian mapping and its geometrical meaning, see [1–3].

Let us denote the image of $\mathcal{F}$ under γ by $\mathcal{F}_\gamma$. The image $\mathcal{F}_\gamma$ is a surface in $\mathbb{R}^{4,1}$. Consider the pseudo-Riemannian space $\mathbb{S}^{3,1} = \{a \in \mathbb{R}^{4,1} : \langle a; a \rangle = 1\}$. Note that

(1.2) $$\langle Y; Y \rangle = 1,$$

i.e., $\mathcal{F}_\gamma \subset \mathbb{S}^{3,1}$.

Let us complexify $\mathbb{R}^{4,1} \supset \mathbb{S}^{3,1}$. It is easy to verify that the parametrization of $\mathcal{F}_\gamma$ by the vector $Y = Y(z, \bar{z})$ defined in (1.1) is conformal, i.e.,

(1.3) $$\langle Y_z; Y_z \rangle = \langle Y_{\bar{z}}; Y_{\bar{z}} \rangle = 0.$$

Let us consider the moving frame $\{Y, Y_z, Y_{\bar{z}}, X, Z\}$, where the real vector $Z \in \mathbb{R}^{4,1}$ completes the set $Y, Y_z, Y_{\bar{z}}, X$ to a basis of $\mathbb{C}^5$. We impose on Z the following restrictions

(1.4) $$\langle Z; Z \rangle = \langle Z; Y \rangle = \langle Z; Y_z \rangle = \langle Z; Y_{\bar{z}} \rangle = 0, \quad \langle Z; X \rangle = 1,$$

which determine the vector Z uniquely.

Using the explicit formula (1.1) for the Gaussian map and relations (1.2)–(1.4), let us write down the Gauss–Weingarten system for the frame $\{Y, Y_z, Y_{\bar{z}}, X, Z\}$:

(1.5)
$$\begin{pmatrix} Y \\ Y_z \\ Y_{\bar{z}} \\ X \\ Z \end{pmatrix}_z = \begin{pmatrix} 0 & 1 & 0 & 0 & 0 \\ 0 & E_z/E & 0 & A & B \\ -E & 0 & 0 & H_\gamma & 0 \\ 0 & 0 & -B/E & C & 0 \\ 0 & -H_\gamma/E & -A/E & 0 & -C \end{pmatrix} \begin{pmatrix} Y \\ Y_z \\ Y_{\bar{z}} \\ X \\ Z \end{pmatrix},$$

$$\begin{pmatrix} Y \\ Y_z \\ Y_{\bar{z}} \\ X \\ Z \end{pmatrix}_{\bar{z}} = \begin{pmatrix} 0 & 0 & 1 & 0 & 0 \\ -E & 0 & 0 & H_\gamma & 0 \\ 0 & 0 & E_{\bar{z}}/E & \overline{A} & \overline{B} \\ 0 & -\overline{B}/E & 0 & \overline{C} & 0 \\ 0 & -\overline{A}/E & -H_\gamma/E & 0 & -\overline{C} \end{pmatrix} \begin{pmatrix} Y \\ Y_z \\ Y_{\bar{z}} \\ X \\ Z \end{pmatrix},$$

where

(1.6) $$A = \langle Y_{zz}; Z \rangle, \quad B = \langle Y_{zz}; X \rangle, \quad C = \langle X_z; Z \rangle,$$
$$H_\gamma = \langle Y_{z\bar{z}}; Z \rangle, \quad E = \langle Y_z; Y_{\bar{z}} \rangle.$$

Let $u_{\mathcal{F}}$ and $Q_{\mathcal{F}}$ be the metric and the Hopf differential induced by $\mathcal{F}$ on $\mathbb{C}_z$:

$$\langle f_z; f_{\bar{z}} \rangle_{\mathbb{R}^4} = 2 \exp\{u_{\mathcal{F}}\}, \quad Q_{\mathcal{F}} = \langle f_{zz}; n \rangle_{\mathbb{R}^4}.$$

Then we can write

(1.7) $$Y_{z\bar{z}} = -Q_{\mathcal{F}} \overline{Q}_{\mathcal{F}} e^{-u_{\mathcal{F}}} Y/2 + (H_{z\bar{z}} + H Q_{\mathcal{F}} \overline{Q}_{\mathcal{F}} e^{-u_{\mathcal{F}}}/2) X,$$
$$\langle Y_z; Y_{\bar{z}} \rangle = Q_{\mathcal{F}} \overline{Q}_{\mathcal{F}} e^{-u_{\mathcal{F}}}/2,$$

i.e.,

(1.8) $$E = Q_{\mathcal{F}} \overline{Q}_{\mathcal{F}} e^{-u_{\mathcal{F}}}/2, \quad H_\gamma = H_{z\bar{z}} + H Q_{\mathcal{F}} \overline{Q}_{\mathcal{F}} e^{-u_{\mathcal{F}}}/2 = H_{z\bar{z}} + HE.$$

Let us denote the principal curvatures of $\mathcal{F}$ at f by k_1 and k_2. Then

$$((k_1 - k_2)/2)^2 = Q_{\mathcal{F}} \overline{Q}_{\mathcal{F}} \exp\{-2u_{\mathcal{F}}\}/4 \stackrel{\text{def}}{=} \omega^2.$$

We can rewrite (1.8) as

$$E = 2\omega^2 e^{u_{\mathcal{F}}}, \quad H_\gamma = H_{z\bar{z}} + 2\omega^2 e^{u_{\mathcal{F}}} H.$$

Equations (1.5) determine a two-dimensional manifold in the group $O(4, 1)$. Now we shall consider another representation of this group: the one in terms of matrices from $SL(4, \mathbb{C})$ (see [9]).

Let us consider $\mathbb{C}^4$ and some fixed matrix $\widehat{\Psi} \in SL(4, \mathbb{C})$ (we put $\widehat{\Psi} = I$). Let $\{e_1, e_2, e_3, e_4\}$ be the basis of $\mathbb{C}^4$ formed by the rows of the matrix $\widehat{\Psi}$. On $\mathbb{C}^4$ consider the bilinear antisymmetric operation of external product, denoted by $\wedge$:

$$\mathbb{C}^4 \times \mathbb{C}^4 \xrightarrow{\wedge} \Lambda^2(\mathbb{C}^4) \stackrel{\text{def}}{=} \{a \wedge b : a, b \in \mathbb{C}^4\}.$$

$\Lambda^2(\mathbb{C}^4)$ can be naturally immersed in the linear space $\mathbb{C}^6$ that consists of all linear combinations of the form $\sum_j \alpha_j a_j \wedge b_j$, where $\alpha_j \in \mathbb{C}$, $a_j, b_j \in \mathbb{C}^4$.

We choose the set $\{\widehat{E}^1, \widehat{E}^2, \widehat{E}^3, \widehat{E}^4, \widehat{E}^5, \widehat{E}^6\}$, where

$$\widehat{E}^1 = e_1 \wedge e_2, \quad \widehat{E}^2 = e_3 \wedge e_4, \quad \widehat{E}^3 = e_1 \wedge e_3,$$
$$\widehat{E}^4 = e_4 \wedge e_2, \quad \widehat{E}^5 = e_2 \wedge e_3, \quad \widehat{E}^6 = e_1 \wedge e_4,$$

as the basis of this space. Define the action of complex 4×4 matrices on $\mathbb{C}^6$ by the formula

$$S\left(\sum_i \alpha_i a_i \wedge b_i\right) \stackrel{\text{def}}{=} \sum_i \alpha_i (a_i S) \wedge (b_i S).$$

Define the scalar product $\langle \cdots ; \cdots \rangle_\wedge$ on the elements of $\Lambda^2(\mathbb{C}^4)$ as follows:

$$\langle a_1 \wedge b_1 ; a_2 \wedge b_2 \rangle_\wedge = \det(a_1 b_1 a_2 b_2),$$

where $(a_1 b_1 a_2 b_2)$ is the matrix consisting of the rows a_1, b_1, a_2, b_2. We extend $\langle \cdots ; \cdots \rangle_\wedge$ to all of $\mathbb{C}^6$ by linearity. It is easy to see that matrices from $SL(4, \mathbb{C})$ preserve this scalar product; moreover, this construction gives us the isomorphism $SL(4, \mathbb{C}) \cong O(6)$.

Let $\widehat{E} \in \mathbb{C}^6$ be some fixed vector, and $\mathbf{S}$ be the following subgroup of $SL(4, \mathbb{C})$:

$$\mathbf{S} = \{S \in SL(4, \mathbb{C}) : S\widehat{E} = \widehat{E}\},$$

$\mathbf{S} \cong O(5) \subset O(6)$. Thus we have constructed a representation of $O(5)$ by matrices from $\mathbf{S} \subset SL(4, \mathbb{C})$, which naturally act on $\widetilde{\mathbf{Y}}$,

$$\widetilde{\mathbf{Y}} \stackrel{\text{def}}{=} \{Y \in \mathbb{C}^6 : \langle Y ; \widehat{E} \rangle_\wedge = 0\} \cong \mathbb{C}^5.$$

Let us immerse $\mathbb{R}^{4,1}$, which contains our surface $\mathcal{F}_\gamma$, into $\widetilde{\mathbf{Y}} \subset \mathbb{C}^6$ and define the moving frame of $\mathbb{C}^6$ $\{E^1, E^2, E^3, E^4, E^5, E^6\}$, which is connected with $\{Y, Y_z, Y_{\bar{z}}, X, Z\}$ (i.e., with a point of $\mathcal{F}_\gamma$), by setting

(1.9)
$$E^1 - E^2 = \sqrt{2}\widehat{E}/i = \text{const}, \quad E^1 + E^2 = \sqrt{2} Y, \quad E^3 = Y_z/\sqrt{E},$$
$$E^4 = Y_{\bar{z}}/\sqrt{E}, \quad E^5 = X, \quad E^6 = Z.$$

Since $\langle E^i ; E^j \rangle_\wedge = \langle \widehat{E}^i ; \widehat{E}^j \rangle_\wedge$ and $E^1 - E^2 = \widehat{E}^1 - \widehat{E}^2 = \text{const}$, there exists a rotation $\Psi \in O(5)$ of the space $\widetilde{\mathbf{Y}} \subset \mathbb{C}^6$ taking $\{\widehat{E}^k\}_{k=1}^6$ to $\{E^k\}_{k=1}^6$. Suppose it is described by the matrix $\Psi_o \in SL(4, \mathbb{C})$. We set

$$\{\widehat{E}^k\}_{k=1}^6 \big|_{z=0} = \{E^k\}_{k=1}^6 \big|_{z=0}, \quad \text{i.e.,} \quad \Psi_o(z, \bar{z}) \big|_{z=0} = I.$$

The matrix Ψ_o describes the rotation of the moving frame to the given one, therefore we can write the Gauss–Weingarten system in the new representation:

(1.10) $$\Psi_{o_z} = U_1 \Psi_o, \quad \Psi_{o_{\bar{z}}} = U_2 \Psi_o,$$

where
$$U_1 = \begin{pmatrix} \partial \log E/4 - C/2 & A/\sqrt{E} & 0 & 0 \\ 0 & -\partial \log E/4 + C/2 & \sqrt{E}/2 & 0 \\ 0 & 0 & \partial \log E/4 + C/2 & B/\sqrt{E} \\ -\sqrt{E}/2 & 0 & -H_\gamma/\sqrt{E} & -\partial \log E/4 - C/2 \end{pmatrix}$$

and

$$U_2 = \begin{pmatrix} -\bar{\partial} \log E/4 - \overline{C}/2 & H_\gamma/\sqrt{E} & 0 & \sqrt{E}/2 \\ -\overline{B}/\sqrt{E} & \bar{\partial} \log E/4 + \overline{C}/2 & 0 & 0 \\ 0 & -\sqrt{E}/2 & -\bar{\partial} \log E/4 + \overline{C}/2 & 0 \\ 0 & 0 & -\overline{A}/\sqrt{E} & \bar{\partial} \log E/4 - \overline{C}/2 \end{pmatrix}.$$

Note that
a) $\mathcal{F}_\gamma$ is real and
b) $\mathcal{F}_\gamma \subset \widetilde{\mathbf{Y}} \cong \mathbb{C}^5 \subset \mathbb{C}^6$;
in terms of the matrices Ψ_o and $U_{o1,2}$ this means:

(1.11) $$\overline{\Psi}_o = J_c^{-1} \Psi_o J_c, \qquad \overline{U}_{o1} = J_c^{-1} U_{o2} J_c.$$

This is the reality condition "a)". Next,

(1.12) $$\Psi_o J_s \Psi_o^T = J_s, \qquad (U_{o1,2} J_s)^T = U_{o1,2} J_s.$$

This is the property[2] "b)": $\mathcal{F}_\gamma \subset \widetilde{\mathbf{Y}}$. In these formulas

$$J_c = \begin{pmatrix} 0 & \sigma_2 \\ \sigma_2 & 0 \end{pmatrix}, \quad J_s = \begin{pmatrix} \sigma_2 & 0 \\ 0 & -\sigma_2 \end{pmatrix}, \quad \sigma_2 = \begin{pmatrix} 0 & -i \\ i & 0 \end{pmatrix}.$$

Let us consider another system, namely

(1.13) $$\Psi_z = U(\lambda_1, \lambda_2)\Psi, \qquad \Psi_{\bar{z}} = V(\lambda_1, \lambda_2)\Psi$$

with

$$U(\lambda_1, \lambda_2) = \begin{pmatrix} \partial \log X/2 & A_H & 0 & 0 \\ 0 & -\partial \log X/2 & \lambda_2/\sqrt{2} & 0 \\ 0 & 0 & \partial \log \overline{X}/2 & B_H \\ -\lambda_1/\sqrt{2} & 0 & -H_H/\overline{X} & -\partial \log X/2 \end{pmatrix},$$

$$V(\lambda_1, \lambda_2) = \begin{pmatrix} 0 & \overline{H}_H & 0 & E/(\sqrt{2}\lambda_1) \\ -\overline{B}_H/X & 0 & 0 & 0 \\ 0 & -E(\sqrt{2}\lambda_2) & 0 & 0 \\ 0 & 0 & -\overline{A}_H/X & 0 \end{pmatrix},$$

where $A_H = A/\sqrt{c}$, $B_H = B\sqrt{c}$, $X = E\sqrt{c/\bar{c}}$, $H_H = H_\gamma/\sqrt{c}$, and c is any solution of the equation

(1.14) $$C = -\partial \log c/2.$$

The system (1.13) is equivalent to (1.10) when $\lambda_1 = \lambda_2 = 1$ because $\Psi_o = d \cdot \Psi|_{\lambda_1=\lambda_2=1}$, d being the diagonal matrix:

$$d = \mathrm{diag}((E/\bar{c})^{1/4}, (E/\bar{c})^{-1/4}, (E\bar{c})^{1/4}, (E\bar{c})^{-1/4}).$$

---

[2]Note that (1.12) means that $\Psi \subset Sp(2, \mathbb{C})$, but we shall not discuss this here.

This system satisfies the following reductions:

(1.15)
$$\overline{\Psi(\lambda_1; \lambda_2)} = \chi^{-1} J_c^{-1} \Psi(1/\bar\lambda_1; 1/\bar\lambda_2) J_c \chi_0,$$
$$\overline{U(\lambda_1; \lambda_2)} = (\chi^{-1})_{\bar z} \chi + \chi^{-1} J_c^{-1} V(1/\bar\lambda_1; 1/\bar\lambda_2) J_c \chi,$$

where $\chi = \mathrm{diag}(\overline{X}^{-1/2}, \overline{X}^{1/2}, X^{-1/2}, X^{1/2})$ and $\chi_0 = \chi|_{z=0}$. This is the reality condition, and

(1.16)
$$\Psi(\lambda_1; \lambda_2) J_s \Psi(\lambda_2; \lambda_1)^T = J_s,$$
$$(U(\lambda_1; \lambda_2) J_s)^T = U(\lambda_2; \lambda_1) J_s, \quad (V(\lambda_1; \lambda_2) J_s)^T = V(\lambda_2; \lambda_1) J_s$$

is the second reduction.

The compatibility condition for this system is:

(1.17)
$$\begin{cases} \partial \bar\partial \log X = -E + 2 A_H \overline{B_H}/X, \\ \bar\partial B_H = 0, \\ \bar\partial A_H = \partial(\overline{H_H}/X) X, \\ \lambda_1 \lambda_2 H_H = H_H, \text{ i.e., } \lambda_1 \lambda_2 = 1, \text{ or } H_H \equiv 0. \end{cases}$$

In the case of a minimal surface[3] ($H_H = H_\gamma \equiv 0$), the compatibility condition does not contain the parameters $\lambda_{1,2}$ and therefore gives the Gauss–Peterson–Codazzi equations for the conformal Gaussian image of $\mathcal{F}$, i.e., for the surface $\mathcal{F}_\gamma \subset \mathbb{S}^{4,1}$.

It is not difficult to restore f (i.e., $\mathcal{F}$) from the matrix Ψ.

THEOREM 1. *Let Ψ be the solution of the (G–W) system for some functions A_H, B_H, X, H_H that satisfy the Gauss–Peterson–Codazzi equations (GPC); then we can reconstruct the function f by the formulas:*

(1.18)
$$f = (f'_1, f'_2, f'_3, f'_4)/f'_5,$$
$$f'_1 = \Delta^{12} + \Delta^{34}, \quad f'_2 = \Delta^{13} - \Delta^{24}, \quad f'_3 = (\Delta^{13} + \Delta^{24})/i,$$
$$f'_4 = \Delta^{23} + \Delta^{14}, \quad f'_5 = \Delta^{23} - \Delta^{14},$$

where $\Delta^{m,n} = \det \begin{vmatrix} \psi_2^m & \psi_2^n \\ \psi_3^m & \psi_3^n \end{vmatrix}$, and ψ_k^m are elements of Ψ.

PROOF. It is sufficient to find the components of the vector $X = E^5$. We have

$$E^5 = \Psi_o \widehat{E}^5 = e_2 \Psi_o \wedge e_3 \Psi_o = \sqrt{\bar c^{-1}} e_2 \Psi \wedge e_3 \Psi.$$

To determine the coefficients, let us choose our basis $\widehat{E}^k$ so that

$$\check e^1 = (\widehat{E}^1 + \widehat{E}^2)/\sqrt{2}, \quad \check e^2 = (\widehat{E}^3 + \widehat{E}^4)/\sqrt{2}, \quad \check e^3 = i(\widehat{E}^3 - \widehat{E}^4)/\sqrt{2},$$
$$\check e^4 = (\widehat{E}^5 - \widehat{E}^6)/\sqrt{2}, \quad \check e^5 = (\widehat{E}^5 + \widehat{E}^6)/\sqrt{2}.$$

Now it is not difficult to determine X:

$$X = (f_1, f_2, f_3, f_4, 1) = \sqrt{\bar c^{-1}}(f'_1, f'_2, f'_3, f'_4, f'_5),$$

where $f'_k = \varepsilon_k \langle e_2 \Psi \wedge e_3 \Psi; \check e^k \rangle$. Consequently $\sqrt{\bar c} = -\langle e_2 \Psi \wedge e_3 \Psi; \check e^5 \rangle$. This completes the proof. □

---

[3]Actually system (1.13) depends only of the product $\lambda_1 \lambda_2$, so the case $\lambda_1 = \lambda_2^{-1}$ is not interesting.

§2. The Toda lattice and the Gauss–Weingarten equations for $\mathcal{F}_\gamma$

Let us consider a Willmore surface[4] $\mathcal{F}$. Its conformal Gaussian image $\mathcal{F}_\gamma$ is a minimal ($H_H = H_\gamma \equiv 0$) surface in $\mathbb{S}^{3,1}$. Recall that c is defined up to an anti-holomorphic function (see formula (1.14)), consequently A_H and B_H^{-1} are defined up to a common holomorphic multiplier. In the case when $H_\gamma = 0$, the GPC-equation on the function A_H is $\bar{\partial} A_H = 0$, so we can demand:

$$A_H = B_H.$$

Moreover, we can change the variable z in a such way that

$$A_H = B_H \equiv 1/\sqrt{2}.$$

In this case the Gauss–Weingarten system (1.13) has the form

(2.19)
$$\Psi_z = \frac{1}{2}\begin{pmatrix} \partial \log X & \sqrt{2} & 0 & 0 \\ 0 & -\partial \log X & \sqrt{2}\lambda_2 & 0 \\ 0 & 0 & \partial \log \overline{X} & \sqrt{2} \\ -\lambda_1\sqrt{2} & 0 & 0 & -\partial \log \overline{X} \end{pmatrix} \Psi = U\Psi,$$

$$\Psi_{\bar{z}} = \frac{1}{2}\begin{pmatrix} 0 & 0 & 0 & \sqrt{2}E/\lambda_1 \\ -\sqrt{2}/X & 0 & 0 & 0 \\ 0 & -\sqrt{2}E/\lambda_2 & 0 & 0 \\ 0 & 0 & -\sqrt{2}/\overline{X} & 0 \end{pmatrix} \Psi = V\Psi.$$

Note that in this case the compatibility condition (Gauss–Peterson–Codazzi equation) can be written as a single equation on a single (complex-valued) function X:

(2.20) $$\partial \bar{\partial} \log X = -|X| + X^{-1}.$$

Let us put $\lambda_2 = 1$, $\lambda \stackrel{\text{def}}{=} -\lambda_1$ and introduce the new notations

$$z_1 = z/\sqrt{2}, \quad z_2 = -\bar{z}/\sqrt{2},$$
$$x_0 = \log X/2, \quad x_1 = \log X^{-1}/2, \quad x_2 = \log \overline{X}/2, \quad x_3 = \log \overline{X}^{-1}/2.$$

Let us consider the following infinite system of linear equations:

(2.21) $$\psi_{n z_1} = \left(\frac{\partial}{\partial z_1} x_n\right)\psi_n + \psi_{n+1}, \quad \psi_{n z_2} = \exp\{x_n - x_{n-1}\}\psi_{n-1},$$

for which the compatibility condition is the well-known Toda lattice equation[5]

(2.22) $$\frac{\partial^2 x_n}{\partial z_1 \partial z_2} = \exp\{x_n - x_{n-1}\} - \exp\{x_{n+1} - x_n\}.$$

It is not difficult to see that equations (2.19), (2.20) coincide with equations (2.21), (2.22) if

(2.23) $$x_{n+4} = x_n,$$
(2.24) $$\psi_{n+4}/\psi_n = \lambda_1 = \text{const},$$

---

[4]For Willmore surfaces and their conformal Gaussian images see [1, 3–5].
[5]Algebro-geometric solutions for the Toda lattice equation were found by Krichever, see Appendix in [6].

and, additionally,

(2.25) $$x_n = -x_{-n+1} + \text{const},$$
(2.26) $$\overline{x}_n = -x_{-n-1}.$$

We put $X = \exp\{2x_0 - \text{const}\}$. This reduction was considered in [7].

We have obtained the following result [9]:

Any Willmore surface can be locally described as a certain special solution of the Toda lattice equation.

Note that system (2.19) and equation (2.20) are defined only locally, i.e., in the domain of the parametrization z. If our surface cannot be parameterized by one z, we cannot deal with A_H, B_H as constants. In that case they become nontrivial holomorphic differentials of the second kind, and in the regions where two parametrizations exist, we have nontrivial transition functions. The problem of constructing such surfaces is well known, yet unsolved. Here we consider the case when this problem does not arise, i.e., when $\mathcal{F}$ can be described by one parameter z.

Let $\mathcal{F}$ be the torus, and z be the global complex parameter, i.e., $z \in \mathbb{C}_z/\Lambda$, where Λ is a nondegenerate two-dimensional lattice of periods of the function $f(z,\overline{z})$, $\mathbb{C}_z \xrightarrow{f} \mathcal{F} \subset \mathbb{S}^3$. In this case we define on $\mathbb{C}_z$ a double periodic complex-valued function X and a matrix-valued function ψ (in a new notation, x_n and $\{\psi_n^k\}$, the fundamental solution of (2.21)). We look for periodic solutions of the Toda lattice equation.

A rich class of periodic solutions (in some cases (see [10]), all the periodic solutions) is contained in the class of so-called finite-gap (algebro-geometrical) solutions of such equations. The solutions of this type are, in general, quasi-periodic functions, and it is an extra problem to find the periodic ones among them.

Let us construct the finite-gap solutions of the Toda lattice equation.

PROPOSITION 1. *Let Γ be the algebraic curve of genus g, and P_0, P_∞ be two points on it.*

Let v_∞ be a local coordinate in the neighborhood of P_∞, v_0 be a local coordinate in the neighborhood of P_0, $\mathcal{D}$ be a nonspecial divisor with $\deg \mathcal{D} = g$.

Then the functions ψ_n possessing the divisor of poles $\mathcal{D}$ and having the following asymptotic behavior:

(2.27) $$\begin{aligned}\psi_n &\sim (1 + O(v_\infty))v_\infty^{-n}\exp\{z/v_\infty\} &&\text{if } P \sim P_\infty,\\ \psi_n &\sim (\varphi_n + O(v_0))v_0^n \exp\{\varepsilon_\nu \overline{z}/v_0\} &&\text{if } P \sim P_0,\end{aligned}$$

satisfy system (2.21), *where* $x_n = \log \varphi_n$.

We need explicit formulas for this solution. On Γ let us define the following objects:

- $\{\vec{a}, \vec{b}\}$, the canonical basis of $H_1(\Gamma)$ (cycles) on Γ;
- $d\vec{u}$, the normalized ($\oint_{a_k} du_j = 2\pi i \delta_j^k$) basis of holomorphic differentials on Γ;
- B, the matrix of b-periods $d\vec{u}$, $B_j^k \stackrel{\text{def}}{=} \oint_{b_k} du_j$;
- the Riemann Θ-function:

$$\Theta\begin{bmatrix}\alpha\\ \beta\end{bmatrix}(\vec{z}) = \sum_{N \in \mathbb{Z}^g} \exp\left\{\frac{1}{2}\langle B(N+\alpha); N+\alpha\rangle + \langle \vec{z} + 2\pi i\beta; N+\alpha\rangle\right\};$$

- Λ, the lattice of periods $d\vec{u}$, $\Lambda \stackrel{\text{def}}{=} \{\Lambda_{N_a;N_b} \in \mathbb{C}^g : \Lambda_{N_a;N_b} = 2\pi i N_a + B N_b\}$;

- the abelian map $\mathcal{A}(P) \stackrel{\text{def}}{=} \int_{P_\infty}^P d\vec{u}$;
- the normalized (i.e., with zero a-periods) abelian integral $\Omega_\infty \stackrel{\text{def}}{=} \int_{P_\infty}^P d\Omega_\infty$ with a single singularity, i.e., $\Omega_\infty = 1/v_\infty + o(1)$, if $P \sim P_\infty$ (a simple pole at the point[6] P_∞);
- the normalized abelian integral of the second kind $\Omega_0 \stackrel{\text{def}}{=} \int_{P_\infty}^P d\Omega_0$ with a single singularity, i.e., $\Omega_0 = 1/v_0 + \widetilde{\Omega}_0 + o(1)$, if $P \sim P_0$ (a simple pole at the point P_0);
- the normalized abelian integral of the third kind, $\Omega_{0,\infty} \stackrel{\text{def}}{=} \int_{P_\infty}^P d\Omega_{0,\infty}$, where

$$\Omega_{0,\infty} \sim \begin{cases} -\log v_\infty + o(1), & P \sim P_\infty, \\ \log v_0 + \widetilde{\Omega}_{0,\infty} + o(1), & P \sim P_0; \end{cases}$$

- $\vec{V}_k$, the vector b-periods of the integral Ω_k, $k \in \{\text{"}\infty\text{"}, \text{"}0\text{"}, \text{"}0,\infty\text{"}\}$. Moreover, let us denote:

(2.28) $$\vec{Z} \stackrel{\text{def}}{=} \vec{V}_\infty z_1 + \vec{V}_0 z_2, \qquad \vec{Z}_n \stackrel{\text{def}}{=} \vec{Z} + n\vec{V}_{0,\infty},$$

(2.29) $$\Omega \stackrel{\text{def}}{=} z_1 \Omega_\infty + z_2 \Omega_0, \qquad \Omega_n \stackrel{\text{def}}{=} \Omega + n\Omega_{0,\infty}.$$

PROPOSITION 2. *The functions $\psi_n = \psi_n(z_1, z_2, P)$ given by*

(2.30) $$\psi_n(z_1, z_2, P) = \frac{\Theta(D)\Theta(\mathcal{A}(P) - D + \vec{Z}_n)}{\Theta(-D + \vec{Z}_n)\Theta(\mathcal{A}(P) - D)} \exp\{\Omega_n\}$$

satisfy system (2.21). *Here $D \in \mathbb{C}$ is any complex vector in general position, i.e., the corresponding divisor $\mathcal{D}$, $\deg \mathcal{D} = g$, is nonspecial, $D = \mathcal{A}(\mathcal{D}) - \mathcal{K}$, where $\mathcal{K}$ is the vector of Riemann constants.*

PROOF. The proof is standard for the theory of the Baker–Akhiezer functions (see [6]).

The solution of (2.22) can be obtained from the asymptotic expansion of (2.30) in the neighborhood of P_0.

(2.31) $$\varphi_n(z_1, z_2) = \frac{\Theta(D)\Theta(\mathcal{A}(P_0) - D + \vec{Z}_n)}{\Theta(\mathcal{A}(P_0) - D)\Theta(-D + \vec{Z}_n)} \cdot \exp\{z_1 \widetilde{\Omega}_\infty + z_2 \widetilde{\Omega}_0 + n\widetilde{\Omega}_{0,\infty}\},$$

where $\varphi_n \stackrel{\text{def}}{=} \exp\{x_n\}$, $\widetilde{\Omega}_\infty = \Omega_\infty(P_0)$.

We are looking for solutions that are periodic with respect to n, $\varphi_{n+4} = \varphi_n$. It is known [6] that solutions of this type are constituted by surfaces on which there exist functions with divisors of zero-poles $4P_0 - 4P_\infty$. Denote such a function by λ. Let $\lambda = \tilde{\lambda}^2$, i.e., suppose that on Γ, besides λ, there exists a function $\tilde{\lambda}$ with only a double zero and a double pole. It is not difficult to see that equation (2.20) becomes a sinh-Laplace or a cosh-Laplace equation in this case. These cases have been investigated in detail (see [1, 11, 12]) and we shall not consider them here.

Let Γ be a surface on which the function λ exists, but the function $\tilde{\lambda}$ does not, i.e., $\lambda = \infty$ and $\lambda = 0$ are two nonsingular branch points P_∞ and P_0, where all four sheets are attached.

---

[6]We define integrals at singular points as the analytical continuation of the function $\oint_{P_\infty}^P (d\Omega - d\text{"the principal part of }\Omega\text{"}) + \text{"the principal part of }\Omega\text{"}$.

The vector
$$\begin{bmatrix} \Delta_2 \\ \Delta_1 \end{bmatrix} \stackrel{\text{def}}{=} 4\widetilde{V}_{0,\infty} = 4\mathcal{A}(P_0)$$
is the vector of the lattice Λ, and by formula (2.30) we have
$$\psi_{n+4} = \psi_n \exp\{\langle -\mathcal{A}(P); \Delta_1 \rangle + 4\Omega_{0,\infty}(P)\}.$$

PROPOSITION 3. *The function* $\vec{\psi} \stackrel{\text{def}}{=} (\psi_0, \psi_1, \psi_2, \psi_3)^T$, *where* ψ_k *is defined by* (2.30), *satisfies the system*

(2.32)
$$\vec{\psi}_{z_1} = \begin{pmatrix} \partial_{z_1} x_0 & 1 & 0 & 0 \\ 0 & \partial_{z_1} x_1 & 1 & 0 \\ 0 & 0 & \partial_{z_1} x_2 & 1 \\ \lambda & 0 & 0 & \partial_{z_1} x_3 \end{pmatrix} \vec{\psi},$$

$$\vec{\psi}_{z_2} = \begin{pmatrix} 0 & 0 & 0 & e^{x_0-x_3}/\lambda \\ e^{x_1-x_0} & 0 & 0 & 0 \\ 0 & e^{x_2-x_1} & 0 & 0 \\ 0 & 0 & e^{x_3-x_2} & 0 \end{pmatrix} \vec{\psi},$$

where $\lambda = \lambda(P) = \exp\{\langle -\mathcal{A}(P); \Delta_1 \rangle + 4\Omega_{0,\infty}(P)\}$.

Note that $\lambda(P)$ is a function with only one zero and one pole, both of the fourth order. For the construction of the Willmore tori, we must ensure that reductions (2.25) and (2.26) are satisfied. We shall do this in the next section.

§3. The main example: surfaces with two symmetries

We need the solutions of the Toda lattice equation with two reductions (holomorphic and anti-holomorphic[7]), so it is natural to consider surfaces with two additional symmetries. Let us prove the following theorem.

THEOREM 2. *Suppose that on a Riemann surface* Γ *we are given*
1) *a meromorphic function* $\lambda(P)$, $(\lambda) = 4P_0 - 4P_\infty$;
2) *a holomorphic involution* σ *such that* $\sigma P_0 = P_0$ *and* $\sigma P_\infty = P_\infty$;
3) *an anti-involution* τ *such that* $\tau P_0 = P_\infty$ *and* $\tau P_\infty = P_0$;
4) *a meromorphic function* $v(P)$ *with the following properties*:
   a) $\deg(v)_{\text{poles}} = g + 1$,
   b) v *has a simple pole at* P_0,
   c) $v(P)\overline{v(\tau P)} = \varepsilon_v = \text{const}$;
5) *an abelian differential of the second kind* $d\check{\Omega}_0$ *with the following properties*:
   a) $(d\check{\Omega}_0)_{\text{poles}} = 2P_0$,
   b) $\dfrac{d\check{\Omega}_0(P)}{\overline{d\check{\Omega}_0(\tau P)}} v(P) v(\sigma P) \lambda(P) = \text{const}.$

Then for
$$v_\infty = v(P), \quad v_0 = 1/v(P), \quad z_1 = z/\sqrt{2}, \quad z_2 = \varepsilon_v \bar{z}/\sqrt{2}, \quad \mathcal{D} = (v)_{\text{poles}} - P_0$$

the function $\psi \stackrel{\text{def}}{=} (\psi_0, \psi_1, \psi_2, \psi_3)$ *(formula (2.30)) satisfies the system of equations*
$$\psi_z = U\psi, \qquad \psi_{\bar{z}} = -\varepsilon_v V\psi,$$

---

[7]A similar problem for other reductions was considered in [13]; see also [14, 15].

where U, V are defined by (2.19) for $\lambda_1 = -\lambda(P)$, $\lambda_2 = 1$ (P is an arbitrary point of Γ), $X = \varphi_0^2/C_\Omega$, $C_\Omega = d\check{\Omega}_0/dv|_{P_\infty}$, $E = |X|$.

PROOF. (i) For the reduction (1.12), it is not difficult to see that $(d\check{\Omega}_0) = \mathcal{D} + \sigma\mathcal{D} - 2P_0$ and the action of σ on the local parameters v_∞ and v_0 is given by:

$$\sigma_* v_0 = -v_0(1 + o(1)), \qquad \sigma_* v_\infty = -v_\infty(1 + o(1)).$$

Note that $\varepsilon_v \in \mathbb{R}$, so we can assume that $\varepsilon_v \in \{+1, -1\}$.

The function $\psi_n(P)$ has the following asymptotics:

(3.34)
$$\begin{aligned} \psi_n &\sim (1 + O(v_\infty))v_\infty^{-n}\exp\{z/v_\infty\} &\text{if } P \sim P_\infty, \\ \psi_n &\sim (\varphi_n + O(v_0))v_0^n \exp\{\varepsilon_v \bar{z}/v_0\} &\text{if } P \sim P_0. \end{aligned}$$

Consider the abelian differential $\psi_n(P)\psi_{-n+1}(\sigma P)\,d\check{\Omega}_0(P)$. It has only two singular points, poles at P_0 and P_∞, with the following principal parts:

$$\begin{aligned} \psi_n\psi_{-n+1}(\sigma P)d\check{\Omega}_0(P) &\sim (-1)^{n+1}v_\infty^{-1}dv_\infty \,\text{const}_\infty &\text{if } P \sim P_\infty, \\ \psi_n\psi_{-n+1}(\sigma P)d\check{\Omega}_0(P) &\sim (-1)^n \varphi_n\varphi_{-n+1}v_0^{-1}dv_0 \,\text{const}_0 &\text{if } P \sim P_0. \end{aligned}$$

The sum of residues must be zero, and therefore

(3.35) $$\varphi_n \varphi_{-n+1} = \text{const}.$$

The calculation of this constant gives $\varphi_n\varphi_{-n+1} = d\check{\Omega}_0/dv|_{P_\infty} \stackrel{\text{def}}{=} C_\Omega$.

(ii) For the reduction (1.11), let us consider the auxiliary functions ψ_n^+:

1) the divisor of the poles ψ_n^+ is the divisor $\mathcal{D}^+$ such that $\mathcal{D} + \mathcal{D}^+$ is the divisor of zeros of the abelian differential of the third kind $d\check{\Omega}_{0\infty}$ with simple poles at P_∞ and P_0,

2) the function ψ_n has the asymptotics:

$$\begin{aligned} \psi_n^+ &\sim (\varphi_n^+ + O(v_\infty))v_\infty^n \exp\{-z/v_\infty\} &\text{if } P \sim P_\infty, \\ \psi_n^+ &\sim (1 + O(v_0))v_0^{-n}\exp\{-\varepsilon_v \bar{z}/v_0\} &\text{if } P \sim P_0. \end{aligned}$$

Consider the differential $\psi_n \psi_n^+ d\check{\Omega}_{0\infty}$. It has only two singularities, namely, the poles with residues ψ_n and $-\psi_n^+$; consequently $\psi_n = \psi_n^+$.

Consider the differential $\psi_n^+(P)\overline{\psi}_{-n-1}(\tau P) v(P)\,d\check{\Omega}_{0\infty}$. It has only two singularities, namely, a simple pole at P_∞ with residue $\varepsilon_v^{-n-1}\varphi_n^+\overline{\varphi}_{-n-1}$ and a simple pole at P_0 with residue $-\varepsilon_v^{-n-1}$, consequently

(3.36) $$\varphi_n^+ \overline{\varphi}_{-n-1} = \varphi_n \overline{\varphi}_{-n-1} = 1.$$

The relations (3.35), (3.36) are equivalent to the desired reductions. The theorem is proved. $\square$

THEOREM 3. *Let the conditions of the previous theorem hold. Then*

(3.37) $$\lambda(\tau P) = 1/\bar{\lambda}.$$

PROOF. Notice that $\lambda(P)\overline{\lambda}(\tau P) = \text{const}$. Let us denote this constant by ε_λ; one can see that $\varepsilon_\lambda \in \mathbb{R}$. Multiplying $\lambda(P)$ by an arbitrary number does not change the argument of ε_λ, so we can assume that $\varepsilon_\lambda \in \{+1, -1\}$. Set $\lambda(P)v^4(P)|_{P_\infty} = 1$; then $\lambda(P)v^4(P)|_{P_0} = \varepsilon_\lambda$.

Let us consider the function
$$\frac{d\check{\Omega}_0(P)}{d\overline{\check{\Omega}}_0(\tau P)} v(P) v(\sigma P) \lambda(P).$$

By the assumption of the theorem it is a constant. Let us calculate this constant at the points P_∞ and P_0:
$$\frac{d\check{\Omega}_0(P)}{d\overline{\check{\Omega}}_0(\tau P)} v(P) v(\sigma P) \lambda(P) \sim \begin{cases} C_\Omega \varepsilon & \text{if } P \sim P_\infty, \\ \varepsilon\varepsilon_\lambda/\overline{C}_\Omega & \text{if } P \sim P_0. \end{cases}$$

Consequently, $|C_\Omega|^2 = \varepsilon_\lambda$, i.e., $|C_\Omega| = 1$ and $\varepsilon_\lambda = 1$. □

Conditions 4) and 5) of Theorem 2 are not constructive. Let us reformulate them in terms of the vector $D = \mathcal{A}(\mathcal{D}) - \mathcal{K}$, where $\mathcal{D}$ is the divisor of the poles of the Ψ-function.

Let $\{\vec{a}; \vec{b}\} \in H_1(\Gamma)$ be a basis of cycles such that

(3.38) $$\sigma\vec{a} = S\vec{a} \quad \text{and} \quad \tau\vec{a} = T\vec{a}.$$

Then $S^2 = T^2 = I$, $\sigma\vec{b} = S^T\vec{b} + \Phi_\sigma\vec{a}$, and $\tau\vec{b} = -T^T\vec{b} + \Phi_\tau\vec{a}$, where Φ_σ is an antisymmetric matrix and Φ_τ is a symmetric matrix. Set
$$\kappa_\sigma \stackrel{\text{def}}{=} \text{diag}(S^T\Phi_\sigma), \qquad \kappa_\tau \stackrel{\text{def}}{=} \text{diag}(T^T\Phi_\tau)$$

and define

(3.39) $$\sigma\mathbf{z} = S^T\mathbf{z}, \qquad \tau\mathbf{z} = -T^T\overline{\mathbf{z}}$$

for any $\mathbf{z} \in \mathbb{C}^g$. It is not difficult to prove that

(3.40) $$\overline{\Theta(\mathbf{z})} = \Theta(\tau\mathbf{z} + \pi i \kappa_\tau),$$
(3.41) $$\Theta(\mathbf{z}) = \Theta(\sigma\mathbf{z} + \pi i \kappa_\sigma),$$
(3.42) $$\tau \begin{bmatrix} \alpha \\ \beta \end{bmatrix} = \begin{bmatrix} -T\alpha \\ T^T\beta + \Phi_\tau\alpha \end{bmatrix},$$
(3.43) $$\sigma \begin{bmatrix} \alpha \\ \beta \end{bmatrix} = \begin{bmatrix} S\alpha \\ S^T\beta + \Phi_\sigma\alpha \end{bmatrix},$$

where

(3.44) $$\begin{bmatrix} \alpha \\ \beta \end{bmatrix} \stackrel{\text{def}}{=} B\alpha + 2\pi i\beta \in \mathbb{C}^g.$$

THEOREM 4. *The conditions 4) and 5) of Theorem 3 are equivalent to the compatibility condition of the following linear system*:

(3.45) $$\begin{cases} D + S^T D \equiv 2 \begin{bmatrix} \Delta_1/4 \\ \Delta_2/4 \end{bmatrix} - \pi i \kappa_\sigma, \\ D + T^T \overline{D} \equiv \pi i \kappa_\tau. \end{cases}$$

PROOF. (i) We can rewrite condition 4) as

$$\mathcal{D} + P_0 - \tau\mathcal{D} - P_\infty \equiv 0. \tag{3.46}$$

It is not difficult to prove that $\tau\mathcal{K} \equiv \mathcal{K} + \pi i \kappa_\tau + (g-1)\mathcal{A}(\tau P_\infty)$. For $D = \mathcal{A}(\mathcal{D}) - \mathcal{K}$ this gives the desired result, namely $D + T^T \overline{D} \equiv \pi i \kappa_\tau$.

(ii) Now let us consider the holomorphic involution σ.

Notice that 5) is equivalent to the relation

$$\mathcal{D} + \sigma\mathcal{D} - 2P_0 \equiv \mathcal{C}, \tag{3.47}$$

where $\mathcal{C}$ is the canonical class of Γ. Condition (3.47) means that there exists a differential with one singularity, the double pole at P_0, and the divisor of zeros $\mathcal{D} + \sigma\mathcal{D}$.

The involution σ acts on points $\mathbf{z}$ of $\mathbb{C}^g \supset \mathcal{J}(\Gamma)$ according to the rule $\sigma \mathbf{z} = S^T \mathbf{z}$. Further $\sigma P_\infty = P_\infty$, and for $\mathcal{K}$ we have $\sigma \mathcal{K} \equiv \mathcal{K} + \pi i \kappa_\sigma$. Finally, since $\mathcal{A}(P_0) = \begin{bmatrix} \Delta_1/4 \\ \Delta_2/4 \end{bmatrix}$, using $\mathcal{C} = 2\mathcal{K}$, we can rewrite (3.47) as

$$D + S^T D \equiv 2\begin{bmatrix} \Delta_1/4 \\ \Delta_2/4 \end{bmatrix} - \pi i \kappa_\sigma. \qquad \square$$

In order to construct the Willmore surfaces, we need the constants from conditions 4c) and 5b) of Theorem 2. So some additional constructions and calculations are required.

Let $\mathcal{D}_\infty$ be a small neighborhood of P_∞ and $\mathcal{L}$ be a path going from P_∞ to P_0 that does not intersect the basis cycles a_k, b_k (i.e., the same path as in the abelian transformation $\mathcal{A}(P_0)$). Choose $\mathcal{L}$ inside $\mathcal{D}_\infty$ so that $\lambda(P'_\infty) > 0$ if $P'_\infty \in \mathcal{L} \cap \mathcal{D}_\infty$. In the neighborhood of P_0 let $\mathcal{L}$ satisfy

$$\tau(\mathcal{L} \cap \mathcal{D}_\infty) = \mathcal{L} \cap (\tau \mathcal{D}_\infty). \tag{3.48}$$

In $\mathcal{D}_\infty$ we can define $\sqrt[4]{\lambda(P)}$. Choose a branch of $\sqrt[4]{\lambda}$ such that $\sqrt[4]{\lambda(P'_\infty)} > 0$ if $P'_\infty \in \mathcal{L} \cap \mathcal{D}_\infty$. In the neighborhood of $\mathcal{L}$ (and, consequently, in $\mathcal{D}_\infty$ and $\mathcal{D}_0 \stackrel{\text{def}}{=} \tau \mathcal{D}_\infty$ as well) we define[8] the function $v_\infty(P)$ as the analytic continuation of $(\sqrt[4]{\lambda(P)})^{-1}$ along $\mathcal{L}$:

$$v_\infty(P) = (\sqrt[4]{\lambda(P)})^{-1}, \qquad v_\infty(P) > 0, \quad P \in \mathcal{L} \cap \mathcal{D}_\infty.$$

Denote

$$v_0(P) \stackrel{\text{def}}{=} 1/v_\infty(P). \tag{3.49}$$

We have defined $v_0(P)$, $v_\infty(P)$, $\mathcal{L}$ so that

$$v_0(\tau P'_\infty) = \varepsilon \overline{v_\infty(P'_\infty)}, \tag{3.50}$$

where

$$\varepsilon \stackrel{\text{def}}{=} \arg v_0(P'_0) = -\arg v_\infty(P'_0) \quad \text{if } P'_0 \in \mathcal{L} \cap \mathcal{D}_0. \tag{3.51}$$

---

[8] The definitions of v_∞ and v_0 in this section are different from the definitions in §3. They can have additional factors, some $\sqrt[4]{1}$, but the solution does not depend on this factor (the factor ε in the definition of z_2 (3.58) compensates this difference).

Let us consider the cycle $\check{\mathcal{L}} \stackrel{\text{def}}{=} \mathcal{L} + \tau \mathcal{L}$. We can assume, by (3.48), that $\check{\mathcal{L}}$ does not go through the neighborhoods $\mathcal{D}_\infty$ and $\mathcal{D}_0$. Let it circle n_τ times around P_0. Then

$$\varepsilon = \exp\left\{\frac{2\pi i n_\tau}{8}\right\} = \exp\left\{\frac{1}{8}\oint_{\check{\mathcal{L}}} \frac{d\lambda}{\lambda}\right\}.$$

We know that
$$\mathcal{A}(P_0) = \begin{bmatrix} \Delta_1/4 \\ \Delta_2/4 \end{bmatrix} = \int_{\mathcal{L}} d\vec{u} = \oint_{\bar{b}} d\Omega_{\infty,0}.$$

Consequently,
$$d\Omega_{\infty,0}(P) = \frac{1}{4}\frac{d\lambda}{\lambda} + \left\langle \frac{\Delta_1}{4}; d\vec{u}\right\rangle.$$

Therefore for Δ_1, Δ_2 we have:

$(-\Delta_1)_k = \frac{1}{2\pi i}\oint_{a_k} d\lambda/\lambda$ is the number of revolutions of the cycle a_k around P_0,
$(\Delta_2)_k = \frac{1}{2\pi i}\oint_{b_k} d\lambda/\lambda$ is the number of revolutions of the cycle b_k around P_0.

Notice that
$$\oint_{\check{\mathcal{L}}} d\vec{u} = \int_{\check{\mathcal{L}}} d\vec{u} + \int_{\tau\mathcal{L}} d\vec{u} = \begin{bmatrix}\Delta_1/4\\\Delta_2/4\end{bmatrix} + \tau\begin{bmatrix}\Delta_1/4\\\Delta_2/4\end{bmatrix} = \begin{bmatrix}(I-T)\Delta_1/4\\(I+T^T)\Delta_2/4 + \Phi_\tau\Delta_1/4\end{bmatrix},$$

and consequently (in $H_1(\Gamma)$)
$$\check{\mathcal{L}} = \langle (I+T^T)\Delta_2/4 + \Phi_\tau\Delta_1/4; \vec{a}\rangle + \langle (I-T)\Delta_1/4; \vec{b}\rangle \in H_1(\Gamma).$$

Set
$$M_\tau \stackrel{\text{def}}{=} (I+T^T)\Delta_2/4 + \Phi_\tau\Delta_1/4 \in \mathbb{Z}^g, \qquad N_\tau \stackrel{\text{def}}{=} (I-T)\Delta_1/4 \in \mathbb{Z}^g.$$

Let us calculate ε. We know the representation of $\check{\mathcal{L}}$ in $H_1(\Gamma)$, but this is not sufficient for calculating n_τ, because there are cycles which are homological to zero (in $H_1(\Gamma)$) but integrals along them are nonzero. Examples of such cycles are l_∞ and l_0, the small cycles around P_∞ and P_0:

$$\oint_{l_\infty}\frac{d\lambda}{\lambda} = -\oint_{l_0}\frac{d\lambda}{\lambda} = -8\pi i.$$

For the differential $d\lambda/\lambda$, we take $H_1(\Gamma\setminus\{P_\infty; P_0\})$. The basis of $H_1(\Gamma\setminus\{P_\infty; P_0\})$ is $\{\vec{a}; \vec{b}; l_\infty\}$. Let

$$\check{\mathcal{L}} = \langle M_\tau; \vec{a}\rangle + \langle N_\tau; \vec{b}\rangle + m_\tau l_\infty \in H_1(\Gamma\setminus\{P_\infty; P_0\}).$$

Then
$$\varepsilon = (-1)^{m_\tau}\exp\{-\pi i\langle T\Delta_1/4; 2\Delta_2 + T^T\Phi_\tau\Delta_1\rangle/4\}.$$

LEMMA 1. *Let $z_1 = z$, $z_2 = \varepsilon\bar{z}$, and $D \in \mathbb{C}^g$ be a vector such that*

(3.52) $$-T^T(-\overline{D}) \equiv -D + \kappa_\tau.$$

Then φ_n satisfies the equation

(3.53) $$\overline{\varphi}_n\varphi_{-n-1} = C_\tau \varepsilon_\tau^{-2n-1},$$

where

(3.54) $$\varepsilon_\tau \stackrel{\text{def}}{=} \exp\{\pi i\langle T\Delta_1/4; 2\Delta_2 + \Phi_\tau\Delta_1\rangle/4\},$$

and
$$C_\tau = \exp\left\{\left\langle N_D - \Delta_1/4; \Re\begin{bmatrix}\Delta_1/4\\\Delta_2/4\end{bmatrix}\right\rangle + \langle N_\tau; \Re D\rangle\right\}$$

is some positive constant; here

$$\tau(-D) + \kappa_\tau + (-D) = \begin{bmatrix}N_D\\M_D\end{bmatrix} \in \Lambda, \quad N_D, M_D \in \mathbb{Z}^g.$$

PROOF. We can obtain the desired formula by directly calculating the product $\overline{\varphi}_n \varphi_{-n-1}$ using formula (3.40); this yields

$$\widetilde{\Omega}_{\infty,0} = \lim_{P\to P_0}(\Omega_{\infty,0}(P) - \log v_0(P)) = \left\langle \Delta_1/4; \begin{bmatrix}\Delta_1/4\\\Delta_2/4\end{bmatrix}\right\rangle,$$

$$z\widetilde{\Omega}_\infty + \varepsilon\bar{z}\widetilde{\Omega}_0 + \overline{z\widetilde{\Omega}_\infty + \varepsilon\bar{z}\widetilde{\Omega}_0} = \oint_{\tilde{\mathcal{L}}}(z\,d\Omega_\infty + \varepsilon\bar{z}\,d\Omega_0) = \langle \vec{Z}(z,\bar{z}); N_\tau\rangle,$$

$$\tau\begin{bmatrix}\Delta_1/4\\\Delta_2/4\end{bmatrix} = -\begin{bmatrix}\Delta_1/4\\\Delta_2/4\end{bmatrix} + \oint_{\tilde{\mathcal{L}}} d\vec{u} = -\begin{bmatrix}\Delta_1/4\\\Delta_2/4\end{bmatrix} + \begin{bmatrix}N_\tau\\M_\tau\end{bmatrix}. \quad \square$$

Notice that

(3.55)
$$\varepsilon\varepsilon_\tau = \exp\{\pi i\,(m_\tau + \langle \Phi_\tau N_\tau; N_\tau\rangle/2)\}$$
$$= \exp\{\pi i\,(m_\tau + \langle (T-I)\Delta_1/4; \Phi_\tau \Delta_1/4\rangle)\} \stackrel{\text{def}}{=} \tilde{\varepsilon} \in \{+1; -1\}.$$

We defined $\tilde{\varepsilon}$ as $\varepsilon\varepsilon_\tau$.

We collect all these statements in the following theorem.

THEOREM 5. *Let Γ be a four-sheeted covering of the $\mathbb{C}_\lambda$-plane with at least two branch-points of the fourth order (all four sheets are glued): at P_∞ ($\lambda = \infty$) and at P_0 ($\lambda = 0$). On Γ suppose there exists an involution σ such that $\sigma P_\infty = P_\infty$, $\sigma P_0 = P_0$ and the anti-involution $\tau: \lambda \to 1/\bar\lambda$. Suppose the basis $\{\vec{a}, \vec{b}\} \in H_1(\Gamma)$ satisfies conditions (3.38) and the vector D is the solution of system (3.45). Suppose in addition that*

(3.56)
$$\tilde{\varepsilon} = -1.$$

Then the functions ψ_n ($n = 0, \ldots, 3$) from (3.20) satisfy the Gauss–Weingarten system (2.19) ($\lambda_1 = \lambda(P) = -1$, $\lambda_2 = 1$) for

(3.57)
$$X = \varepsilon_\tau \frac{\Theta(D + \begin{bmatrix}\Delta_1/4\\\Delta_2/4\end{bmatrix})\Theta^2(-D + \vec{Z} + \begin{bmatrix}\Delta_1/4\\\Delta_2/4\end{bmatrix})}{\Theta(D - \begin{bmatrix}\Delta_1/4\\\Delta_2/4\end{bmatrix})\Theta^2(-D + \vec{Z})} \exp\{\widetilde{\Omega}\},$$

where

(3.58)
$$\vec{Z} = z\vec{V}_\infty/\sqrt{2} + \varepsilon\bar{z}\vec{V}_0/\sqrt{2}$$

and

(3.59)
$$\widetilde{\Omega} = z\widetilde{\Omega}_\infty/\sqrt{2} + \varepsilon\bar{z}\widetilde{\Omega}_0/\sqrt{2}.$$

§4. Real double coverings over hyperelliptic surfaces

We are investigating surfaces Γ that are real four-sheeted coverings of the complex plane $\mathbb{C}_\lambda$ with an additional involution σ; σ has at least two fixed points P_∞ and P_0.

It is inconvenient to work with the canonical basis of cycles on the four-sheeted covering of the complex plane because the corresponding picture is too complicated. We shall use the noncanonical basis $\{\vec{a}, \vec{f}\}$.

LEMMA 2. *Let $a_k, f_k \in H_1(\Gamma)$, $1 \leq k \leq g$, be cycles such that $\{a_i \circ a_j\} = 0$, $\{a_i \circ f_j\} = \delta_{i,j}$. Then $\{\vec{a}; \vec{b}\}$ is the canonical basis of $H_1(\Gamma)$, where*

$$\vec{b} = \vec{f} - F^+ \vec{a}, \qquad F_{ij}^+ \stackrel{\text{def}}{=} \begin{cases} \{f_i \circ f_j\} & \text{if } i < j, \\ 0 & \text{otherwise.} \end{cases}$$

PROOF. We need only verify that $\{b_i \circ b_j\} = 0$. Let $i < j$; then

$$\{b_i \circ b_j\} = \{(f_i - F_{ik}^+ a_k) \circ (f_j - F_{jm}^+ a_m)\}$$
$$= \{f_i \circ f_j\} - F_{jm}^+ \{f_i \circ a_m\} - F_{ik}^+ \{a_k \circ f_j\}$$
$$= F_{ij}^+ + F_{ji}^+ - F_{ij}^+ = 0. \qquad \square$$

LEMMA 3. *Suppose the automorphisms σ and τ on Γ act on the cycles in the following way (compare with (3.38)):*

(4.60) $\qquad \tau \vec{a} = T \vec{a}, \qquad \sigma \vec{a} = S \vec{a},$

(4.61) $\qquad \tau \vec{b} = -T^T \vec{b} + \Phi_\tau \vec{a}, \qquad \sigma \vec{b} = S^T \vec{b} + \Phi_\sigma \vec{a}.$

Then,

(4.62) $\qquad \Phi_\tau = \widetilde{\Phi}_\tau - (F^+ T + (F^+ T)^T), \quad$ where $(\widetilde{\Phi}_\tau)_{jk} \stackrel{\text{def}}{=} \{\tau f_j \circ f_k\}$,

(4.63) $\qquad \Phi_\sigma = \widetilde{\Phi}_\sigma - (F^+ S - (F^+ S)^T), \quad$ where $(\widetilde{\Phi}_\sigma)_{jk} \stackrel{\text{def}}{=} \{\sigma f_j \circ f_k\}$.

PROOF. Formula (4.62) follows from a simple calculation:

$$\Phi_{ik} = \{\tau b_i \circ b_k\} = \{\tau(f_i - F_{is}^+ a_s) \circ (f_k - F_{km}^+ a_m)\}$$
$$= \{\tau f_i \circ f_k\} - F_{km}^+ \{\tau f_i \circ (T_{ms} \tau a_s)\} - F_{is}^+ T_{sk}$$
$$= \{\tau f_i \circ f_k\} - (F_{km}^+ T_{mi} + F_{is}^+ T_{sk}).$$

Formula (4.63) can be proved in the same way. $\qquad \square$

Let us consider the quotient Γ/σ. It is not difficult to see that this is a two-sheeted covering of the λ-plane, i.e., a hyperelliptic surface. Let us denote it by Γ_h, and its hyperelliptic involution by σ_h. Let π_σ be the projection of Γ to Γ_h. Consider $\pi_\sigma \circ \tau \stackrel{\text{def}}{=} \tau_h$. This is an anti-holomorphic involution on Γ_h, hence Γ_h is a real hyperelliptic surface. Note that on each Γ_h there exist two anti-involutions: τ_h and $\sigma_h \tau_h$. We will consider them separately, each Γ_h twice (we shall try to lift both anti-involutions from Γ_h to Γ).

Let Γ be a two-sheeted ramified (at least at the two points P_∞ and P_0) covering of Γ_h. We shall consider the surface Γ_h with a system of cuts $\mathcal{L}$ on it as a model of Γ, i.e.,
$$\Gamma = \{\Gamma_h; \mathcal{L}\}.$$
Two systems of cuts $\mathcal{L}_1$ and $\mathcal{L}_2$ on Γ_h are said to be *equivalent* if they define isomorphic surfaces.

DEFINITION. Let $\mathcal{L}_1 - \mathcal{L}_2$ be the curve $\check{l}$ consisting of all points of $\mathcal{L}_1$ and $\mathcal{L}_2$ with their multiplicities taken in account; here we set the multiplicity of the internal points to be 1 and the multiplicity of the endpoints to be $1/2$.

Let us consider the group $H_1(\Gamma_h, \mathbb{Z}_2)$; for $l \in H_1(\Gamma_h)$ we have
$$l \stackrel{\mathbb{Z}_2}{=} 0 \iff \exists m \in H_1(\Gamma_h) \ l = 2m \ (\text{in } H_1(\Gamma)).$$
Notice that the cycles in $H_1(\Gamma_h, \mathbb{Z}_2)$ are nonoriented, i.e., $l \stackrel{\mathbb{Z}_2}{=} -l$.

PROPOSITION 4. *If two cuts $\mathcal{L}_1$ and $\mathcal{L}_2$ have the same endpoints (the corresponding coverings have the same branch points), then $\mathcal{L}_1 - \mathcal{L}_2$ determines a class in $H_1(\Gamma_h, \mathbb{Z}_2)$.*

THEOREM 6. *Let Γ_h be a real hyperelliptic surface, and τ_h be an anti-involution on it. Then the anti-involution τ_h can be lifted to $\{\Gamma_h, \mathcal{L}\}$ if and only if the set of endpoints of $\mathcal{L}$ is fixed under τ_h, and*

(4.64)
$$\mathcal{L} - \tau_h \mathcal{L} \stackrel{\mathbb{Z}_2}{=} 0.$$

PROOF. It is easy to see that if $\mathcal{L} - \tau_h \mathcal{L} \stackrel{\mathbb{Z}_2}{=} 0$, then τ_h can be lifted to Γ. We need only prove that the relation $\mathcal{L} - \tau_h \mathcal{L} \stackrel{\mathbb{Z}_2}{=} 0$ holds provided that τ_h can be lifted to $\{\Gamma_h, \mathcal{L}\}$. Consider the surfaces $\{\Gamma_h, \tau_h \mathcal{L}\}$ and $\{\Gamma_h, \mathcal{L}\}$. If τ_h can be lifted, then they are isomorphic; we denote the corresponding isomorphism by $\tilde{\kappa}$. Consider $P \in \{\Gamma_h, \mathcal{L}\}$ and $P' = \tilde{\kappa}P \in \{\Gamma_h, \tau\mathcal{L}\}$. Let $P_\pi = \pi_\sigma P = \pi_\sigma \tilde{\kappa}^{-1} P' \in \Gamma_h$ be the projection of the points P and P' on Γ_h. Let P_π run along a loop $\check{l}_\pi$ on Γ_h which begins and ends at some fixed point. The preimages of $\check{l}_\pi$ on $\{\Gamma_h, \mathcal{L}\}$ and on $\{\Gamma_h, \tau_h \mathcal{L}\}$ are either both closed cycles or both nonclosed lines whose two endpoints lie on different copies of Γ_h. Consequently, *any closed line $\check{l}_\pi$ on Γ_h intersects $\mathcal{L}$ and $\tau\mathcal{L}$ the same number of times* (mod 2), or, in other words, $\mathcal{L}$ and $\tau\mathcal{L}$ have the same endpoints, and any closed line on Γ_h intersects the cycle $\mathcal{L} - \tau\mathcal{L}$ an even number of times. This is equivalent to the statement of the theorem: such a cycle (with an arbitrary orientation) has only even coefficients in its expansion over any canonical basis $\{\vec{a}, \vec{b}\} \in H_1(\Gamma_h)$. $\square$

Let us denote the anti-involution lifted to $\Gamma = \{\Gamma_h, \mathcal{L}\}$ by τ. Notice that there is no guarantee that $\tau^2 = 1$! We shall check this in each case individually.

How many different real coverings can one construct with one Γ_h?

Let us fix some system $\mathcal{L}_0$ which connects the branch points Γ (over Γ_h). All other systems $\mathcal{L}$ can be written as
$$\mathcal{L} = \mathcal{L}_0 + \check{l}, \quad \text{where } \check{l} \subset H_1(\Gamma_h; \mathbb{Z}_2).$$

By Theorem 6 the condition "τ_h *can be lifted to* $\{\Gamma_h, \mathcal{L}\} = \{\Gamma_h, \mathcal{L}_0 + \check{l}\}$" can be written in the form

(4.65)
$$\tau \mathcal{L} - \mathcal{L} \stackrel{\mathbb{Z}_2}{=} 0, \text{ or } \tau \check{l} - \check{l} \stackrel{\mathbb{Z}_2}{=} \tau \mathcal{L}_0 - \mathcal{L}_0 \in H_1(\Gamma, \mathbb{Z}_2).$$

We can choose any basis of $H_1(\Gamma)$ $\{\vec{a};\vec{b}\}$, and rewrite (4.65) as a linear system for the coefficients (mod 2) of the vector $\vec{l}$. Each solution of this system gives us the surface $\{\Gamma_h;\mathcal{L}\}$ with an anti-involution on it.

If system (4.65) is noncompatible, then the anti-involution τ_h cannot be lifted to Γ, and there is no real surfaces with such a Γ_h.

Let us introduce some notation. We call the cuts along which we glue two copies of Γ_h the γ-*cuts*, and the corresponding branch-points will be called the γ-*branch points*. The cuts along which we glue two copies of $\mathbb{C}_\lambda$ (to obtain Γ_h) will be called h-*cuts*, and the branch points of Γ_h over $\mathbb{C}_\lambda$ will be called the h-*branch points*.

Let $\tau_h P_\infty = P_0$, where P_0 and P_∞ are some h-branch points. There are two possible types of symmetries: $\lambda \to 1/\bar{\lambda}$ and $\lambda \to -1/\bar{\lambda}$. We know (Theorem 3) that only the first type of symmetry is compatible with the conditions of Theorem 2, and therefore we consider only such surfaces here.

§5. Classification of real double-coverings over hyperelliptic surfaces. Examples

Our classification is based on the properties of the anti-involution τ_h. On each real Γ_h there exists two anti-involutions. If there are real h-branch points (on the unit circle), we consider them one at a time, i.e., we choose different $\mathcal{L}_0$ for each of these anti-involutions. We consider the cases when there is (is not) real oval separately.

On the figures below the following notation is used:

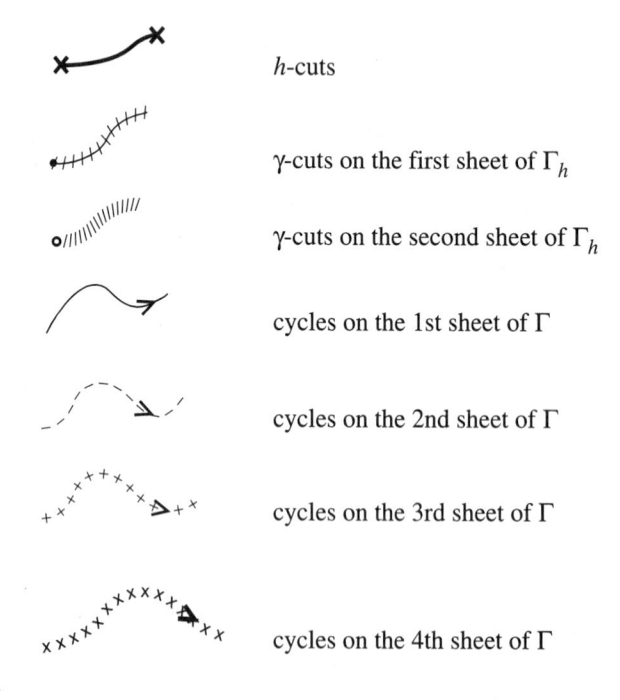

h-cuts

γ-cuts on the first sheet of Γ_h

γ-cuts on the second sheet of Γ_h

cycles on the 1st sheet of Γ

cycles on the 2nd sheet of Γ

cycles on the 3rd sheet of Γ

cycles on the 4th sheet of Γ

I. τ_h **has a real oval.** Choose $\mathcal{L}_0$ as in Figure 1: $\mathcal{L}_0$ connects the fixed γ-branch points along the real oval[9]. We connect conjugate pairs of branch points by symmetric lines which intersect the unit circle at points of the real oval.

---
[9]It is not difficult to prove that on each real oval there is an even number of γ-branch points. In the converse case no real covering exist.

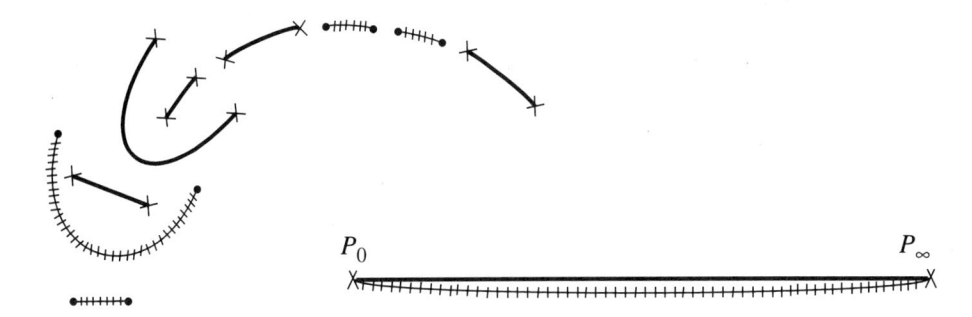

FIGURE 1

All other cuts (the solutions of the equation $\tau \mathcal{L} \stackrel{\mathbb{Z}_2}{\equiv} \mathcal{L}$) can be written in the form $\mathcal{L} = \mathcal{L}_0 + \check{l}_\Re + \check{l}_\Im + \check{l}^{(m)}$, where $\check{l}_\Re$ is any linear combination of real ovals, $\check{l}_\Im$ is any linear combination of imaginary ovals, and $\check{l}^{(m)}$ is any linear combination of $a^{(m)}$-cycles (the symmetrical cycles around conjugated pairs of h-branch points).

II. τ_h **has no real oval.** In this case there are no h-branch points on the unit circle. We denote by $n^{(m)}$ the number of pairs of h-branch points and by $\tilde{n}^{(m)}$ the number of pairs of γ-branch points (P_0 and P_∞ are included in these sets).

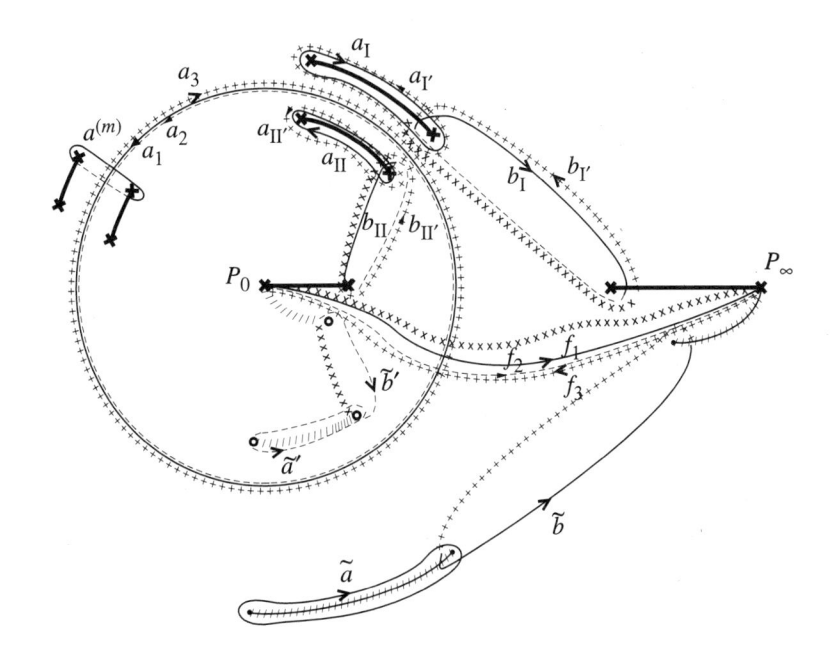

FIGURE 2

a) $n^{(m)}$ **and** $\tilde{n}^{(m)}$ **are even.** Let $\mathcal{L}_0$ be as in Figure 2, i.e., symmetric with respect to τ_h; here we regard $\mathcal{L}_0$ as a set of points (without orientation). The other cuts $\mathcal{L}$ are the solutions of (4.65), i.e., $\mathcal{L} = \mathcal{L}_0 + \check{l}^{(m)} + \varepsilon_a O$, where $\check{l}^{(m)}$ is any linear combination of the $a^{(m)}$-cycles (i.e., the symmetric cycles around conjugate pairs of h-branch points as in the previous case) and O is a simple cycle over the unit circle ($\varepsilon_a \in \{0; 1\}$).

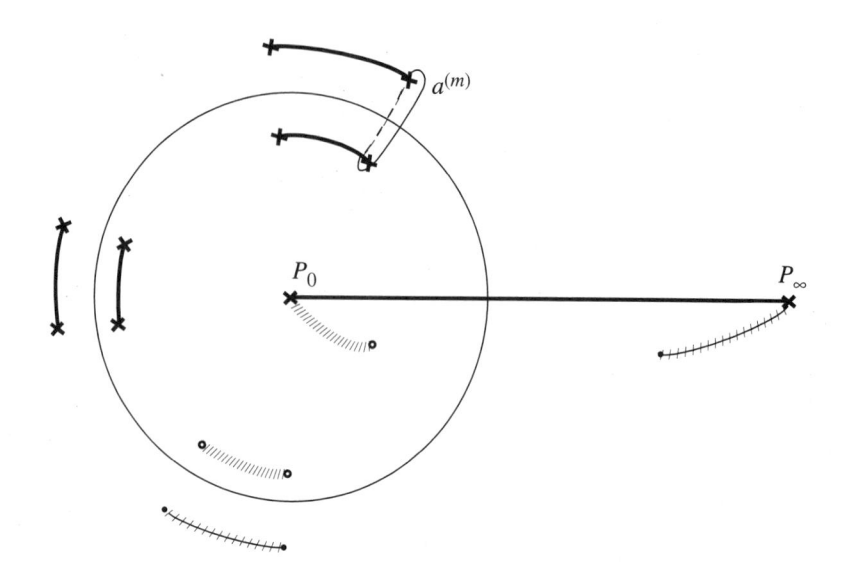

FIGURE 3

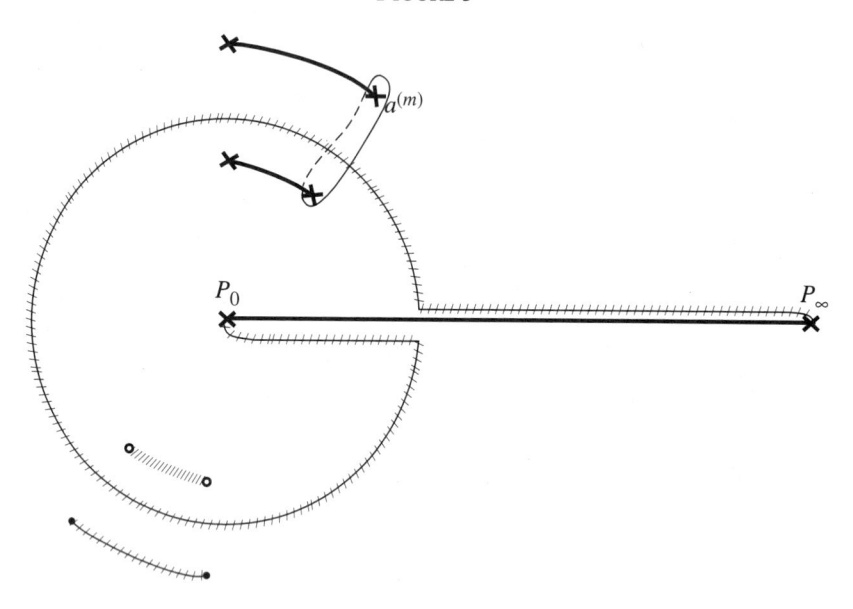

FIGURE 4

b) $n^{(m)}$ **is even and** $\tilde{n}^{(m)}$ **is odd.** Then there are no real coverings.

c) $n^{(m)}$ **is odd and** $\tilde{n}^{(m)}$ **is even.** Let $\mathcal{L}_0$ be as in Figure 3. The cut $\mathcal{L}_0$ is symmetric with respect to τ_h, and $\mathcal{L} = \mathcal{L}_0 + \breve{l}^{(m)}$, where $\breve{l}^{(m)}$ is a linear combination of the $a^{(m)}$-cycles.

d) $n^{(m)}$ **is odd and** $\tilde{n}^{(m)}$ **is odd.** Let $\mathcal{L}_0$ be as shown in Figure 4. The difference $\mathcal{L}_0 - \tau \mathcal{L}_0$ is the cycle over the unit circle (which is homologous to zero in this case). The other systems of $\mathcal{L}$ are $\mathcal{L} = \mathcal{L}_0 + \breve{l}^{(m)}$, where $\breve{l}^{(m)}$ is a linear combination of the $a^{(m)}$-cycles.

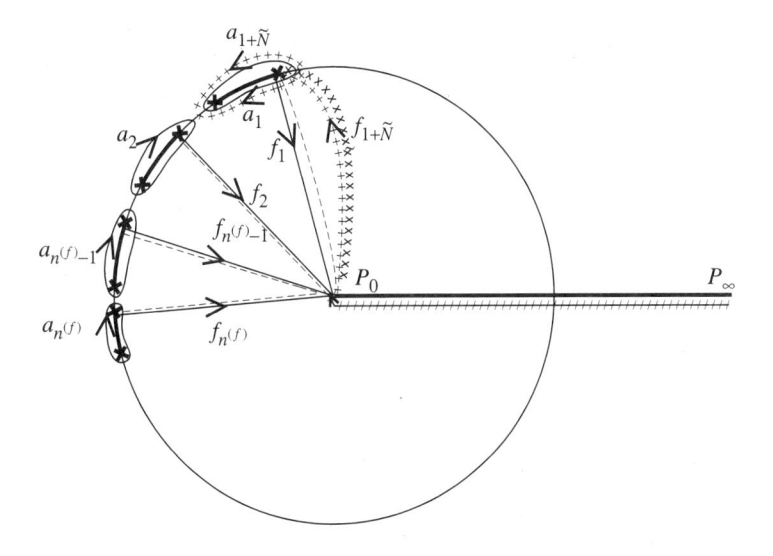

FIGURE 5

5.1. Examples. Let us consider surfaces with the basic system of cuts $\mathcal{L}_0$ from the families I and II.a) as examples. Before turning to concrete examples, we note that in the case when some γ-branch points coincide with h-branch points (i.e., when all the four sheets are glued), it is possible to choose a basis of $H_1(\Gamma)$ that has the same symmetry properties as the one considered in the case of distinct branch points (except for P_∞ and P_0). This means that without loss of generality we can assume that all the γ- and h-branch points are distinct (except for P_∞ and P_0).

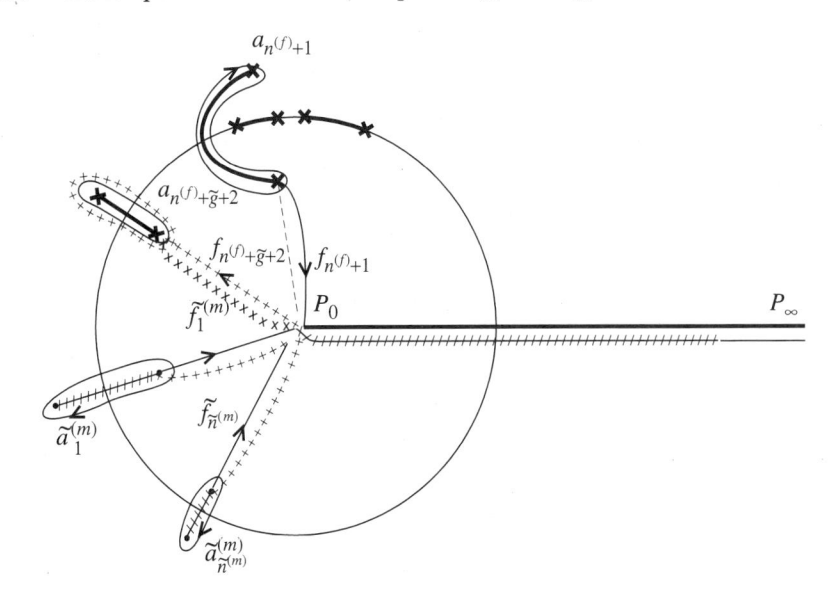

FIGURE 6

I. τ_h **has a real oval.** We choose the basis of cycles $\{\vec{a}, \vec{f}\}$ as in Figures 5, 6, 7 (one picture with all types of cycles is too complicated and we present different kinds

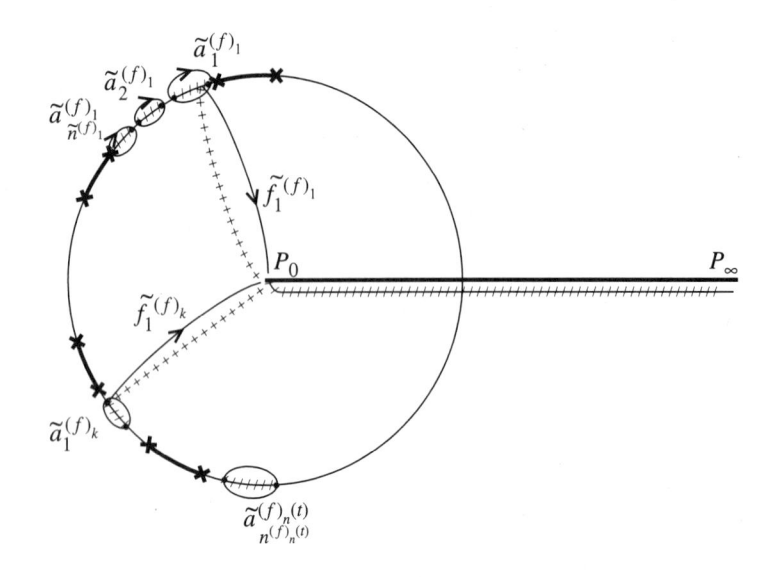

FIGURE 7

of cycles on different figures). The first $n^{(f)}$ cycles (where $n^{(f)}$ is the number of pairs of h-branch points on the unit circle) go around the fixed h-cuts (Figure 5). The next $n^{(m)}$ cycles (where $n^{(m)}$ is the number of pairs of h-branch points symmetric with respect to the unit circle) go around the corresponding h-cuts (Figure 6). Notice that the sum $n^{(f)} + n^{(m)} = g_h$ is the genus of Γ_h. The next group of a-cycles (marked by a tilde) go around the γ-cuts. The cycles $\tilde{a}_i^{(f)_k}$, $i \in \{1, \ldots, \tilde{n}^{(f)_k}\}$, $\sum_k \tilde{n}^{(f)_k} = \tilde{n}^{(f)}$, go around the γ-cuts that lie on the kth τ_h-real oval (Figure 7). The corresponding f-cycle is denoted by $\tilde{f}_i^{(f)_k}$, $k \in \{1, \ldots, n^{(f)}\}$. The total number of γ-cuts on the real ovals is denoted by $\tilde{n}^{(f)}$.

The cycles $\tilde{a}_j^{(m)}$, $j \in \{1, \ldots, \tilde{n}^{(m)}\}$, go around γ-cuts (Figure 6) through the real h-oval ($\tilde{n}^{(m)}$ is the number of pairs of γ-branch points symmetric with respect to the unit circle). The total number of pairs of γ-branch points (except P_∞ and P_0) is denoted by $\tilde{g}$, so that $\tilde{g} = \tilde{n}^{(f)} + \tilde{n}^{(m)}$. The last g_h a-cycles are the same cycles as the first ones, but on the second copy of Γ_h (Figure 5) and have the opposite direction, $a_{k+\tilde{g}} = -\sigma a_k$, where $k = \{1, \ldots, g_h\}$. Finally, the cycles f_i go from the corresponding cut to P_0 and back on another sheet (see Figures 5–7).

The matrix F^+ for this basis is

$$(5.66) \qquad F^+ = \begin{pmatrix} (0\backslash 1) & -A^T & 0 \\ 0 & (0\backslash 1) & A \\ 0 & 0 & (0\backslash 1) \end{pmatrix},$$

where

$$(0\backslash 1)_{nm} = \begin{cases} 1 & \text{if } n < m, \\ 0 & \text{if } n \geqslant m, \end{cases}$$

and A is the intersection matrix of the f- and $\tilde{f}$-cycles, which is given by

$$\{\tilde{f}_i^{(f)k_1} \circ f_{k_2}^{(f)}\} = \begin{cases} 0 & \text{if } k_1 \geqslant k_2, \\ 1 & \text{if } k_1 < k_2, \end{cases}$$

$$\{\tilde{f}_{k_1}^{(m)} \circ f_{k_2}^{(f)}\} = 0, \qquad \{\tilde{f}_{k_1}^{(m)} \circ f_{k_2}^{(m)}\} = 0.$$

Note that a-cycles cannot circle around P_0, so $\Delta_1 = 0$ and the number of revolutions of a b-cycle is equal to the number of revolutions of the corresponding f-cycle. As the result, we have

$$\Delta_2 = (-\vec{1}_{g_h}, -2\vec{1}_{\tilde{g}}, \vec{1}_{g_h}), \qquad \vec{t}_k \stackrel{\text{def}}{=} (t, \ldots, t) \in \mathbb{R}^k.$$

Let us consider the involution σ. In our basis

$$S = -\begin{pmatrix} 0 & 0 & I_{g_h} \\ 0 & I_{\tilde{g}} & 0 \\ I_{g_h} & 0 & 0 \end{pmatrix},$$

where g_h and $\tilde{g}$ are the dimensions of the corresponding unit matrices. Further

$$\Phi_\sigma = \begin{pmatrix} 0 & -A^T & 0 \\ A & 0 & -A \\ 0 & A^T & 0 \end{pmatrix}, \qquad \kappa_\sigma = 0.$$

Consider the anti-involution. For τ, which does not transpose the copies of Γ_h, we have $\tilde{\varepsilon} = 1$ (see formula (3.55)). This means that in spite of the solvability of system (3.45), this case gives us no Willmore surfaces. The case of Willmore surfaces corresponds to the other anti-involution $\tau_1 = \sigma\tau$, which permutes copies of Γ_h. For this anti-involution, we have:

$$T_1 = \begin{pmatrix} 0 & 0 & I_{g_h} \\ 0 & I_{\tilde{g}} & 0 \\ I_{g_h} & 0 & 0 \end{pmatrix} = -S, \qquad \Phi_{\tau_1} = -\begin{pmatrix} 0 & 0 & (1\backslash 1)_{n^{(f)}} \\ 0 & (1\backslash 1)_{\tilde{n}^{(f)}} & 0 \\ (1\backslash 1)_{n^{(f)}} & 0 & 0 \end{pmatrix},$$

$$\kappa_{\tau_1} = \pi i (\vec{1}_{n^{(f)}}, \vec{0}_{n^{(m)}}, \vec{1}_{\tilde{n}^{(f)}}, \vec{0}_{\tilde{n}^{(m)}}, \vec{1}_{n^{(f)}}, \vec{0}_{n^{(m)}}),$$

where $(1\backslash 1)_{n^{(f)}} \in \mathbb{Z}^{g_h \times g_h}$ and $(1\backslash 1)_{\tilde{n}^{(f)}} \in \mathbb{Z}^{\tilde{g}_h \times \tilde{g}_h}$ are given by

$$((1\backslash 1)_k)_{nm} = \begin{cases} 1 & \text{if } n < m, \\ 1 & \text{if } n = m \leqslant k, \\ 0 & \text{if } n = m > k, \\ 1 & \text{if } n > m. \end{cases}$$

The vector Z in the argument of the Θ-function is $\vec{V}_\infty z - \vec{V}_0 \bar{z}$. The system (3.45) has two classes of solutions. The first is:

$$D = \begin{bmatrix} \delta/2, \vec{1}_{n^{(m)}}/2; & 0, \vec{1}_{\tilde{n}^{(m)}}/2; & \delta/2, \vec{1}_{n^{(m)}}/2 \\ d^\beta; & \tilde{d}^\beta; & d^\beta + \vec{1}_{g_h}/2 \end{bmatrix},$$

where $\delta \in \mathbb{Z}^{n^{(f)}}$, $d^\beta \in \mathbb{R}^{g_h}$, $\tilde{d}^\beta \in \mathbb{R}^{\tilde{g}}$ with the following additional restrictions:
a) $\sum_{k=1}^{g_h} \delta_k = n^{(m)} \pmod{2}$;
b) $n^{(f)} \neq 0$ if $\tilde{n}^{(m)}$ is odd and $n^{(f)} = 0$ if $\tilde{n}^{(m)}$ is even.

The second solution is

$$D = \begin{bmatrix} \delta/2, \ \vec{1}_{n^{(m)}}/2; & \vec{0}_{\tilde{g}}; & \delta/2, \ \vec{1}_{n^{(m)}}/2 \\ d^\beta & \tilde{d}^\beta & d^\beta + \vec{1}_{g_h}/2 \end{bmatrix}$$

with the additional restrictions:
a) $\sum_{k=1}^{g_h} \delta_k = n^{(m)} \pmod 2$;
b) $\tilde{n}^{(f)} = 0$ (in this case $\tilde{g} = \tilde{n}^{(m)}$).

Let us consider the second example, namely the surface of type II.a). Choose the basis of cycles as in Figure 2. It is natural to separate the basis cycles into seven groups:

1) The first three a-cycles lie over the unit circle on different sheets. The corresponding f-cycles run from P_0 to P_∞ and back on the fourth sheet (on which there is no a-cycle).

The other cycles have a canonical matrix of intersection, so we do not need to introduce f-cycles and present the canonical basis $\{a, b\}$.

2) $\Rightarrow$ 3) $a_I b_I$-cycles and $a_{II} b_{II}$-cycles are the basis cycles of Γ_h (see Figure 2). They lie on the first copy of Γ_h. The dimension of these vectors is $(g_h - 1)/2$, where g_h is the genus of Γ_h.

4) $\Rightarrow$ 5) $a_{I'} b_{I'}$-cycles and $a_{II'} b_{II'}$-cycles are the same as $a_I b_I$ and $a_{II} b_{II}$, but lie on another copy of Γ_h and have opposite direction.

6) $\tilde{a}$-cycles go around the γ-cuts outside the unit circle; the $\tilde{b}$-cycles are similar (see Figure 2).

7) $\tilde{a}'$-cycles go around the γ-cuts inside the unit circle; the $\tilde{b}'$-cycles are similar (see Figure 2).

For this basis, the nontrivial part of the matrix F^+ (the first diagonal block 3×3, which corresponds to the group 1) has the form

$$(F^+)_{(1)} = \begin{pmatrix} 0 & -1 & 1 \\ 0 & 0 & 1 \\ 0 & 0 & 0 \end{pmatrix}.$$

For this basis, the matrix S and the nontrivial block Φ_σ have the form

$$S = \begin{pmatrix} & 1 & & & & & & \\ 1 & 1 & -1 & & & & & \\ 1 & & & & & & & \\ & & & I & & & & \\ & & & & I & & & \\ & & & & & I & & \\ & & & & & & I & \\ & & & & & & & I \end{pmatrix}, \quad (\Phi_\sigma)_{(1)} = \begin{pmatrix} 0 & -1 & 0 \\ 1 & 0 & -1 \\ 0 & 1 & 0 \end{pmatrix},$$

$$\begin{bmatrix} \Delta_1/4 \\ \Delta_2/4 \end{bmatrix} = \frac{1}{4} \begin{bmatrix} 0, & -1, & 2, 0, \ldots, 0 \\ 1, & 1, & -1, 0, \ldots, 0 \end{bmatrix}, \quad \kappa_\sigma = (1, 0, 1, 0, \ldots, 0).$$

First let us consider an anti-involution τ that does not permute the copies of Γ_h. The matrix T has the form:

$$T = \begin{pmatrix} 1 & & & & & & & & \\ & 1 & & & & & & & \\ & 1 & 1 & -1 & & & & & \\ & & & & I & & & & \\ & & & & & I & & & \\ & & & & & & I & & \\ & & & & & & & I & \\ & & & & & & & & I \\ & & & & & & & I & \end{pmatrix}.$$

The nontrivial part of the matrix Φ_τ (first diagonal block) and the vector κ_τ are

$$(\Phi_\tau)_{1)} = \begin{pmatrix} 0 & -1 & 0 \\ -1 & 0 & 0 \\ 0 & 0 & 0 \end{pmatrix}, \quad \kappa_\tau = (1, 1, 0, 0, \dots, 0).$$

Let us consider the other anti-involution $\tau_1 = \sigma\tau$. It permutes the copies of Γ_h. In this case:

$$T_1 = \begin{pmatrix} -1 & -1 & 1 & & & & & & \\ & & -1 & & & & & & \\ & -1 & & & & & & & \\ & & & & I & & & & \\ & & & & & I & & & \\ & & & & & & I & & \\ & & & & I & & & & \\ & & & & & & & & I \\ & & & & & & & I & \end{pmatrix}.$$

The nontrivial part of the matrix Φ_{τ_1} and the vector κ_{τ_1} are:

$$(\Phi_{\tau_1})_{(1)} = \begin{pmatrix} 0 & 0 & 0 \\ 0 & 0 & 1 \\ 0 & 1 & 0 \end{pmatrix}, \quad \kappa_{\tau_1} = (0, 1, 1, 0, \dots, 0).$$

The subsequent computations are trivial and we only describe the result, namely the vectors D, the solution of the system (3.45), and $\vec{Z}$, the argument of the Θ-function (2.28). For the anti-involution τ, we have

$$D = \begin{bmatrix} \alpha_0 & -\alpha_0 & \alpha_0 + 1/2 & \alpha_1 & -\alpha_1 & \alpha_1 & -\alpha_1 & \alpha_2 & -\alpha_2 \\ \beta_0 & \beta_0 + \alpha_0 + (1-\delta_0)/2 & \delta/2 & \beta_1 & \beta_1 & \beta_1 & \beta_1 & \beta_2 & \beta_2 \end{bmatrix},$$

$$\varepsilon = -i, \; \tilde{\varepsilon} = -1, \quad \text{and} \quad \vec{Z} = \vec{V}_\infty z - i\vec{V}_0 \bar{z}.$$

For the anti-involution τ_1, we have

$$D = \begin{bmatrix} \delta_1/2 & (\delta_1+1)/2 & (\delta_1+1)/2 & \alpha_1 & \alpha_1 & \alpha_1 & \alpha_1 & \alpha_2 & \alpha_2 \\ \delta_2/2 & \beta_0 & (\delta_1+\delta_2)/2 - \beta_0 & \beta_1 & -\beta_1 & \beta_1 & -\beta_1 & \beta_2 & -\beta_2 \end{bmatrix},$$

$$\varepsilon = i, \; \vec{Z} = \vec{V}_\infty z + i\vec{V}_0 \bar{z} \text{ and } \tilde{\varepsilon} = -1.$$

Here $\alpha_0, \beta_0 \in \mathbb{R}$, $\alpha_1, \beta_1 \in \mathbb{C}^{(g_h-1)/2}$, $\alpha_2, \beta_2 \in \mathbb{C}^{\tilde{g}}$, and $\delta_k \in \mathbb{Z}$.

References

1. A. I. Bobenko, *Constant mean curvature surfaces and integrable equations*, Uspekhi Mat. Nauk **46** (1991), no. 4, 3–42; English transl., Russian Math. Surveys **46** (1991), no. 4, 1–45.
2. G. Thomsen, *Über Konforme Geometrie* I: *Grundlagen der Konformen Flachentheorie*, Abh. Math. Sem. Univ. Hamburg (1923), 31–56.
3. T. J. Willmore, *Note on embedded surfaces*, An. Ştiinţ. Univ. "Al. I. Cuza" Iaşi Secţ. I a Mat. **11** (1965), 493–496.
4. J.-H. Eschenburg, *Willmore surfaces and Moebius geometry* (to appear).
5. B. Palmer, *The conformal Gauss map and the stability of Willmore surfaces* (to appear).
6. B. A. Dubrovin, *Theta-functions and non-linear equations*, Uspekhi Mat. Nauk **36** (1981), no. 2, 11–80; English transl., Russian Math. Surveys **36** (1991), no. 2, 11–92.
7. A. P. Fordy and J. Gibbons, *Integrable nonlinear Klein-Gordon equations and Toda lattices*, Comm. Math. Phys. **77** (1980), 21–30.
8. E. D. Belokolos, A. I. Bobenko, V. Z. Enolskiĭ, A. R. Its, and V. B. Matveev, *Algebro-geometric approach to nonlinear integrable equations* (to appear).
9. D. Ferus, F. Pedit, U. Pinkall, and I. Sterling, *Minimal tori in S^4*, Preprint TU-Berlin (1991).
10. U. Pinkall and I. Sterling, *On the classification of constant mean curvature tori*, Ann. of Math. **130** (1989), 407–451.
11. _____, *Willmore surfaces*, Math. Intelligencer **9** (1987), 38–43.
12. M. Babich and A. Bobenko, *Willmore tori with umbilic lines and minimal surfaces in hyperbolic space*, Duke Math. J. **72** (1993), 151–185.
13. I. Cherdantsev and R. Shapirov, *Finite-gap solutions of the Bullough–Dodd–Shabat equation*, Teoret. Mat. Fiz. **81** (1990), no. 1, 155–160; English transl. in Theoret. and Math. Phys. **81** (1990).
14. A. Veselov and S. Novikov, *Finite-gap two-dimensional Schrödinger operators. Explicit formulas and evolution equations*, Dokl. Akad. Nauk SSSR **279** (1984), no. 1, 20–24; English transl. in Soviet Math. Dokl. **30** (1985).
15. _____, *Finite-gap two-dimensional Schrödinger operators. Potential operators*, Dokl. Akad. Nauk SSSR **279** (1984), no. 4, 784–788; English transl. in Soviet Math. Dokl. **30** (1985).

Translated by THE AUTHOR

POMI, FONTANKA 27, ST. PETERSBURG 191011, RUSSIA
E-mail address: babich@lomi.spb.su

On the Birkhoff Standard Form of Linear Systems of ODE

A. A. Bolibruch

Dedicated to Boarding School #45 on the occasion of its 30th anniversary

ABSTRACT. In this paper we present sufficient conditions for an irreducible reducible system of differential equations to be analytically transformed to Birkhoff standard form.

§1. Introduction

Consider a linear system of differential equations

$$(1) \qquad x\frac{dy}{dx} = C(x)y,$$

where $C(x)$ is a matrix of size (p, p) of the form

$$(2) \qquad C(x) = x^r \sum_{n=0}^{\infty} C_n x^{-n}, \qquad C_0 \neq 0, \ r \geq 0,$$

x is a complex variable, and the power series converges in some neighborhood of ∞.

Under the transformation

$$(3) \qquad z = \Gamma(x)y$$

system (1) is transformed to the system

$$(4) \qquad x\frac{dz}{dx} = B(x)z,$$

where

$$(5) \qquad B(x) = x\frac{d\Gamma}{dx}\Gamma^{-1} + \Gamma C(x)\Gamma^{-1}.$$

If $\Gamma(x)$ is holomorphically invertible in some neighborhood of ∞, then such a transformation is called *analytic*.

1991 *Mathematics Subject Classification*. Primary 34A30; Secondary 34A20.

This work was partially supported by Grant M3C000 from the International Science Foundation.

©1996, American Mathematical Society

If the matrix $B(x)$ in (4) is a polynomial in x of degree r, then (4) is called *the Birkhoff standard form for* (1).

Birkhoff [**Bi**] claimed that each system (1) can be analytically transformed to the Birkhoff standard form, but Gantmacher [**Ga**] presented a counterexample to this statement. It turned out that Birkhoff's proof was valid only for systems with diagonalizable monodromy matrices. But later it was found that the essential obstacle for system (1) to be analytically transformed to the Birkhoff standard form involves its reducibility.

A system (1) is called *reducible* if there exists a matrix $\Gamma(x)$, holomorphically invertible in some neighborhood of ∞, such that under the transformation (3) system (1) is transformed to system (4) with a lower triangular block matrix

$$(6) \qquad B(x) = \begin{pmatrix} B' & 0 \\ * & B'' \end{pmatrix}$$

(or equivalently, with an upper triangular block matrix). Otherwise a system (1) is called *irreducible*. In [**Bo5, Bo6, Bo7**] it was proved that *each irreducible system* (1) *of p equations can be analytically transformed to the Birkhoff standard form*. (For $p = 2$ a similar result was proved by Jurkat, Lutz, and Peyerimhoff [**JLP**], and for $p = 3$, by Balser [**Ba1**].)

For a reducible system (1) consider an analytic (in some neighborhood of infinity) transformation (3) that takes the system to a system (4) of the form

$$(7) \qquad B(x) = \begin{pmatrix} B_1 & W_{12} & \cdots & W_{1m} \\ 0 & B_2 & \cdots & W_{2m} \\ \vdots & \vdots & \ddots & \vdots \\ 0 & 0 & \cdots & B_m \end{pmatrix},$$

where each system with the coefficient matrix B_i, $i = 1, \ldots, m$ of size s_i determines an irreducible system. Denote by $\widehat{B}$ the matrix

$$\begin{pmatrix} B_2 & \cdots & W_{2m} \\ \vdots & \ddots & \vdots \\ 0 & \cdots & B_m \end{pmatrix}.$$

Below we consider fundamental matrices for systems (4), (7) of the form

$$(8) \qquad Z(x) = \begin{pmatrix} Z_1 & * \\ 0 & Z_2 \end{pmatrix}$$

only, where Z_1 is a fundamental matrix to the system with matrix B_1, and Z_2 is a fundamental matrix to the system with matrix $\widehat{B}$.

The following theorem gives sufficient conditions for a reducible system to be analytically transformed to the Birkhoff standard form.

THEOREM 1. *Let the monodromy matrix G of a fundamental matrix $Z(x)$ of form* (8) *for the system* (4), (7) *be expressed as*

$$(9) \qquad G = \begin{pmatrix} G' & 0 \\ 0 & G'' \end{pmatrix},$$

where the size l of the matrix G' does not exceed s_1. If the system with the matrix $\widehat{B}$ can be analytically transformed to Birkhoff standard form, then so does the system (4), (7) (and so does the original system (1)).

COROLLARY 1. *Let the system* (1) *be analytically transformed to* (4) *with the matrix $B(x)$ of the form* (7) *with two diagonal blocks and one off-diagonal block only. If the monodromy matrix G of a fundamental matrix $Z(x)$ of the form* (8) *for the system* (4), (7) *has block diagonal form, then the system can be analytically transformed to the Birkhoff standard form.*

COROLLARY 2. *If the monodromy matrix G of a fundamental matrix $Z(x)$ of the form* (8) *for the system* (4), (7) *has the form*

$$G = \mathrm{diag}(G_1, \ldots, G_m),$$

where the sizes of G_i coincide with those of the corresponding B_i from (7), *then the system can be analytically transformed to the Birkhoff standard form.*

COROLLARY 3 ([**Bi**]). *If the monodromy matrix of system* (1) *is diagonalizable, then the system can be analytically transformed to the Birkhoff standard form.*

Now let transformation (3) be meromorphic. This means that this transformation is holomorphically invertible in some punctured neighborhood of ∞ and is only meromorphic at ∞. Is it possible to transform the arbitrary system (1) to the Birkhoff standard form with the help of a meromorphic transformation (recall that such a transformation must not increase the order r of pole of $C(x)$ at ∞, which is called *the Poincaré rank of singularity*)?

In 1963 Turritin [**Tu**] proved the following statement.

THEOREM 2. *If all eigenvalues of the matrix C_0 in* (2) *are distinct, then system* (1) *can be meromorphically transformed to the Birkhoff standard form* (*without increasing the Poincaré rank r*).

In the general case the question mentioned above is still open. Up to this moment only for dimension $p = 2$ in [**JLP**] and for dimension $p = 3$ in [**Ba2**] a positive answer for this problem was obtained.

In §3 we present a new version of the proof of Turritin's theorem. It turns out that this theorem follows from the theorem on the analytic transformation of an irreducible system to the Birkhoff standard form.

§2. Preliminary information

In what follows we need the following improvement of Theorem 5 from [**Bo6**] (Theorem 1 in [**Bo7**]).

PROPOSITION 1. *Let system* (1) *be irreducible and let its monodromy matrix G be of form* (9) *and upper triangular. Then for any real d the system can be analytically transformed to the Birkhoff standard form* (4) *such that in some neighborhood of zero the corresponding fundamental matrix $Z(x)$ of* (4) *has the form*

(10) $$Z(x) = U_0(x) x^{A_0} x^E,$$

where U_0 is holomorphically invertible at zero, $A_0 = \mathrm{diag}(\lambda_1, \ldots, \lambda_p)$, $\lambda_k \in \mathbb{Z}$, $\lambda_1 > \cdots > \lambda_l$, $\lambda_{l+1} > \cdots > \lambda_p > d$, $E = (2\pi i)^{-1} \log G$.

PROOF. Consider a fundamental matrix $Y(x)$ for the system (1) with the following factorization

(11) $$Y(x) = M(x) x^E,$$

where $M(x)$ is a single-valued matrix function with nonvanishing determinant in some punctured neighborhood O of ∞, E is the same as above. This factorization determines a vector bundle on $P^1(\mathbb{C})$ with coordinate neighborhoods O, $\mathbb{C}$ and the transition function $M(x)$. Denote this bundle by F and consider all its *admissible extensions* F^Λ, i.e., any extension determined by the transition function $M(x) x^{-\Lambda}$, where $\Lambda = \text{diag}(f_1, \ldots, f_p)$ and

(12) $$f_1 \geqslant \cdots \geqslant f_l, \qquad f_{l+1} \geqslant \cdots \geqslant f_p.$$

Consider the decomposition of F^Λ into the direct sum of line bundles (see [**OSS**])

(13) $$F^\Lambda \cong \mathcal{O}(-k_1^\Lambda) \oplus \cdots \oplus \mathcal{O}(-k_p^\Lambda),$$

where $k_1^\Lambda \geqslant \cdots \geqslant k_p^\Lambda$.

The numbers $(k_1, \ldots, k_p)$ (we omit the index Λ here) are said to form a *splitting type* of F^Λ.

LEMMA 1. *If system* (1) *is irreducible, then the family of vector bundles F^Λ has the finiteness property, i.e., the following inequalities hold for the splitting type of an arbitrary admissible extension F^Λ of F*:

(14) $$k_i - k_{i+1} \leqslant r, \qquad i = 1, \ldots, p - 1.$$

PROOF. Using a matrix $T(x)$ holomorphically invertible in some neighborhood of ∞, we transform the matrix $M(x) x^{-\Lambda}$ to the form

(15) $$T(x) M(x) x^{-\Lambda} = x^K U(x),$$

where K is a splitting type of F^Λ, $U(x)$ is holomorphically invertible on the complex plane. (The existence of such a transformation is equivalent to the existence of the decomposition (13).)

Assume that for some $m = 1, \ldots, p - 1$ we have

(16) $$k_m - k_{m+1} > r.$$

Consider the system with a fundamental matrix

(17) $$Y'(x) = x^K U(x) x^\Lambda x^E.$$

This system has only two singular points 0 and ∞ on the whole Riemann sphere (since $U(x)$ is holomorphically invertible everywhere except ∞) and its coefficient matrix

$$C' = x \frac{dY'}{dx} (Y')^{-1}$$

has a pole of order r at ∞. The latter statement follows from the fact that this system is obtained from the original system (1) by the transformation $T(x)$ which is analytic in some neighborhood of ∞.

On the other hand, from (17) it follows that

(18) $$C'(x) = K + x^K \left[\frac{dU}{dx} U^{-1} + U(\Lambda + L) U^{-1} \right] x^{-K},$$

where $L = x^\Lambda E x^{-\Lambda}$. It follows from (12) and the fact that E has the block and upper triangular form (9), that L is holomorphic on the whole complex plane. Therefore, the matrix in brackets in (18) is holomorphic everywhere except at the point ∞. Denote this matrix by $W(x)$. Then

(19) $$C'(x) = K + x^K W(x) x^{-K}.$$

Since the element c'_{ij} of the matrix C' and the element w_{ij} of the matrix $W(x)$ are connected as follows,

$$c'_{ij} = x^{k_i - k_j} w_{ij}, \quad i \neq j,$$

we obtain from (16) that $k_i - k_j > r$ for $i = 1, \ldots, m$, $j = m+1, \ldots, p$. Therefore, for every pair of such i, j, the element c'_{ij} has a zero of order $s > r$ at 0 while the order of its pole at ∞ is at most r. This means that

$$c'_{ij} \equiv 0, \quad i = 1, \ldots, m, \ j = m+1, \ldots, p$$

and therefore, the original system (1) is reducible. But this contradicts the assumption of the lemma. This contradiction means that relations (14) hold. The lemma is proved. $\square$

Let us continue the proof of the proposition. Consider a matrix Λ from (12) such that

(20) $$f_i - f_{i+1} > r(p-1), \quad i = 1, \ldots, l-1, l+1, \ldots, p-1,$$
$$f_p - d > r(p-1)$$

and consider the corresponding matrices $T(x)$ and K from (15). By a certain technical lemma (which is called Kimura's lemma in [**Si**], we see a similar statement in [**Bo6**] and [**Bo2**]), applied to the matrix $x^K U(x)$ from (17), there exists a matrix $\Gamma(x)$ holomorphically invertible everywhere except the point ∞ such that

(21) $$\Gamma(x) x^K U(x) = V(x) x^D,$$

where $D = \mathrm{diag}(d_1, \ldots, d_p)$ is obtained by some permutation of the diagonal elements of the matrix K, $V(x)$ is holomorphically invertible in O.

Under the analytic (in some neighborhood of ∞) transformation

(22) $$z = \Gamma(x) T(x) y$$

our original system (1) is transformed to system (4) with a fundamental matrix

(23) $$Z(x) = \Gamma(x) T(x) Y(x) = V(x) x^{D+\Lambda} x^E,$$

where $V(x)$ is a matrix holomorphically invertible on the whole Riemann sphere except at the point ∞.

It follows from (21) and Lemma 1 that for elements d_j of the integer-valued diagonal matrix D the following inequalities hold:

(24) $$|d_j - d_{j+1}| \leqslant r(p-1), \quad j = 1, \ldots, p-1.$$

Indeed,

$$|d_j - d_{j+1}| \leqslant \max_j d_j - \min_i d_i = k_1 - k_p$$
$$= (k_1 - k_2) + (k_2 - k_3) + \cdots + (k_{p-1} - k_p) \leqslant r(p-1),$$

since D is obtained by some permutation of diagonal elements of K.

Thus, from (24) and (20) we see that the first l and the last $p-l$ diagonal elements of the matrix $D+\Lambda$ form nonincreasing sequences (separately for each group). Since the matrix E is upper triangular and has the block form (9), we see again that

$$L' = x^{D+\Lambda} E x^{-D-\Lambda}$$

is an entire matrix function (i.e., all its entries are entire functions). Therefore,

$$B(x) = x\frac{dZ}{dx} Z^{-1} = x\frac{dV}{dx} V^{-1} + V(D+\Lambda+L') V^{-1}$$

is an entire matrix function too. Since all transformations are analytic at infinity, we conclude that $B(x)$ has a pole of the same order there as the original system. So $B(x)$ is a polynomial of degree r. Moreover $d_p + f_p > d$. This completes the proof of the proposition.

REMARK 1. Of course from the proposition (in the case $l = p$) it follows that each irreducible system can be analytically transformed to the Birkhoff standard form (see [**Bo6**, Theorem 5], [**Bo7**, Theorem 1]). Note that this proof is similar to the proof of the positive solvability of the Riemann–Hilbert problem for an irreducible case presented in [**Bo3**]. It is based on the finiteness property of some family of vector bundles as well as the proof mentioned above.

The following technical lemma is a simplified version of Lemma 1 from [**Bo6**] (see also [**Bo3**]).

LEMMA 2. *Suppose that the matrix $W(x)$ of size $(l, p-l)$ is meromorphic, and the matrix $V(x)$ of size $(p-l, p-l)$ is holomorphically invertible in a neighborhood O of the point 0. Then there exists a matrix function $\Gamma(x)$, meromorphic on the whole Riemann sphere and holomorphically invertible everywhere except the point 0, such that*

$$(25) \qquad \Gamma(x)\begin{pmatrix} W(x) \\ V(x) \end{pmatrix} = \begin{pmatrix} W'(x) \\ V(x) \end{pmatrix},$$

where the matrix $W'(x)$ is holomorphic in O and $W'(0) = 0$.

PROOF. If the matrix $W(x)$ has a pole of order m at zero, then it can be presented as follows (in O):

$$(26) \qquad x^m W(x) = Q^1(x) + x^{m+1} H^1(x),$$

where $Q^1(x)$ is a matrix polynomial of degree m and the matrix $H^1(x)$ is holomorphic at zero. Decompose $V(x)$ in a similar way:

$$V(x) = Q^2(x) + x^{m+1} H^2(x).$$

Then $\det Q^2(0) \neq 0$ by the assumption of the lemma. Thus there exists a polynomial $R(x)$ in x of degree m such that

$$(27) \qquad R(x) Q^2(x) = -Q^1(x) + x^{m+1} H^3(x)$$

with holomorphic $H^3(x)$. Indeed, let $Q^i(x) = \sum_{j=0}^{m} Q_j^i x^j$, $i = 1, 2$, then $R(x) = \sum_{j=0}^{m} R_j x^j$, where R_j is uniquely determined by the lower triangular system

$$\begin{cases} R_0 Q_0^2 = -Q_0^1, \\ \quad \ldots\ldots\ldots\ldots\ldots \\ \sum_{j=0}^{m} R_j Q_{m-j}^2 = -Q_m^1 \end{cases}$$

with $\det Q_0^2 = \det Q^2(0) \neq 0$. Consider the matrix

$$\Gamma(x) = \begin{pmatrix} I^l & x^{-m} R(x) \\ 0 & I^{p-l} \end{pmatrix}.$$

It follows from (26), (27) that (25) holds. The lemma is proved. $\square$

§3. Proof of Theorem 1

Now we have all we need to prove Theorem 1.

PROOF OF THEOREM 1. It follows from the proof of Lemma 1 (formula (18)) that each reducible system can be analytically transformed to the form (7), where each system with coefficient matrix B_i, $i = 1, \ldots, m$ of size s_i defines an irreducible system, and $B(x)$ has a pole at zero, a pole of the order r at ∞, and is holomorphic on the whole Riemann sphere except at these two points. Moreover the system (4) with the matrix $B(x)$ is regular at zero.

Let us analytically transform the system with the matrix $\widehat{B}(x)$ to the Birkhoff standard form with the help of a change Γ_2. Denote the matrix of the obtained system again by $\widehat{B}(x)$. Consider a fundamental matrix $Z_2(x)$ of the system with the factorization

(29) $$Z_2(x) = U_2(x) x^{A_2} x^{E_2},$$

where $U_2(x)$ is holomorphically invertible in $\mathbb{C}$, $A_2 = \operatorname{diag}(\mu_1, \ldots, \mu_s)$ with $\mu_k \in \mathbb{Z}$, $s = s_2 + \cdots + s_m$, E_2 is the block of the matrix E corresponding to the system under consideration, and

(30) $$x^{A_2} E_2 x^{-A_2}$$

is holomorphic at zero. (Each system which is Fuchsian at zero admits such a factorization with $\det U_2(0) \neq 0$, see [Ga], [Le] or [Bo1]. In our case $U(x)$ is even holomorphically invertible in $\mathbb{C}$, since $\widehat{B}(x)$ is holomorphic in $\mathbb{C}$.)

Let $d = \max_{i=1,\ldots,s} \mu_i$. Apply Proposition 1 to the system with matrix $B_1(x)$. Transform it analytically with the help of a change Γ_1 to the form (10) with the fundamental matrix

(31) $$Z_1(x) = U_1(x) x^{A_1} x^{E_1},$$

where $U_1(x)$ is holomorphically invertible in $\mathbb{C}$, A_1 is a matrix of the same form as A_0 from (10) (with $p = s_1$), E_1 is the block of E corresponding to the system with matrix B_1. It follows from the form (9) of the matrix E that E_1 has the same form (9) (with G' replaced by E' and G'' by the smaller block E_1'' of size $s_1 - l$). Thus

(32) $$x^{A_1} E_1 x^{-A_1}$$

is holomorphic at zero.

Transform our system (4), (7) using

$$\Gamma = \begin{pmatrix} \Gamma_1 & 0 \\ 0 & \Gamma_2 \end{pmatrix},$$

denoting the fundamental matrix of the constructed system by $Z(x)$ again. Then by construction

(33) $$Z(x) = U(x) x^A x^E,$$

where

$$U(x) = \begin{pmatrix} U_1(x) & W(x) \\ 0 & U_2(x) \end{pmatrix}, \quad A = \operatorname{diag}(A_1, A_2), \quad E = \operatorname{diag}(E', E''),$$

and the matrix $W(x)$ is meromorphic at zero and holomorphic in $\mathbb{C} \setminus \{0\}$.

Denote by A'' the matrix $\operatorname{diag}(\lambda_{l+1}, \ldots, \lambda_{s_1}, A_2)$. Since

$$\lambda_{l+1} \geqslant \cdots \geqslant \lambda_{s_1} > d \geqslant \mu_i \quad \text{for all } i = 1, \ldots, s$$

and because of (30) we see that the matrix

(34) $$x^{A''} E'' x^{-A''}$$

is holomorphic at zero.

Since $A = \operatorname{diag}(A', A'')$, where $A' = \operatorname{diag}(\lambda_1, \ldots, \lambda_l)$, it follows from (32) and (34) that the matrix

(35) $$x^A E x^{-A}$$

is holomorphic at zero (and on the whole set $\mathbb{C}$).

Apply to the matrix

$$\begin{pmatrix} W(x) \\ U_2(x) \end{pmatrix}$$

the procedure of Lemma 2 (replacing $V(x)$ by $U_2(x)$). Consider the matrix Γ' from this lemma such that

$$\Gamma' \begin{pmatrix} W(x) \\ U_2(x) \end{pmatrix} = \begin{pmatrix} W'(x) \\ U_2(x) \end{pmatrix}$$

with a matrix $W'(x)$ holomorphic in $\mathbb{C}$.

Transform our system (4), (7) using $\Gamma'(x) \Gamma(x)$. As the result, we obtain a system with the factorization (33), the matrix $U(x)$ being replaced by

$$U'(x) = \begin{pmatrix} U_1(x) & W'(x) \\ 0 & U_2(x) \end{pmatrix},$$

which is already holomorphically invertible in $\mathbb{C}$.

Having in mind the holomorphy condition of (35), one can see that the matrix $B(x)$ of the constructed system is equal to

$$B(x) = x \frac{dZ(x)}{dx} Z^{-1} = x \frac{dU'(x)}{dx} (U'(x))^{-1} + U'(x)[A + x^A E x^{-E}](U'(x))^{-1},$$

and thus it determines an entire function in $\mathbb{C}$. Since all transformations were analytic at infinity, we see that $B(x)$ has a pole of the same order there as the original system. So $B(x)$ is a polynomial of degree r. The theorem is proved.

REMARK 2. Of course, the statement of the theorem does not depend on the location of the block $\widehat{B}$. If we denote by $\widehat{B}$ the matrix obtained from B by deleting the last s_m rows and columns and if we replace the phrase "*the size l of the matrix G' does not exceed s_1*" by the following one "*the size l of the matrix G'' does not exceed s_m*", then we obtain a theorem analogous to Theorem 1. In order to prove it, all we need is to replace the condition

$$\lambda_1 > \cdots > \lambda_l, \qquad \lambda_{l+1} > \cdots > \lambda_p > d$$

in (10) by

$$d > \lambda_1 > \cdots > \lambda_l, \qquad \lambda_{l+1} > \cdots > \lambda_p,$$

to replace the relation $d = \max_{i=1,\ldots,s} \mu_i$ by $d = \min_{i=1,\ldots,s} \mu_i$ with $s = s_1 + \cdots + s_{m-1}$, and to replace the inequality $f_p - d > r(p-1)$ in (19) by the following one: $d > f_1 + r(p-1)$.

PROOF OF COROLLARY 1. Under the assumptions of the corollary, the monodromy matrix G is of the form (9), where either the size of G' does not exceed s_1, or the size of G'' does not exceed s_2. Since the matrices B_1, B_2 are the coefficient matrices of the corresponding irreducible systems, we are under the conditions of Theorem 1 or Remark 2 respectively.

PROOF OF COROLLARY 2. It follows by induction with respect to the number m of blocks of B in (7), using Theorem 1 at each step of the induction.

Corollary 3 is a particular case of Corollary 2.

PROOF OF THEOREM 2. If the Poincaré rank r is equal to zero, then the assumption of the theorem means that the monodromy matrix of system (1) (which is Fuchsian at infinity for $r = 0$) is diagonalizable. Thus, in this case the theorem follows from Corollary 3.

Let r be greater than zero. By Remark 1, it is necessary to prove the theorem only for reducible systems. As it was already mentioned at the beginning of the proof of Theorem 1, it follows from the proof of Lemma 1 (formula (18)) that each reducible system can be analytically transformed to the form (4) with the matrix $B(x)$ of the form (7), where each system with the coefficient matrix B_i, $i = 1, \ldots, m$ of size s_i determines an irreducible system, and $B(x)$ has a pole at zero, a pole of order r at ∞, and is holomorphic on the whole Riemann sphere outside of these two points.

Moreover, due to Remark 1 (or to [**Bo7**, Theorem 1]), we can assume that systems with matrices B_i, $i = 1, \ldots, m$ already have the Birkhoff standard forms, the matrices W_{kl} are holomorphic off zero and infinity, with a pole of order r at infinity and a pole of order k at zero.

Transform our system using (3), where

(36) $\quad \Gamma(x) = \mathrm{diag}\,(\underbrace{0,\ldots,0}_{s_1}, \underbrace{-k,\ldots,-k}_{s_2}, \ldots, \underbrace{-k(m-1),\ldots,-k(m-1)}_{s_m}).$

As the result, we obtain a system (4) with a matrix $B(x)$ (we keep the same notation for the coefficient matrix of the new system) of the form (7), with all elements already holomorphic on $\mathbb{C}$. The matrices B_i are still polynomials of degree r, but W_{kl} are polynomials of degree $r + t$, $t = k(m-1)$. Note that all transformations used above did not change the eigenvalues of the matrix C_0 of the original system. Indeed, the analytic transformations could not change them, because C_0 has a multiplier x^r, $r > 0$, while $x(d\Gamma/dx)\Gamma^{-1}$ from (5) is holomorphic at infinity for analytic Γ. Since the transformation (36) changes the elements lying over blocks B_i only (and adds a constant diagonal matrix to B_i), it does not change the eigenvalues of the leading term B^0 of the matrix

$$\begin{pmatrix} B_1 & \cdots & 0 \\ \vdots & \ddots & \vdots \\ 0 & \cdots & B_m \end{pmatrix}.$$

Let us transform each leading term B_i^0 of B_i to the diagonal form by a matrix S_i. Then the transformation (3) with $\Gamma = S = \mathrm{diag}(S_1, \ldots, S_m)$ transforms our system to system (4) with the matrix $B(x)$ of form (7) such that the leading term B^0 of the matrix B' (we use the old notation for these new matrices; in fact they are equal to SB^0S^{-1} and $SB'S^{-1}$ respectively) is equal to

$$(37) \qquad B^0 = \mathrm{diag}(b_1, \ldots, b_p), \qquad b_i \neq b_j, \; i \neq j.$$

Let us show that with help of a meromorphic transformation, which is holomorphically invertible off infinity, the order of poles of matrices W_{kl} (matrices lying above diagonal blocks B_i in (7)) can be decreased to $r - 1$ (without increasing the order of the poles of B_i). This will complete the proof.

Consider an element $w_{kl} = a_{kl}x^{r+t} + o(x^{r+t})$ of the matrix B lying in some W_{ij} and consider the matrix

$$(38) \qquad \Gamma_1 = I + C_{kl},$$

where C_{kl} is the matrix whose elements equal zero except only one element

$$c_{kl} = \frac{a_{kl}}{b_k - b_l} x^t.$$

It is obvious that $\Gamma_1^{-1} = I - C_{kl}$. Under the transformation (3) with $\Gamma = \Gamma_1$ our system is changed to the system (4) with the same matrix B'. Only the matrices W_{ij} are changed under this transformation. Note that the element w'_{kl} of the new matrix B will have a pole of order less than $r + t$ and that this transformation will not change the leading terms of other elements of W_{ij}. Note also that it does not create new terms with poles of order $r + t$ in W_{ij} (all these facts immediately follow from the form of the transformation and from (5)). Therefore, with the help of transformations of form (38), we can decrease the order of the poles of all the matrices W_{ii+1}, $i = 1, \ldots, p - 1$ to $r - 1$, then we can do the same with W_{ii+2}, $i = 1, \ldots, p - 2$, etc. After a finite number of steps, we obtain the required form (4). The theorem is proved.

Note that some results presented in this work can be used for the investigation of normal forms for families of systems of ODE with bifurcation of the poles (see [Ko]).

Acknowledgement. This work was prepared during my stay at the University of Nice in June 1994. I am grateful to all the staff of the Department of Mathematics of the University of Nice for their hospitality.

References

[Ba1] W. Balser, *Analytic transformation to Birkhoff standard form in dimension three*, Funkcial. Ekvac. **33** (1990), no. 1, 59–67.

[Ba2] _____, *Meromorphic transformation to Birkhoff standard form in dimension three*, J. Fac. Sci. Univ. Tokyo Sect. IA Math. **36** (1989), 233–246.

[Bi] G. D. Birkhoff, *A theorem on matrices of analytic functions*, Math. Ann. **74** (1913), 122–133.

[Bo1] A. A. Bolibrukh, *The Riemann–Hilbert problem*, Uspekhi Mat. Nauk **45** (1990), no. 2, 3–47; English transl., Russian Math. Surveys **45** (1990), no. 2, 1–47.

[Bo2] _____, *Construction of a Fuchsian equation from a monodromy representation*, Mat. Zametki **48** (1990), no. 5, 22–34; English transl., Math. Notes **48** (1990), no. 5, 1090–1099.

[Bo3] _____, *On sufficient conditions for the positive solvability of the Riemann–Hilbert Problem*, Mat. Zametki **51** (1992), no. 2, 9–19; English transl., Math. Notes **51** (1992), no. 2, 110–117.

[Bo4] _____, *Fuchsian systems with reducible monodromy and the Riemann–Hilbert Problem*, Nelineynye Operatory v Global'nom Analize, Voronezh Gos. Univ., 1991, pp. 3–19; English transl., Lecture Notes in Math., vol. 1520, pringer-Verlag, Heidelberg and Berlin, 1992, pp. 139–155.

[Bo5] _____, *On analytic transformation to Birkhoff standard form*, preprint MPI/93-64 (1993), Max-Planck-Institut für Mathematik, Bonn.

[Bo6] _____, *Fuchsian systems and holomorphic vector bundles on the Riemann sphere*, Preprint 347 (1994), University of Nice.

[Bo7] _____, *On analytic transformation to Birkhoff standard form*, Dokl. Akad. Nauk Rossii **334** (1994), 553-555; English transl. in Proc. Russian Acad. Sci. **49** (1994).

[Ga] F. R. Gantmacher, *Theory of matrices*, Vol. II, Chelsea, New York, 1959.

[JLP] W. B. Jurkat, D. A. Lutz, and A. Peyerimhoff, *Birkhoff invariants and effective calculations for meromorphic differential equations*, Part I, J. Math. Anal. Appl. **53** (1976), 438–470; Part II, Houston J. Math. **2** (1976), 207–238.

[Ko] V. K. Kostov, *Normal forms of unfoldings of non-fuchsian systems*, C. R. Acad. Sci. Paris Sér. I Math **318** (1994), 623–642.

[Le] A. H. M. Levelt, *Hypergeometric functions*, Nederl. Akad. Wetensch. Proc. Ser. A **64** (1961).

[OSS] G. Okonek, H. Schneider, and H. Spindler, *Vector bundles on complex projective spaces*, Birkhäuser, Boston, MA, 1980.

[Si] Y. Sibuya, *Linear differential equations in the complex domain: problems of analytic continuation*, Amer. Math. Soc., Providence, RI, 1990.

[Tu] H. L. Turritin, *Reduction of ordinary differential equations to the Birkhoff canonical form*, Trans. Amer. Math. Soc. **107** (1963), 485–507.

Translated by THE AUTHOR

STEKLOV MATHEMATICAL INSTITUTE OF RUSSIAN ACADEMY OF SCIENCES UL. VAVILOVA 42, 117966, MOSCOW GSP-1 RUSSIA

E-mail address: bolibr@class.mian.su

NC Solving of a System of Linear Ordinary Differential Equations in Several Unknowns

D. Yu. Grigoriev

In acknowledgement to the teachers and in memory of unforgetable years spent at Boarding School #45

Introduction

We study the solvability of a system of linear ordinary differential equations in several unknowns

$$\sum_j L_{ij} u_j = b_i,$$

where

$$b_i \in \mathbb{Q}(X) \quad \text{and} \quad L_{ij} = \sum_k f_k \frac{d^k}{dx^k}$$

are linear ordinary differential operators with rational coefficients $f_k \in \mathbb{Q}(X)$. We study the solvability of the system in the unknowns u_j in the differential closure of $\mathbb{C}(X)$ (in fact, since we deal with linear operators, this is equivalent to the solvability in the Picard–Vessiot closure of $\mathbb{C}(X)$) (see [K]), in which any (in particular any linear) ordinary differential equation has a solution. In other words, the solvability in the Picard–Vessiot closure means that the system cannot be brought to a contradiction by equivalent transformations over the ring $\mathcal{R} = \mathbb{C}(X)[d/dX]$ of linear differential operators, or more precisely, that the ideal in the ring of differential polynomials in $\{u_j\}$ generated by the differential polynomials $\sum_j \{L_{ij} u_j - b_i\}$, differs from the unit one.

Note that this problem is a special case of the solvability problem (over the differential closure) of a system of nonlinear ordinary differential equations in several unknowns (more generally, the quantifier elimination problem for these systems), for

---

1991 *Mathematics Subject Classification.* Primary 68Q99; Secondary 34A99.
Key words and phrases. Differential equations, solvability problem, NC solving.

©1996, American Mathematical Society

which an algorithm with elementary complexity (more precisely, double-exponential) was designed in [**G87**]. In the present paper we deal with a linear fragment of this general problem and for it describe an algorithm (see the theorem in §4 below) of complexity class NC, i.e., with polynomial time and polylog depth (parallel time); moreover, the algorithm produces a "triangular" basis for the space of solutions of the system.

A problem close to the one under consideration is that of solving a linear system over the ring $\mathcal{R}$ of differential operators, for which an algorithm was designed in [**G91**] (even for the case of the differential operators with coefficients in many variables from $\mathbb{Q}(X_1, \ldots, X_n)$). The problem considered in the present paper is more subtle from the point of view of the allowed transformations of the system, since in the definition of the equivalence we are not allowed to multiply the equations of the system by differential operators, as we could in the case of linear systems over $\mathcal{R}$ (see [**G91**]).

Therefore, we must carry out elementary transformations of the system with a matrix over $\mathcal{R}$ (see §1 below), in order to reduce the matrix to a standard basis form, which is a special case of the differential standard basis [**G, O, C**] for partial differential operators. However, the bounds necessary for our complexity purposes are unknown. Since the ring $\mathcal{R}$ is noncommutative (some of its properties can be found, say, in [**B**]), difficulties arise in estimating the standard basis form of the matrix over $\mathcal{R}$, in contrast to the case of matrices over the (Euclidean) rings of integers or univariate polynomials, because for the latter, one exploits the notion of determinant. But still we are able (see Lemma 4) to bound the size of the quasi-inverse of a matrix over $\mathcal{R}$ (for an invertible matrix a similar bound follows from the bound in [**O**] obtained for the more general situation of nonlinear operators) and define the rank of a matrix over $\mathcal{R}$ (see [**J**], also Lemma 5 below). To replace the notion of determinant, we consider (see §2 below) the order [**R**] of a system of linear differential operators, i.e., of a matrix over $\mathcal{R}$, as the dimension over $\mathbb{C}(X)$ of the quotient of the free $\mathcal{R}$-module over the submodule generated by the rows of the matrix. Despite the fact that the order of a prime ideal in the ring of differential polynomials was studied in [**R, Ko**], the author could not find certain necessary properties for the linear case, and included them in §2; they are applied to get the main complexity result and, hopefully, are of some independent interest. We prove that the order is additive with respect to the product of the square matrices (Lemmas 6, 7). Relying on Lemma 7, on the analog of the Bezout theorem for differential equations [**R, Ko**] (see also Lemma 9 below), and on a bound on the quasi-inverse (see Lemma 4 in §1), we estimate in §3 the size of the standard basis form of the matrix (see Lemma 10). In the concluding §4 we give an algorithm from the NC for constructing the standard basis form of a matrix, using the bounds obtained in §3. This provides the desired algorithm from the class NC for testing the solvability of a system of linear ordinary differential equations and producing a "triangular" basis for the space of solutions of a system (see the theorem at the end of the paper).

Let us also mention that the problem of solving a single linear ordinary differential equation in one unknown leads to the problem of factoring the equation, while for the latter problem an algorithm was proposed in [**G88**]. A slight generalization of this is solving a first-order system of linear ordinary differential equations, and an algorithm for reducing a matrix of this system to block-triangular form was exhibited in [**G90**]. The relationship of first-order linear systems with general linear systems considered in the present paper is discussed in §4 below.

§1. Transformations and the rank of matrices over the ring of linear differential operators

Denote $D = d/dX$, $\mathcal{R} = \mathbb{C}(X)[D]$, and let $\mathcal{F}$ be the Picard–Vessiot closure (see [**K**]), i.e., any linear differential equation

$$L = \left(\sum_{0 \leqslant i \leqslant n} f_i D^i \right) u = 0$$

with coefficients $f_i \in \mathcal{F}$ and the leading coefficient $\ell c(L) = f_n \neq 0$ has n solutions in $\mathcal{F}$ that are linearly independent over $\mathbb{C}$. Furthermore, the subfield of constants of $\mathcal{F}$ (i.e., the elements $c \in \mathcal{F}$ such that $Dc = 0$) coincides with $\mathbb{C}$.

We consider the solvability problem in $\mathcal{F}$ for a system of linear ordinary differential equations in several unknowns

(1) $$\sum_{1 \leqslant j \leqslant s} L_{ij} u_j = b_i, \quad 1 \leqslant i \leqslant k,$$

where $L_{ij} \in \mathbb{Q}(X)[D]$, $b_i \in \mathbb{Q}(X)$ and the solutions $u_1, \ldots, u_s$ should be in $\mathcal{F}^s$. For an operator

$$L = \sum_{0 \leqslant i \leqslant n} f_i D^i \in \mathcal{R} \quad \text{with } \ell c(L) = f_n \neq 0,$$

denote $n = \operatorname{ord} L$ and by $\deg L$ denote $\sum_{0 \leqslant i \leqslant n} \deg_X f_i$. Consider the $k \times s$ matrix $\mathcal{L} = (L_{ij})$ and assume that $\operatorname{ord} \mathcal{L} \leqslant r$, $\deg \mathcal{L} \leqslant d$, $\deg(b_i) \leqslant d$, i.e., $\operatorname{ord} L_{ij} \leqslant r$, $\deg L_{ij} \leqslant d$ for all i, j. Assume also that the bit-size of each (rational) coefficient of L_{ij}, b_i does not exceed M.

Now consider the $k \times s$ matrix $A = (A_{ij})$ with entries $A_{ij} \in \mathcal{R}$, and assume that $\operatorname{ord}(A_{ij}) \leqslant r$. Since the ring $\mathcal{R}$ is left-Euclidean, making elementary transformations over $\mathcal{R}$ with the rows, one can reduce A to the following standard basis form, see [**J**]; this is a special case of a characteristic set considered in [**R**] in the nonlinear case, or of the differential standard basis ([**C, G, O**])

(2) $$Q = \begin{pmatrix} 0 & \cdots & 0\, Q_{1 p_1} & & & & & \\ 0 & \cdots & \cdots & & 0\, Q_{2 p_2} & & & \\ 0 & \cdots & \cdots & & \cdots & 0\, Q_{3 p_3} & * & \\ \vdots & & & & & & \ddots & \\ 0 & \cdots & \cdots & & \cdots & \cdots & & 0\, Q_{\ell, p_\ell} & \cdots \\ & & 0 & & & & & 0 & \end{pmatrix},$$

where $p_1 < \cdots < p_\ell$, all the rows starting with $(\ell + 1)$th vanish. Let us admit also as an elementary transformation the multiplication (from the left) of a row by a nonzero element from $\mathbb{C}(X)$. In other words, there is a $k \times k$ matrix $B = (B_{ij})$ over $\mathcal{R}$ which is the product of elementary matrices and satisfies $BA = Q$. The rows of Q provide a triangular basis of the left $\mathcal{R}$-module $\mathcal{R}^k A \subset \mathcal{R}^s$ generated by the rows of the matrix A.

The next lemma and the corollary can be deduced from the results of [**J**].

LEMMA 1. *A square $k \times k$ matrix A over $\mathcal{R}$ is invertible from the left if and only if A is equal to a product of elementary matrices.*

COROLLARY. *A square matrix is invertible from the left if and only if it is invertible from the right. The left inverse is unique and coincides with the right inverse. Thus, one can speak simply about invertible matrices.*

We say that a $k \times s$ matrix A is *quasi-invertible from the left* if there exists an $s \times k$ matrix G over $\mathcal{R}$ such that

$$GA = \begin{pmatrix} C_1 & \cdots & 0 \\ \vdots & \ddots & \vdots \\ 0 & \cdots & C_s \end{pmatrix}$$

is a diagonal matrix with nonzero diagonal elements $C_1, \ldots, C_s$ (in a similar way one can define quasi-invertibility from the right).

LEMMA 2. *A is quasi-invertible from the left if and only if $\dim_{\mathbb{C}(X)}(\mathcal{R}^s/\mathcal{R}^k A)$ is finite.*

PROOF. If

$$GA = \begin{pmatrix} C_1 & \cdots & 0 \\ \vdots & \ddots & \vdots \\ 0 & \cdots & C_s \end{pmatrix}$$

and $\operatorname{ord} C_1 = r_1, \ldots, \operatorname{ord} C_s = r_s$, then the vectors

$$\Pi(e_i^{(j)}) = \Pi(\underbrace{0, \ldots, 0, D^j, 0, \ldots, 0}_{i}) \in \mathcal{R}^s/\mathcal{R}^k A \quad \text{for } 1 \leqslant i \leqslant s, \ 0 \leqslant j < r_i$$

constitute a generating set over $\mathbb{C}(X)$ of the $\mathcal{R}$-module $\mathcal{R}^s/\mathcal{R}^k A$, where $\Pi : \mathcal{R}^s \to \mathcal{R}^s/\mathcal{R}^k A$ is the natural projection, hence $\dim_{\mathbb{C}(X)}(\mathcal{R}^s/\mathcal{R}^k A) \leqslant r_1 + \cdots + r_s$ (better inequalities are given below in Lemmas 9 and 10).

Let $\dim_{\mathbb{C}(X)}(\mathcal{R}^s/\mathcal{R}^k A) < \infty$. Then one can reduce A by elementary transformations of the rows to the standard basis form (2) and if $\ell < s$, then the vectors of the infinite family $\Pi(e_p^{(0)}), \Pi(e_p^{(1)}), \ldots$, where $1 \leqslant p \leqslant s$ is distinct from $p_1, \ldots, p_\ell$, are independent over $\mathbb{C}(X)$ and we get a contradiction. Therefore, $\ell = s$. One can show that there exists an $s \times k$ matrix G over $\mathcal{R}$ such that

$$GQ = \begin{pmatrix} C_1 & \cdots & 0 \\ \vdots & \ddots & \vdots \\ 0 & \cdots & C_s \end{pmatrix}$$

with nonzero $C_1, \ldots, C_s$. Indeed, multiply the first row of Q by a suitable element $0 \neq \alpha_1 \in \mathcal{R}$ such that $\alpha_1 Q_{12} = \beta_1 Q_{22}$ for a certain $\beta_1 \in \mathcal{R}$ (this is possible since $\mathcal{R}$ is an Ore domain [B]), then from the first row subtract the second one multiplied by β_1, so that the entry with coordinates $(1, 2)$ will vanish. Continuing in a similar way, we can make all the entries in the first row (except the diagonal entry) equal to zero. Then we proceed to the second row and so on. As the result, we obtain a diagonal matrix, which shows that A is quasi-invertible from the left.

Observe that when $\dim_{\mathbb{C}(X)}(\mathcal{R}^s/\mathcal{R}^k A) < \infty$, this dimension coincides with the order of the system $Au = 0$ [R]. In [R], the order was introduced for a prime ideal in the ring of differential polynomials, we use it for a linear ideal generated by the rows of A.

The next lemma was proved in [G91].

LEMMA 3. *A is quasi-invertible from the left iff there does not exist a vector* $0 \neq v \in \mathcal{R}^s$ *such that* $Av = 0$. *For an* $(s-1) \times s$ *matrix A, one can select a vector* v, $0 \neq v \in \mathcal{R}^s$, *so that* $Av = 0$ *and* $\mathrm{ord}(v) \leq (s-1)r + 1$.

PROOF. If A is quasi-invertible from the left and

$$GA = \begin{pmatrix} C_1 & \cdots & 0 \\ \vdots & \ddots & \vdots \\ 0 & \cdots & C_s \end{pmatrix},$$

then $Av \neq 0$ for any $0 \neq v \in \mathcal{R}^s$ since $\mathcal{R}$ has no divisors of zero ([**B**]).

Conversely, let $Av \neq 0$ for any $0 \neq v \in \mathcal{R}^s$. Let us show that in the standard basis form (2), $\ell = s$. Suppose $\ell < s$. Consider the $\mathbb{C}(X)$-space $\mathcal{R}^{s;N}$ of vectors $(\alpha_1, \ldots, \alpha_s) \in \mathcal{R}^s$ for which $\mathrm{ord}(\alpha_1), \ldots, \mathrm{ord}(\alpha_s) < N$. Let $\mathrm{ord}(Q_{ij}) \leq R$ for all i, j. Then the composition of the mapping $Q: v \to Qv$ with the restriction to the first ℓ coordinates (notice that the others are zeros, see (2)) is the map $\bar{Q}: \mathcal{R}^{s;N} \to \mathcal{R}^{\ell;N+R}$. Since $\dim_{\mathbb{C}(X)} \mathcal{R}^{s;N} = sN$, we obtain

$$\dim_{\mathbb{C}(X)} \mathcal{R}^{s;N} > \dim_{\mathbb{C}(X)} \mathcal{R}^{\ell;N+R} \quad \text{for } N = \left[\frac{\ell R}{s-\ell}\right] + 1,$$

and therefore there exists a vector $0 \neq v \in \mathcal{R}^{s;N}$ such that $Qv = 0$, hence $BAv = 0$ and thus $Av = 0$ because B is a product of elementary matrices (cf. Lemma 1). This contradiction with the assumption justifies the equality $\ell = s$. Then one can show that A is quasi-invertible from the left. This proves the first statement of the lemma. For the second statement, we follow the previous proof, but instead of $\bar{Q}$ we consider the mapping $\bar{A}: \mathcal{R}^{s;M} \to \mathcal{R}^{s-1;M+\mathrm{ord}(A)}$ for $M = (s-1)\mathrm{ord}\, A + 1$.

The next lemma was proved in [**G91**].

LEMMA 4. *A square $s \times s$ matrix A is quasi-invertible from the left iff A is quasi-invertible from the right. In this case there exists a G for which*

$$GA = \begin{pmatrix} C_1 & & 0 \\ & \ddots & \\ 0 & & C_s \end{pmatrix} \quad \text{with } \mathrm{ord}(G) \leq (s-1)r + 1.$$

PROOF. Let A be quasi-invertible from the left. Then for an appropriate matrix B, chosen as the product of elementary matrices, we have

$$BA = \begin{pmatrix} Q_{11} & Q_{12} & \\ & \ddots & \ddots \\ 0 & & Q_{ss} \end{pmatrix},$$

where $Q_{11} \cdots Q_{ss} \neq 0$ (see (2) and the proof of Lemma 2). Let us show that for any vector $0 \neq w \in \mathcal{R}^s$ we have $wA \neq 0$ (this would imply that A is quasi-invertible from the right because of Lemma 3). Assume that $0 = wA$. Then $0 = wA = (wB^{-1}Q)$ and we get a contradiction, which justifies that A is quasi-invertible from the right.

In order to prove the necessary bound on G, consider for each $1 \leq j \leq s$ the matrix $A^{(j)}$ obtained from A by deleting its jth column. Lemma 3 shows that there exists a vector $0 \neq g^{(j)} \in \mathcal{R}^s$ such that $g^{(j)} A^{(j)} = 0$ and $\mathrm{ord}\, g^{(j)} \leq (s-1)r + 1$. For the matrix G, take a matrix with jth row equal to $g^{(j)}$.

Note that when A is invertible, Lemma 4 follows from Theorem 6 in [**O**], where a similar bound was proved in the much more general situation of an invertible nonlinear differential map.

Thus, for a square matrix A, we can say that it is quasi-invertible without specifying from the left or from the right. Note (see also [**G91**]) that a square matrix A is quasi-invertible iff its Dieudonné determinant ([**A**]) does not vanish.

Define the rank $\mathrm{rk}(A)$ as the maximal ℓ such that there exists an $\ell \times \ell$ quasi-invertible submatrix of A (cf. [**J**]). The following lemma can be deduced from the results in [**J**].

LEMMA 5. $\mathrm{rk}(A)$ *coincides with*
(a) *ℓ in the standard basis form* (2);
(b) *the maximal number of $\mathcal{R}$-linearly independent columns of A;*
(c) *the maximal number of $\mathcal{R}$-linearly independent rows of A.*

§2. Some properties of the order of a system of linear differential operators

For brevity we adopt the notation $\dim(\mathcal{R}^s/\mathcal{R}^k A) = \dim_{\mathbb{C}(X)}(\mathcal{R}^s/\mathcal{R}^k A)$.

LEMMA 6. *For any $m \times k$ matrix B and any $k \times s$ matrix A over $\mathcal{R}$, we have*

$$\dim(\mathcal{R}^s/\mathcal{R}^m BA) \leqslant \dim(\mathcal{R}^k/\mathcal{R}^m B) + \dim(\mathcal{R}^s/\mathcal{R}^k A).$$

If A is quasi-invertible from the right, then this inequality turns into an equality.

PROOF. Consider the natural projections

$$\Pi_1: \mathcal{R}^s \to \mathcal{R}^s/\mathcal{R}^k A, \quad \Pi_2: \mathcal{R}^k \to \mathcal{R}^k/\mathcal{R}^m B, \quad \Pi_3: \mathcal{R}^s \to \mathcal{R}^s/\mathcal{R}^m BA.$$

Let $u_1, \ldots, u_\gamma \in \mathcal{R}^k$ be such that $\Pi_2(u_1), \ldots, \Pi_2(u_\gamma)$ form a basis of $\mathcal{R}^k/\mathcal{R}^m B$ over $\mathbb{C}(X)$, and $v_1, \ldots, v_\delta \in \mathcal{R}^s$ be such that $\Pi_1(v_1), \ldots, \Pi_1(v_\delta)$ constitute a basis of $\mathcal{R}^s/\mathcal{R}^k A$ over $\mathbb{C}(X)$ (note that γ or δ can be infinite). Let us prove that

$$\Pi_3(v_1), \ldots, \Pi_3(v_\delta), \Pi_3(u_1 A), \ldots, \Pi_3(u_\gamma A)$$

generate $\mathcal{R}^s/\mathcal{R}^m BA$ over $\mathbb{C}(X)$ and constitute a basis when A is quasi-invertible from the right. Indeed, for some

$$f_1, \ldots, f_\delta, g_1, \ldots, g_\gamma \in \mathbb{C}(X) \quad \text{and} \quad (\beta_1, \ldots, \beta_m) \in \mathcal{R}^m$$

let us have

$$f_1 v_1 + \cdots + f_\delta v_\delta + (g_1 u_1 + \cdots + g_\gamma u_\gamma) A = (\beta_1, \ldots, \beta_m) BA;$$

then $f_1 = \cdots = f_\delta = 0$. If A is quasi-invertible from the right, then

$$g_1 u_1 + \cdots + g_\gamma u_\gamma = (\beta_1, \ldots, \beta_m) B$$

by virtue of Lemma 3, hence $g_1 = \cdots = g_\gamma = 0$.

On the other hand, for any vector $w \in \mathcal{R}^s$ there exists $f_1, \ldots, f_\delta \in \mathbb{C}(X)$ and a vector $v \in \mathcal{R}^k$ for which $w = f_1 v_1 + \cdots + f_\delta v_\delta + vA$. Then $v = g_1 u_1 + \cdots + g_\gamma u_\gamma + uB$ for suitable $g_1, \ldots, g_\gamma \in \mathbb{C}(X), u \in \mathcal{R}^m$. Therefore

$$w = f_1 v_1 + \cdots + f_\delta v_\delta + g_1 u_1 A + \cdots + g_\gamma u_\gamma A + uBA,$$

i.e., $\dim(\mathcal{R}^s/\mathcal{R}^m BA) \leqslant \gamma + \delta = \dim(\mathcal{R}^k/\mathcal{R}^m B) + \dim(\mathcal{R}^s/\mathcal{R}^k A)$.

We can reformulate what was proved above by saying that we have the following exact sequence of $\mathbb{C}(X)$-vector spaces

$$\mathcal{R}^k/\mathcal{R}^m B \xrightarrow{\alpha} \mathcal{R}^s/\mathcal{R}^m BA \xrightarrow{\pi} \mathcal{R}^s/\mathcal{R}^k A \to 0,$$

where $\alpha(v + \mathcal{R}^m B) = vA + \mathcal{R}^m BA$ and $\pi(w + \mathcal{R}^m BA) = w + \mathcal{R}^k A$. In the case of a quasi-invertible A, the following sequence is exact:

$$0 \to \mathcal{R}^k/\mathcal{R}^m B \xrightarrow{\alpha} \mathcal{R}^s/\mathcal{R}^m BA \xrightarrow{\pi} \mathcal{R}^s/\mathcal{R}^k A \to 0.$$

LEMMA 7. *If a matrix A is square, then*

$$\dim(\mathcal{R}^s/\mathcal{R}^m BA) = \dim(\mathcal{R}^s/\mathcal{R}^m B) + \dim(\mathcal{R}^s/\mathcal{R}^s A).$$

PROOF. If A is quasi-invertible (see Lemma 4), then we use Lemma 6.
If A is not quasi-invertible, then $\dim(\mathcal{R}^s/\mathcal{R}^m BA) \geq \dim(\mathcal{R}^s/\mathcal{R}^s A) = \infty$.

Note that in the following example

$$\dim\left(\mathcal{R}^2/\mathcal{R}^2 \begin{pmatrix} 1 & 1 \\ 1 & D \end{pmatrix}\right) = 1, \quad \dim\left(\mathcal{R}/\mathcal{R}^2 \begin{pmatrix} 1 \\ 1 \end{pmatrix}\right) = 0,$$

and hence for the product of these matrices we have $\dim\left(\mathcal{R}/\mathcal{R}^2 \begin{pmatrix} 2 \\ 1+D \end{pmatrix}\right) = 0$, so that the inequality in Lemma 6 for rectangular matrices can be strict.

LEMMA 8. (a) *For a triangular $k \times s$ (where $k \geq s$) matrix*

$$C = \begin{pmatrix} C_1 & & * \\ & \ddots & \\ 0 & & C_s \end{pmatrix},$$

we have $\dim(\mathcal{R}^s/\mathcal{R}^k C) = \operatorname{ord} C_1 + \cdots + \operatorname{ord} C_s$, *provided that* $C_1 \cdots C_s \neq 0$.
(b) $\dim(\mathcal{R}^s/\mathcal{R}^k A) < \infty$ iff $\ell = s$ *in the standard basis form* (2). *In this case* $\dim(\mathcal{R}^s/\mathcal{R}^k A) = \operatorname{ord} Q_{11} + \cdots + \operatorname{ord} Q_{ss}$.
(c) *If A is a square matrix, then* $\dim(\mathcal{R}^s/\mathcal{R}^s A) = \dim(\mathcal{R}^s/A\mathcal{R}^s)$, *where in the right side of the equality we regard $\mathcal{R}^s$ as a right $\mathcal{R}$-module.*
(d) *A square matrix A is invertible iff* $\dim(\mathcal{R}^s/\mathcal{R}^s A) = 0$.

PROOF. (a) is obvious.
(b). The first statement can be found in the proof of Lemma 3. The second statement follows from (a) and the equality $\dim(\mathcal{R}^s/\mathcal{R}^k A) = \dim(\mathcal{R}^s/\mathcal{R}^k Q)$.
(c). Because of Lemma 4, both left and right sides of the equality are finite or infinite simultaneously. Assume that they are both finite. Then

$$BA = \begin{pmatrix} Q_{11} & & * \\ & \ddots & \\ 0 & & Q_{ss} \end{pmatrix}$$

(see (2)), where B is the product of elementary matrices and $Q_{11} \cdots Q_{ss} \neq 0$ (see (b)). For any $s \times s$ elementary matrix G, we have $\dim(\mathcal{R}^s/\mathcal{R}^s G) = \dim(\mathcal{R}^s/G\mathcal{R}^s) = 0$, hence by Lemma 7 the same is true for any invertible matrix (cf. Lemma 1), thus $\dim(\mathcal{R}^s/\mathcal{R}^s B) = \dim(\mathcal{R}^s/B\mathcal{R}^s) = 0$. Now (a) implies that for the triangular matrix

$$Q = \begin{pmatrix} Q_{11} & & * \\ & \ddots & \\ 0 & & Q_{ss} \end{pmatrix},$$

we have

$$\dim(\mathcal{R}^s/\mathcal{R}^s Q) = \dim(\mathcal{R}^s/Q\mathcal{R}^s) = \operatorname{ord} Q_{11} + \cdots + \operatorname{ord} Q_{ss},$$

so that (c) follows from Lemma 7.

Finally, (d) follows from (b) and Lemma 1.

The following lemma was proved in [R, p. 135] (see also [Ko]) in a more general form (for the order of a prime ideal in the ring of differential polynomials).

LEMMA 9. *If a $k \times s$ matrix A is quasi-invertible from the left, then*

$$\dim(\mathcal{R}^s/\mathcal{R}^k A) \leqslant \max_i \{\operatorname{ord} a_{i1}\} + \cdots + \max_i \{\operatorname{ord} a_{is}\}.$$

§3. Bounds on the standard basis form of a matrix over the ring of differential operators

In this section we estimate $\operatorname{ord}(Q)$, $\operatorname{ord}(B)$ in the standard basis form (2) using the results on the order from §2.

Take any $s \times s$ permutation matrix P such that $P(p_1) = 1, \ldots, P(p_\ell) = \ell$, then

$$BAP = \begin{pmatrix} Q_{1p_1} & & & \\ & \ddots & & \\ & & Q_{\ell p_\ell} & \cdots \\ 0 & & & 0 \end{pmatrix}.$$

Represent AP in the form $(A_1 A_2)$, where the $k \times \ell$ submatrix A_1 consists of the first ℓ columns of AP. Then, by Lemma 5, $\operatorname{rk} A = \operatorname{rk} A_1 = \ell$. Complete A_1 by $(k - \ell)$ columns of the type $(0, \ldots, 0, 1, 0, \ldots, 0)^T$ to a $k \times k$ quasi-invertible matrix $(A_1 A_3)$. Then

$$B(A_1 A_3) = \begin{pmatrix} Q_{1p_1} & & & \\ & \ddots & & \\ & & Q_{\ell p_\ell} & \\ 0 & & & * \end{pmatrix}.$$

Carrying out several elementary transformations with the rows that have indices bigger than ℓ, reduce the matrix at the right to the triangular form

$$B_0(A_1 A_3) = \begin{pmatrix} Q_{1p_1} & & & & * \\ & \ddots & & & \\ & & Q_{\ell p_\ell} & & \\ & & & Q_{\ell+1,\ell+1}^{(0)} & \\ 0 & & & & \ddots \\ & & & & & Q_{kk}^{(0)} \end{pmatrix},$$

here $B = \binom{B_1}{B_2}$, where B_1 is $\ell \times k$ submatrix, $B_0 = \binom{B_1}{B_3}$ and $\dim(\mathcal{R}^k/\mathcal{R}^k B) = \dim(\mathcal{R}^k/\mathcal{R}^k B_0) = 0$ (see Lemma 8 (d)).

Moreover, after elementary transformations with the rows, one can assume without loss of generality that

$$\operatorname{ord}(Q_{ip_j}) < \operatorname{ord}(Q_{jp_j}), \qquad \operatorname{ord}(Q_{ij}^{(0)}) < \operatorname{ord}(Q_{jj}^{(0)}) \quad \text{for all } i < j.$$

By Lemmas 6, 7, 9,

$$\operatorname{ord}(Q_{1p_1}) + \cdots + \operatorname{ord}(Q_{\ell p_\ell}) + \operatorname{ord}(Q_{\ell+1,\ell+1}^{(0)}) + \cdots + \operatorname{ord}(Q_{kk}^{(0)})$$
$$= \dim(\mathcal{R}^k/\mathcal{R}^k A_1 A_3) \leq \max_i\{\operatorname{ord} A_{ip_1}\} + \cdots + \max_i\{\operatorname{ord} A_{ip_\ell}\} \leq \ell r,$$

hence $\operatorname{ord}(Q_{ip_j})$, $\operatorname{ord}(Q_{ij}^{(0)}) \leq \ell r$. By Lemma 4, there exists a $k \times k$ matrix G over $\mathcal{R}$ such that

$$(A_1 A_3) G = \begin{pmatrix} C_1 & & 0 \\ & \ddots & \\ 0 & & C_k \end{pmatrix},$$

where $C_1 \cdots C_k \neq 0$ and $\operatorname{ord}(G) \leq (k-1)r + 1$, hence $\operatorname{ord}(C_i) \leq kr + 1$. But

$$B_0 \begin{pmatrix} C_1 & & 0 \\ & \ddots & \\ 0 & & C_k \end{pmatrix} = \begin{pmatrix} Q_{1p_1} & & * \\ & \ddots & \\ 0 & & Q_{kk}^{(0)} \end{pmatrix} G,$$

and we conclude that $\operatorname{ord}(B_0) \leq (\ell + k - 1)r + 1$.

Observe that $B_0 A$ has a standard basis form similar to (2) (with the same "diagonal" entries $Q_{1p_1}, \ldots, Q_{\ell p_\ell}$ and perhaps different other entries, because we have achieved the conditions $\operatorname{ord} Q_{ip_j} < \operatorname{ord} Q_{jp_j}$, $\operatorname{ord}(Q_{ij}^{(0)}) < \operatorname{ord}(Q_{jj}^{(0)})$, $i < j$). Thus

$$B_0 A = \begin{pmatrix} 0 & \cdots & 0 Q_{1p_1} & & & * \\ 0 & \cdots & & 0 Q_{2p_2} & & \\ \vdots & & & & \ddots & \\ 0 & \cdots & & \cdots & \cdots & 0 Q_{\ell p_\ell} & 0 \\ & & & & & 0 & \end{pmatrix},$$

since $\operatorname{rk} A = \ell$. Therefore, $\operatorname{ord}(Q) \leq (\ell + k)r + 1$. Let us summarize what was proved in the present section in the following lemma.

LEMMA 10. *There exists an invertible matrix B such that $BA = Q$ has the standard basis form* (2) *and moreover*

$$\operatorname{ord}(B) \leq (s + k - 1)r + 1, \qquad \operatorname{ord}(Q) \leq (s + k)r + 1.$$

§4. NC algorithm for finding the standard basis form of a matrix over the ring of differential operators

Let us design an algorithm that finds the standard basis form of a matrix in NC, i.e., in polynomial time and with polylogarithmic depth (parallel complexity).

Join the unit matrix to the matrix A and denote the resulting $k \times (s+k)$ matrix by $\bar{A} = (AE)$. Obviously $\operatorname{rk} \bar{A} = k$ (see Lemma 5). Therefore, the standard basis form of $\bar{A}$ equals

$$B_1 \bar{A} = \begin{pmatrix} 0 & \cdots & 0\, Q_{1p_1} & & & & & \\ \vdots & & & \ddots & & & & \\ 0 & \cdots & & \cdots & 0\, Q_{\ell p_\ell} & * & & \\ \vdots & & & & & \ddots & & \\ 0 & \cdots & & \cdots & & \cdots & 0\, Q_{kp_k} & \cdots \end{pmatrix} = \bar{Q} \quad (\text{see (2)}),$$

where $\dim(\mathcal{R}^k / \mathcal{R}^k B_1) = 0$ (see Lemma 8).

For each $1 \leqslant m \leqslant s+k$ and $0 \leqslant j \leqslant (s+2k)r+1$, the algorithm tests whether there exists a vector $w = (w_1, \ldots, w_k)$ with $\operatorname{ord}(w) \leqslant (s+2k-1)r+1$ (cf. Lemma 10) such that the vector $w\bar{A} = (\underbrace{0, \ldots, 0}_{m}, v, \ldots)$, where $\operatorname{ord} v = j$ and the leading coefficient is $\ell c(v) = 1$. This condition can be written as a linear system $\mathcal{T}_{m,j}$ over $\mathbb{Q}(X)$ with $k((s+2k-1)r+2)$ unknowns that are the coefficients of $w_1, \ldots, w_k$ in the powers of $1, D, \ldots, D^{(s+2k-1)r+1}$ and with at most $(s+k)((s+2k)r+1)$ equations. Since the entries of this linear system are rational functions from $\mathbb{Q}(X)$ with degrees in X not exceeding d and with the size of rational coefficients at most M, the algorithm can solve $\mathcal{T}_{m,j}$ in time $(Md\,skr)^{0(1)}$ with depth (parallel complexity) $\log^{0(1)}(Md\,skr)$ (see [**M**]).

Since the rows of the matrix $\bar{Q}$ constitute a (triangular) basis of the left $\mathcal{R}$-module $\mathcal{R}^k \bar{A}$, the system $\mathcal{T}_{m,j}$ is solvable only for $m = p_1, \ldots, p_k$. For each of these $m = p_i$, take the minimal j_i for which $\mathcal{T}_{p_i,j}$ is solvable. Lemma 10 implies that $\mathcal{T}_{p_i,j}$ is solvable for $j = \operatorname{ord} Q_{ip_i}$, hence $j_i \leqslant \operatorname{ord} Q_{ip_i}$. Take a solution $W^{(i)} = (w_1^{(i)}, \ldots, w_k^{(i)})$ for $\mathcal{T}_{p_i,j_i}$, and denote by W the $k \times k$ matrix with ith row $W^{(i)}$. Then

$$W\bar{A} = \begin{pmatrix} 0 & \cdots & 0\, \widetilde{Q}_{1p_1} & & \\ \vdots & & & * & \\ 0 & \cdots & & \cdots & 0\, \widetilde{Q}_{kp_k} & \cdots \end{pmatrix} = \widetilde{Q},$$

where $\operatorname{ord} \widetilde{Q}_{i,p_i} = j_i$.

Let us prove that $\dim(\mathcal{R}^k / \mathcal{R}^k W) = 0$. Denote by $\bar{A}_0, \bar{Q}_0, \widetilde{Q}_0$ the $k \times k$ matrices obtained from $\bar{A}, \bar{Q}, \widetilde{Q}$, respectively, by taking the submatrices formed by the columns $p_1, \ldots, p_k$. Then $B_1 \bar{A}_0 = \bar{Q}_0$ and $W \bar{A}_0 = \widetilde{Q}_0$.

Lemmas 7, 8 (a) imply

$$0 = \dim(\mathcal{R}^k / \mathcal{R}^k B_1) = \operatorname{ord} Q_{1p_1} + \cdots + \operatorname{ord} Q_{kp_k} - \dim(\mathcal{R}^{s+k} / \mathcal{R}^k \bar{A})$$
$$\geqslant \operatorname{ord} \widetilde{Q}_{1p_1} + \cdots + \operatorname{ord} \widetilde{Q}_{kp_k} - \dim(\mathcal{R}^{s+k} / \mathcal{R}^k \bar{A}) = \dim(\mathcal{R}^k / \mathcal{R}^k W) \geqslant 0,$$

therefore $\dim(\mathcal{R}^k / \mathcal{R}^k W) = 0$ and moreover $\operatorname{ord} \widetilde{Q}_{ip_i} = \operatorname{ord} Q_{ip_i}$, $1 \leqslant i \leqslant k$. Since WA has the desired standard basis form (see (2)), we get the following lemma.

LEMMA 11. *There is an NC-algorithm running in time* $(Md\,skr)^{0(1)}$ *with depth (parallel complexity)* $\log^{0(1)}(Md\,skr)$, *which produces a* $k \times k$ *matrix* W, *invertible over*

$\mathcal{R}$, such that

$$WA = \begin{pmatrix} 0 & \cdots & 0\,\widetilde{Q}_{1p_1} & & & \\ \vdots & & & & * & \\ 0 & \cdots & & \cdots & 0\,\widetilde{Q}_{\ell p_\ell} & \\ & & 0 & & & 0 \end{pmatrix}$$

has the standard basis form.

Now we get a criterion for the solvability of system (1). Namely, apply Lemma 11 to the $k \times (s+1)$ matrix $A = (L_{ij} b_i)_{1 \leq i \leq k, 1 \leq j \leq s}$, so that the last column is $(b_1, \ldots, b_k)^T$. Then the system (1) has a solution in the field $\mathcal{F}$ iff $p_\ell \leq s$ (in other words, iff $p_\ell \neq s+1$), and the standard basis form provides a "triangular" basis of the space of solutions of (1). Let us summarize what we have obtained above in the following main result of the paper.

THEOREM. *One can test the solvability of system* (1) *of linear differential equations in several unknowns in the Picard–Vessiot closure $\mathcal{F}$ and find a "triangular" basis of the space of solutions of* (1) *in the NC complexity class, namely in time* $(Md\,skr)^{0(1)}$ *and with depth (parallel time)* $\log^{0(1)}(Md\,skr)$.

Observe that the space of solutions of the homogeneous system (1) (i.e., when $b_1 = \cdots = b_k = 0$) has finite dimension (over $\mathbb{C}(X)$) if and only if

$$p_1 = 1, \ldots, p_\ell = \ell \quad \text{and} \quad \ell = s$$

(for a $k \times s$ matrix $A = (L_{ij})$, see above). In this case the standard basis form WA of the system can be rewritten in the usual first-order matrix form $DY = HY$ (cf. [**G90**]), where the vector Y has coordinates

$$u_1, Du_1, \ldots, D^{j_1-1}u_1, u_2, \ldots, D^{j_2-1}u_2, \ldots, u_s, \ldots, D^{j_s-1}u_s$$

and $j_i = \operatorname{ord} \widetilde{Q}_{i,p_i}$, $1 \leq i \leq s$, and one can easily get the matrix H over $\mathbb{Q}(X)$ from the matrix WA.

Acknowledgments. The author would like to thank Mike Singer for his attention to this paper.

References

[A] E. Artin, *Geometric algebra*, Interscience, New York and London, 1957.

[B] J.-E. Björk, *Rings of differential operators*, North-Holland, Amsterdam, 1979.

[C] Gll. Carro-Ferro, *Groebner bases and differential ideals*, Lecture Notes Comput. Sci., vol. 356, Springer-Verlag, Berlin and Heidelberg, 1987, pp. 129–140.

[G] A. Galligo, *Some algorithmic questions on ideals of differential operators*, Lecture Notes Comput. Sci., vol. 204, Springer-Verlag, Berlin and Heidelberg, 1985, pp. 413–421.

[G87] D. Grigoriev, *Complexity of quantifier elimination in the theory of ordinary differential equations*, Lecture Notes Comput. Sci., vol. 378, Springer-Verlag, Berlin and Heidelberg, 1989, pp. 11–25.

[G88] _____, *Complexity of factoring and GCD calculating of ordinary linear differential operators*, J. Symbolic Comput. **10** (1990), 7–37.

[G90] _____, *Complexity of irreducibility testing for a system of linear ordinary differential equations*, Proc. Int. Symp. on Symb. Algebr. Comput., ACM (1990), Japan, 225–230.

[G91] _____, *Complexity of solving systems of linear equations over the rings of differential operators*, Proc. Int. Symp. Eff. Meth. in Algebraic Geometry (1990, Italy); Birkhäuser, Boston, 1991, pp. 195–202.

[J] N. Jacobson, *Pseudo-linear transformations*, Ann. Math. **38** (1937), 484–507.

[K] I. Kaplansky, *An introduction to differential algebra*, Hermann, Paris, 1957.

[Ko] E. Kolchin, *Differential algebra and algebraic groups*, Academic Press, New York, 1973.

[M] K. Mulmuley, *A fast parallel algorithm to compute the rank of a matrix over an arbitrary field*, Proc. 18 STOC ACM (1986), 338–339.

[O] F. Ollivier, *Standard bases of differential ideals*, Lecture Notes Comput. Sci., vol. 508, Springer-Verlag, Berlin and Heidelberg, 1991, pp. 304–321.

[R] J. F. Ritt, *Differential algebra*, Amer. Math. Soc. Colloq. Publ., vol. 33, Amer. Math. Soc., NY, 1950.

Translated by THE AUTHOR

DEPARTMENTS OF COMPUTER SCIENCE AND MATHEMATICS PENNSYLVANIA STATE UNIVERSITY, UNIVERSITY PARK, PA 16802 USA, ON LEAVE FROM POMI, FONTANKA 27, ST. PETERSBURG 191011, RUSSIA
E-mail address: dima@math.psu.edu

Hyperreflexivity of Contractions Close to Isometries

V. V. Kapustin

Dedicated to my classmates in School #45

Introduction

In this paper a criterion for the hyperreflexivity of contractions T whose defect index $d_T = \operatorname{rank}(I - T^*T)^{1/2}$ is finite is given in terms of factorizations of the characteristic function of T.

The problem of hyperreflexivity, i.e., the reflexivity of the commutant for contractions of class C_0 was studied in [1] (see also [2, Theorem 4.1.25]). There it was reduced to the case of a scalar inner characteristic function, for which a criterion is obtained in [3]. For weak contractions, the problem reduces to that for C_0-contractions, e.g. see [3, Theorem 2.4]. For contractions whose spectrum fills the unit disc, it is noted in [2, Exercise 4.1.10], that the direct sum $S \oplus M_\theta$ of the unilateral shift and a contraction with scalar inner characteristic function θ is hyperreflexive if and only if M_θ is hyperreflexive.

For a family of operators $\mathcal{A}$, $\operatorname{Lat}\mathcal{A}$ denotes the lattice of subspaces that are invariant under all operators from $\mathcal{A}$; $\operatorname{AlgLat}\mathcal{A} \stackrel{\text{def}}{=} \{A : \operatorname{Lat}\mathcal{A} \subset \operatorname{Lat}\{A\}\}$. It is easily seen that $\mathcal{A} \subset \operatorname{AlgLat}\mathcal{A}$. The family $\mathcal{A}$ is said to be *reflexive* if $\mathcal{A} = \operatorname{AlgLat}\mathcal{A}$. If $\mathcal{A} = \{T\}' \stackrel{\text{def}}{=} \{A : AT = TA\}$, where T is a bounded operator, then $\operatorname{Lat}' T \stackrel{\text{def}}{=} \operatorname{Lat}\{T\}'$ is the lattice of *hyperinvariant* subspaces of T; $\operatorname{AlgLat}' T \stackrel{\text{def}}{=} \operatorname{AlgLat}\{T\}'$. T is called *hyperreflexive* if $\{T\}'$ is a reflexive algebra, i.e., if $\operatorname{AlgLat}' T = \{T\}'$.

By $\mathbb{D}$ and $\mathbb{T}$ we denote the unit disc $\{z : |z| < 1\}$ and the unit circle $\{z : |z| = 1\}$ respectively; m is the Lebesgue measure on $\mathbb{T}$; H^2, H^∞ denote the Hardy spaces; S denotes the operator of multiplication by z in H^2. If θ is an inner function (i.e., $\theta \in H^\infty$, $|\theta| = 1$ a.e. on $\mathbb{T}$), then $K_\theta \stackrel{\text{def}}{=} H^2 \ominus \theta H^2$, P_θ denotes the orthoprojection onto K_θ, M_θ is the operator in K_θ defined by $M_\theta h = P_\theta z h$, $h \in K_\theta$.

T always denotes a contraction on a separable Hilbert space H; $\sigma(T)$ denotes the spectrum of T. The contraction T is called *completely nonunitary* (*c.n.u.*) if it has no invariant subspaces on which it acts as a unitary operator. The defect index of T is defined by $d_T \stackrel{\text{def}}{=} \operatorname{rank}(I - T^*T)^{1/2}$. If $(I - T^*T)^{1/2}$ is of Hilbert–Schmidt class (in

---

1991 *Mathematics Subject Classification.* Primary 47B99.
This research was partially supported by the ISF-grant No. RX64000.

particular, if $d_T < \infty$) and $\sigma(T) \neq \operatorname{clos} \mathbb{D}$, then T is called a *weak* contraction. Some (mainly well-known) results on the structure of contractions are collected in §1. The main result of §1 (Theorem 1.9) is necessary to formulate our main theorem.

Let θ be an inner function. Then there exists a sequence (λ_k), $\sum(1 - |\lambda_k|) < \infty$, a finite singular measure μ on $\mathbb{T}$, and a complex number α, $|\alpha| = 1$, such that

$$(1) \qquad \theta(z) = \alpha \prod \left(\frac{|\lambda_k|}{\lambda_k} \frac{\lambda_k - z}{1 - \bar{\lambda}_k z} \right) \exp \left(\int \frac{z + \xi}{z - \xi} d\mu(\xi) \right), \qquad z \in \mathbb{D}.$$

A set E, $E \subset \mathbb{T}$, is called a *Beurling–Carleson set* if E is closed, $mE = 0$, and $\sum l_k \log(2\pi/l_k) < \infty$, where l_k are the lengths of the complementary arcs of E.

Our main result is the following theorem (for exact definitions see below).

MAIN THEOREM. *Let T be a c.n.u. contraction with $d_T < \infty$, and Θ be its characteristic function. Let $\Theta = \Theta_{10}\Theta_{00}\Theta_{11}\Theta_{01}$ be the canonical factorization of Θ, let θ be the minimal (with respect to the ordering of divisibility in H^∞) inner scalar multiple of Θ_{00}, and let μ be the singular measure defined in* (1).

If at least two of the factors Θ_{10}, Θ_{11}, and Θ_{01} are unitary constants, then T is hyperreflexive if and only if all zeros of θ in $\mathbb{D}$ are simple and $\mu E = 0$ for every Beurling–Carleson set $E \subset \mathbb{T}$.

Otherwise, T is not hyperreflexive.

This theorem seems to remain true for contractions T whose defect operator $D_T = (I - T^*T)^{1/2}$ is of Hilbert–Schmidt class. We discuss this extension in §4.

§1. Preliminary results on the structure of contractions

Let T be a c.n.u. contraction. Define

$$D = (I - T^*T)^{1/2}, \qquad \mathcal{D} = \operatorname{clos} \operatorname{range} D,$$
$$D_* = (I - TT^*)^{1/2}, \qquad \mathcal{D}_* = \operatorname{clos} \operatorname{range} D_*.$$

The function Θ, whose values are operators from $\mathcal{D}$ into $\mathcal{D}_*$,

$$\Theta(\lambda) \stackrel{\text{def}}{=} -T + \lambda D_*(I - \lambda T^*)^{-1} D,$$

is called the *characteristic function* of T. If $\lambda \in \mathbb{D}$, then $\|\Theta(\lambda)\| \leq 1$. Thus, Θ belongs to the space of bounded operator-valued functions $H^\infty(\mathcal{D} \to \mathcal{D}_*)$, and

$$\|\Theta\|_\infty \stackrel{\text{def}}{=} \|\Theta\|_{H^\infty(\mathcal{D} \to \mathcal{D}_*)} \leq 1.$$

Any function $\varkappa_*^* \Theta \varkappa$, where $\varkappa \colon E \to \mathcal{D}$ and $\varkappa_* \colon E_* \to \mathcal{D}_*$ are (constant) unitary operators, is also called a characteristic function of T.

Let E, E_* be Hilbert spaces, $\Theta \in H^\infty(E \to E_*)$. The boundary values of Θ, as well as the operator $f \mapsto \Theta f$, are also denoted by Θ.

The function Θ is called *inner*, if $\Theta^*\Theta \equiv I$; *outer*, if $\operatorname{clos} \Theta H^2(E) = H^2(E_*)$; *$*$-inner*, if $\widetilde{\Theta}$, $\widetilde{\Theta}(z) = [\Theta(\bar{z})]^*$, is inner; *$*$-outer*, if $\widetilde{\Theta}$ is outer.

We write $T \in C_{0*}$, if $T^n h \to 0$ for each $h \in H$; $T \in C_{1*}$, if $T^n h \to 0$ implies $h = 0$; $T \in C_{*0}$, if $T^* \in C_{0*}$; $T \in C_{*1}$, if $T^* \in C_{1*}$. If $\alpha, \beta \in \{0, 1\}$, then $T \in C_{\alpha\beta}$ means that $T \in C_{\alpha*}$ and $T \in C_{*\beta}$.

We write $T \in C_{0*}$ iff its characteristic function Θ is $*$-inner; $T \in C_{*0}$ iff Θ is inner; $T \in C_{1*}$ iff Θ is $*$-outer; $T \in C_{*1}$ iff Θ is outer.

Consider the invariant subspace $H_0 = \{h : T^n H \to 0\}$. Let $H_1 = H \ominus H_0$, and let $T = \begin{pmatrix} T_0 & * \\ 0 & T_1 \end{pmatrix}$ be the matrix of T corresponding to the decomposition $H = H_0 \oplus H_1$. This representation is called the *canonical triangulation* of type $\begin{pmatrix} C_{0*} & * \\ 0 & C_{1*} \end{pmatrix}$. Such a construction for T^* gives the canonical triangulation of type $\begin{pmatrix} C_{*1} & * \\ 0 & C_{*0} \end{pmatrix}$; it corresponds to the inner-outer factorization of Θ. A triangulation of type $\begin{pmatrix} C_{0*} & * \\ 0 & C_{1*} \end{pmatrix}$ corresponds to the factorization of Θ into a product of a $*$-outer and a $*$-inner (operator-valued) function. In general, T admits a triangulation of type

$$\begin{pmatrix} C_{01} & * & * & * & * \\ 0 & C_{00} & * & * & * \\ 0 & 0 & C_{11} & * & * \\ 0 & 0 & 0 & C_{00} & * \\ 0 & 0 & 0 & 0 & C_{10} \end{pmatrix}.$$

For c.n.u. contractions, the H^∞-calculus is defined. The contraction T is said to belong to the class C_0 if there exists a function φ, $\varphi \in H^\infty$, $\varphi \not\equiv 0$, such that $\varphi(T) = 0$. The simplest operator of class C_0 is the operator M_Θ. If $T \in C_0$, then $T \in C_{00}$. All functions φ such that $\varphi(T) = 0$ form an ideal $m_T H^\infty$, where m_T is an inner function called the *minimal annihilator* of T. For $\varphi \in H^\infty$, $\varphi(T) = 0$ if and only if φ is a *scalar multiple* of the characteristic function Θ of T, i.e., there exists a function Ω, $\Omega \in H^\infty(E_* \to E)$, such that $\Omega\Theta = \varphi I$, $\Theta\Omega = \varphi I$. So, the minimal annihilator of T is the "minimal" scalar multiple of its characteristic function.

LEMMA 1.1 [4, pp. 71–72]. *Let $T \in C_{1*}$. Then there exists an isometry $W : K \to K$ and an injection $X : H \to K$ with dense range such that $WX = XT$.*

Such an isometry can be chosen in a canonical way, as in the proof of this fact. It will be called the *canonical isometry* of T. The canonical isometry of an arbitrary contraction T is defined as that of T_1 determined by the canonical triangulation

$$T = \begin{pmatrix} T_0 & * \\ 0 & T_1 \end{pmatrix} \text{ of type } \begin{pmatrix} C_{0*} & * \\ 0 & C_{1*} \end{pmatrix}.$$

COROLLARY 1.2. *If T is a contraction, $T \notin C_{*0}$, then there exist a coisometric operator $V : K \to K$, $K \neq \{0\}$, and an injection $Z : K \to H$ such that $ZV = TZ$.*

PROOF. Consider the canonical triangulation

$$T = \begin{pmatrix} T_1 & * \\ 0 & T_0 \end{pmatrix} \text{ of type } \begin{pmatrix} C_{*1} & * \\ 0 & C_{*0} \end{pmatrix}.$$

Since $T \notin C_{*0}$, the space where T_1 acts is nonzero. Let W, X be the operators constructed in Lemma 1.1 for the C_{1*}-contraction T_1^*. The operators $V = W^*$ and $Z : K \to H$, $Zk \stackrel{\text{def}}{=} X^*k$ have all the desired properties. □

For more detailed information about general contractions, see [4]. As our reference to the theory of contractions of class C_0, we use the book [2].

In [5] the following result is proved.

THEOREM 1.3. *If T is a contraction of class C_{10} with $d_T < \infty$, then the canonical isometry of T is a unilateral shift, i.e., it is unitarily equivalent to the operator $S_{\mathcal{E}}$ of multiplication by z in $H^2(\mathcal{E})$, where $\mathcal{E}$ is an auxiliary Hilbert space.*

We shall use the following corollaries of this result.

COROLLARY 1.4. *Every contraction T of class C_{10} with $d_T < \infty$ is hyperreflexive.*

PROOF. The subspaces $H_\lambda \stackrel{\text{def}}{=} \ker(T^* - \lambda I)$, $\lambda \in \mathbb{D}$, are hyperinvariant under T, and hence they are invariant under A^* if $A \in \operatorname{AlgLat}' T$. Since the eigenvectors of $S_{\mathcal{E}}^*$ form a complete family, it follows from Lemma 1.1 and Theorem 1.3 that $\operatorname{span}_{\lambda \in \mathbb{D}} H_\lambda = H$. Hence $A^* \in \{T\}^*$ and T is hyperreflexive. $\square$

COROLLARY 1.5. *If there exists a triangulation $T = \begin{pmatrix} T_0 & * \\ 0 & T_1 \end{pmatrix}$, where T_1 is a contraction of class C_{10} with $d_{T_1} < \infty$ acting in a nonzero space, then there exists a nonzero operator $X \colon H \to H^2$ such that $SX = XT$.*

PROOF. An operator intertwining T and the shift operator in $H^2(\mathcal{E})$ can be constructed as the superposition of the orthoprojection onto the space where T_1 acts and of the operator defined in Lemma 1.1, taking into account Theorem 1.3. Since the shift operator in $\mathcal{E}$ is unitarily equivalent to $S \oplus S \oplus \cdots \oplus S$, by projecting onto any direct summand we obtain an operator X intertwining T and the operator of simple unilateral shift. $\square$

The following three well-known lemmas give information on the structure of contractions with finite defect index.

LEMMA 1.6. *Let T be a contraction with $d_T < \infty$.*
1. *If the point spectrum of T^* does not fill $\mathbb{D}$, then $d_{T^*} < \infty$.*
2. *If $T = \begin{pmatrix} T_0 & N \\ 0 & T_1 \end{pmatrix}$ is a triangulation of T and $\ker T_0^* = \{0\}$, then $d_{T_1} < \infty$.*

PROOF. (1) If $\ker T^* = \{0\}$, then
$$d_{T^*} = \operatorname{rank} D_{T^*} = \operatorname{rank} T^* D_{T^*} = \operatorname{rank} D_T T^* \leqslant \operatorname{rank} D_T < \infty.$$

If α is a point of the unit disc which is not an eigenvalue of T, then
$$T_\alpha = (T - \alpha I)(I - \bar{\alpha} T)^{-1}$$
is a contraction whose defect index d_{T_α} is finite, and 0 is not an eigenvalue of it. Hence $d_{T_\alpha^*}$ is finite, and this is equivalent to the fact that $d_{T^*} < \infty$.

(2) Obviously, $d_{T_0} < \infty$; according to (1), we have $d_{T_0^*} < \infty$. Since the operator
$$I - T^*T = \begin{pmatrix} I & 0 \\ 0 & I \end{pmatrix} - \begin{pmatrix} T_0^* & 0 \\ N^* & T_1^* \end{pmatrix} \begin{pmatrix} T_0 & N \\ 0 & T_1 \end{pmatrix} = \begin{pmatrix} I - T_0^* T_0 & -T_0^* N \\ -N^* T_0 & I - T_1^* T_1 - N^* N \end{pmatrix}$$
is of finite rank, the operators $T_0^* N$ and $I - T_1^* T_1 - N^* N$ are of finite rank, too. Since $d_T < \infty$, it follows from (1) that T_0^* is a finite rank perturbation of an isometry W; hence WN, $N = W^*WN$, and $I - T_1^* T_1$ are of finite rank. $\square$

LEMMA 1.7 [6, Lemma 3.2]. *Let T be a contraction of class C_{1*} with $d_T < \infty$ and $T = \begin{pmatrix} T_1 & * \\ 0 & T_0 \end{pmatrix}$ be its canonical triangulation of type $\begin{pmatrix} C_{*1} & * \\ 0 & C_{*0} \end{pmatrix}$. Then $T_1 \in C_{11}$.*

LEMMA 1.8. *If T is a contraction of class C_{00} with $d_T < \infty$, then $T \in C_0$.*

THEOREM 1.9 (on the structure of contractions; cf. [7, Lemma 3.2]). *Let T be a c.n.u. contraction, $d_T < \infty$. Then there exists a triangulation*

$$T = \begin{pmatrix} T_{01} & * & * & * \\ 0 & T_{11} & * & * \\ 0 & 0 & T_{00} & * \\ 0 & 0 & 0 & T_{10} \end{pmatrix} \text{ of type } \begin{pmatrix} C_{01} & * & * & * \\ 0 & C_{11} & * & * \\ 0 & 0 & C_{00} & * \\ 0 & 0 & 0 & C_{10} \end{pmatrix}.$$

The defect indices $d_{T_{01}}$, $d_{T_{11}}$, $d_{T_{00}}$, $d_{T_{10}}$ are finite. The characteristic function Θ of T admits a factorization corresponding to the triangulation: $\Theta = \Theta_{10}\Theta_{00}\Theta_{11}\Theta_{01}$, where Θ_{10} is inner, $*$-outer; Θ_{00} is inner, $*$-inner; Θ_{11} is outer, $*$-outer; Θ_{01} is outer, $*$-inner.

T_{00} is of class C_0, and Θ_{00} admits an inner scalar multiple.

PROOF. Consider the canonical triangulation of T,

$$T = \begin{pmatrix} T_1 & * \\ 0 & T_0 \end{pmatrix} \text{ of type } \begin{pmatrix} C_{*1} & * \\ 0 & C_{*0} \end{pmatrix}.$$

For T_1 and T_2 consider the canonical triangulations

$$T_1 = \begin{pmatrix} T_{01} & * \\ 0 & T_{11} \end{pmatrix} \text{ and } T_0 = \begin{pmatrix} T_{00} & * \\ 0 & T_{10} \end{pmatrix} \text{ of type } \begin{pmatrix} C_{0*} & * \\ 0 & C_{1*} \end{pmatrix}.$$

It follows from Lemma 1.6 that the defect indices $d_{T_{01}}$, $d_{T_{11}}$, $d_{T_{00}}$, $d_{T_{10}}$ are finite. By construction, $T_{01} \in C_{01}$, $T_{00} \in C_{00}$, $T_{10} \in C_{10}$. By Lemma 1.7, $T_{11} \in C_{11}$, and the existence of the matrix representation of T follows. Properties of the factors in the corresponding factorization of the characteristic function are obviously satisfied. □

THEOREM 1.10. *Let T be a contraction of class C_{*1} with $d_T < \infty$ and let Θ be the characteristic function of T. Then Θ admits a right scalar multiple (i.e., there exists a bounded operator-valued analytic function Ω such that $\Theta\Omega = \varphi I$ for a nonzero function $\varphi \in H^\infty$). Moreover, the greatest common divisor of the inner parts of all right scalar multiplies of Θ is a constant function.*

This theorem is a reformulation of the theorem on outer functions of [8, p. 21]. □

LEMMA 1.11. *Let T be a contraction of class C_{*0} with $d_T < \infty$, let*

$$T = \begin{pmatrix} T_0 & * \\ 0 & T_1 \end{pmatrix}$$

be the canonical triangulation of T under decomposition $H = H_0 \oplus H_1$ of type $\begin{pmatrix} C_{0*} & * \\ 0 & C_{1*} \end{pmatrix}$, and let θ be the minimal annihilator of T_0 ($T_0 \in C_0$ by Lemma 1.8). Then

1. clos range $\theta(T) \cap H_0 = \{0\}$.
2. $T_\theta \stackrel{\text{def}}{=} T|_{\text{clos range }\theta(T)} \in C_{10}$.

PROOF. Both statements follow from the fact that there exist a Hilbert space $\mathcal{E}$ and a family of operators $\{Y_\alpha\}_{\alpha \in A}$, $Y_\alpha \colon H \to H^2(\mathcal{E}) \oplus H_0$ such that for every nonzero vector $h \in H$ there is an element of the family for which $Y_\alpha h \neq 0$, and $Y_\alpha T = (S_\mathcal{E} \oplus T_0) Y_\alpha$ is satisfied for all elements of this family, where $S_\mathcal{E}$ denotes the shift operator in $H^2(\mathcal{E})$. Indeed, assume that $h \in \operatorname{clos range} \theta(T) \cap H_0$, $h \neq 0$. Choose Y_α so that $Y_\alpha h \neq 0$. Then

$$Y_\alpha h \in \operatorname{clos range} \theta(S_\mathcal{E} \oplus T_0) \cap Y_\alpha H_0 = \theta H^2(\mathcal{E}) \oplus 0 \cap Y_\alpha H_0.$$

Since $T_0 \in C_{0*}$, it is easy to show that $Y_\alpha H_0 \subset 0 \oplus H_0$. Thus $Y_\alpha h = 0$, and this contradiction proves (1).

To prove (2), note that

$$Y_\alpha^0 T_\theta = S_\mathcal{E} Y_\alpha^0,$$

where Y_α^0 is the superposition of Y_α and of the orthoprojection onto the first component in the direct sum $H^2(\mathcal{E}) \oplus H_0$. Since T_θ is the restriction of a contraction of class C_{1*} onto an invariant subspace, T is of class C_{1*} too. For $h \in H^2(\mathcal{E})$ we have $T_\theta^{*n} Y_\alpha^0 h = Y_\alpha^0 (S_\mathcal{E}^*)^n h \to 0$ because of $S_\mathcal{E} \in C_{*0}$. It follows from the properties of the family $\{Y_\alpha\}$ that the vectors $Y_\alpha^0 h$ form a complete family; hence $T_\alpha \in C_{*0}$, and (2) is proved.

Let us construct the operators Y_α in terms of the function model due to Sz.-Nagy and Foiaş. Let $\Theta \in H^\infty(E \to E_*)$ be the characteristic function of T. Then the space H can be identified with $H^2(E_*) \ominus \Theta H^2(E)$, where T acts as $Th = P_H S_{E_*}|_H$. Let $\Theta = \Theta_1 \Theta_0$ be a factorization of Θ, where $\Theta_0 \in H^\infty(E \to F)$ is $*$-inner and $\Theta_1 \in H^\infty(F \to E_*)$ is $*$-outer. Then

$$H_0 = \Theta_1 H^2(F) \ominus \Theta H^2(E) \quad \text{and} \quad H_1 = H \ominus H_0$$

are the canonical subspaces of T. Note that

$$H_0 = \{h \in H : \Delta_* h = 0\}, \quad \text{where } \Delta_* \stackrel{\text{def}}{=} (I - \Theta \Theta^*)^{1/2}.$$

Let

$$T = \begin{pmatrix} T_0 & * \\ 0 & T_1 \end{pmatrix}$$

be the corresponding triangulation. If $\Omega \in H^\infty(E_* \to F)$, $\varphi \in H^\infty$, and $\Omega \Theta_1 = \varphi$, then define

$Y \colon H \to K \oplus H_0$ by $Yh \stackrel{\text{def}}{=} P_H \Theta_1 \Omega h \oplus \Delta_* h$, where $K \stackrel{\text{def}}{=} \operatorname{clos}(\Delta_* H)$;

note that the canonical isometry of T acts on K. It is easy to see that the first component does belong to H_0, and that Y intertwines T and the direct sum of the operator of multiplication by z in K and T_0.

Let us define

$$Z \colon K \oplus H_0 \to H, \quad Z(k \oplus h) \stackrel{\text{def}}{=} k + P_H(\varphi - \Theta_1 \Omega) \Delta_* h, \quad k \in K, \ h \in H_0.$$

Then for $h \in H$,

$$ZYh = P_H \Theta_1 \Omega h + P_H(\varphi - \Theta_1 \Omega) \Delta_*^2 h = P_H(\varphi - \varphi \Theta_1 \Theta_1^* + \Theta_1 \Omega \Theta_1 \Theta_1^*) h$$
$$= P_H(\varphi - \varphi \Theta_1 \Theta_1^* + \Theta_1 \varphi \Theta_1^*) h = P_H \varphi h = \varphi(T) h.$$

It follows from Lemma 1.6 that T_1 is an almost isometric contraction of class C_{10}. By Theorem 1.3, the operator of multiplication by z in K is unitarily equivalent to a unilateral shift $S_\mathcal{E}$, and we shall identify these unitarily equivalent operators. By

Lemma 1.10, there exist families $\{\Omega_\alpha\}|_{\alpha \in A} \subset H^\infty(E_* \to F)$ and $\varphi_\alpha|_{\alpha \in A} \subset H^\infty$ such that $\Omega_\alpha \Theta_1 = \varphi_\alpha$, and the greatest common divisor of the inner parts of the functions φ_α is a constant function. Construct the corresponding operators Y_α and Z_α as described above. Take $h \in H$. If $Y_\alpha h = 0$ for all α, then $\varphi_\alpha(T)h = Z_\alpha Y_\alpha h = 0$ for all α too. Since $\{\varphi : \varphi(T)h = 0\} = \omega H^\infty$ for some inner function ω, it follows from properties of functions φ_α that ω is a constant. Hence $h = 1(T)h = 0$, and the proof is complete. $\square$

§2. Case $\sigma(T) = \operatorname{clos} \mathbb{D}$

In this section we consider a contraction T for which there exists a nonzero operator $X: H \to H^2$ such that $XT = SX$. This implies that $\sigma(T) = \operatorname{clos} \mathbb{D}$. Without loss of generality, we can assume that $\operatorname{clos} \operatorname{range} X = H^2$. Indeed, $\operatorname{clos} \operatorname{range} X$ is a nonzero subspace invariant under S. It is of the form θH^2, where θ is an inner function. Taking the operator $h \mapsto \bar{\theta} X h$ instead of X, we obtain an operator with a dense range.

THEOREM 2.1. *Let T be a c.n.u. contraction, $T \notin C_{*0}$. Let there exist a nonzero operator $X: H \to H^2$ such that $XT = SX$. Then T is not hyperreflexive.*

PROOF. As noted above, we can assume that $\operatorname{clos} \operatorname{range} X = H^2$. Since $T \notin C_{*0}$, according to Corollary 1.2, there exists a coisometric operator $V: K \to K$ and an operator $Z: K \to H$ such that $ZV = TZ$ and $\ker Z = \{0\}$. There exists a vector $k \in K$ for which $Vk \neq 0$. Define a rank one operator A in H, $Ah \stackrel{\text{def}}{=} (Xh)(0) Zk$. We shall show that $A \in \operatorname{AlgLat}' T$, but $A \notin \{T\}'$.

Let $h \in H$, $E_h \stackrel{\text{def}}{=} \operatorname{clos}\{Bh : B \in \{T\}'\}$. We need to show that $Ah \in E_h$. If $Xh = 0$, then $Ah = 0$; let $Xh \neq 0$. Denote by ω the inner factor of the function Xh. There exists a family (Y_α) of operators from H^2 into H such that $Y_\alpha S = TY_\alpha$ and $\operatorname{span}_\alpha \operatorname{range} Y_\alpha = H$. Indeed, such a family exists for the isometric dilation of T, which is an absolutely continuous unitary operator; applying the orthoprojection onto H, we get such a family for T (cf. e.g. [9, Lemma 2.2]). Since $Y_\alpha XT^n \in \{T\}'$, we have

$$\omega(T)H = \omega(T)\operatorname{span}_\alpha Y_\alpha H^2 \subset \operatorname{span}_\alpha \omega(T) Y_\alpha H^2 = \operatorname{span}_\alpha Y_\alpha \omega(S) H^2$$
$$= \operatorname{span}_\alpha Y_\alpha(\omega H^2) = \operatorname{span}_\alpha (Y_\alpha XT^n h) \subset E_h,$$

and it is sufficient to show that $Zk \in \operatorname{range} \omega(T)$. Indeed, since V is a coisometry, $\omega(V)$ is also a coisometry and $\omega(V)K = K$; it follows that $Zk \in Z\omega(V)K = \omega(T)YK \subset \operatorname{range} \omega(T)$. Thus, $Ah \in E_h$ and $A \in \operatorname{AlgLat}' T$.

Let us take $h \in H$ such that $(Xh)(0) \neq 0$. Since $(XTh)(0) = (SXh)(0) = 0$, we get $ATh = 0$. But $TAh = (Xh)(0)TYk = (Xh)(0)ZVk \neq 0$, because of the choice of k and the property $\ker Z = \{0\}$. This means that $AT \neq TA$ and T is not hyperreflexive. $\square$

THEOREM 2.2. *Let T be a contraction of class C_{*0} with $d_T < \infty$; consider its canonical triangulation*

$$T = \begin{pmatrix} T_0 & * \\ 0 & T_1 \end{pmatrix} \quad \text{of type} \quad \begin{pmatrix} C_{00} & * \\ 0 & C_{10} \end{pmatrix}.$$

Then T is hyperreflexive if and only if so is T_0.

This theorem remains true if instead of $d_T < \infty$ we only assume that $D_T \stackrel{\text{def}}{=} (I - T^*T)^{1/2}$ is a Hilbert–Schmidt operator; see Remark 4.2.

LEMMA 2.3. *Let $H_0 \in \text{Lat}' T$. Suppose that $T|_{H_0}$ is a hyperreflexive operator. Let $A \in \text{AlgLat}' T$, $A_0 \stackrel{\text{def}}{=} A|_{H_0}$. Then $A_0 \in \{T_0\}'$.*

PROOF. Let $H_* \subset H_0, H_* \in \text{Lat}' T_0$. Then obviously $H_* \in \text{Lat}' T$ and $AH_* \subset H_*$. This means that $A_0 H_* \subset H_*$ and $A_0 \in \text{AlgLat}' T_0 = \{T_0\}'$. □

COROLLARY 2.4. *Let there exist a family of subspaces $(H_\alpha) \subset \text{Lat}' T$ which span the whole space H and are such that $T_\alpha \stackrel{\text{def}}{=} T|_{H_\alpha}$ is hyperreflexive for each α. Then T is hyperreflexive.*

PROOF. Let $A \in \text{AlgLat}' T$. By Lemma 2.3, for every α, $A_\alpha \stackrel{\text{def}}{=} (A|_{H_\alpha})$ commutes with T_α. $A \in \{T\}'$ follows from $\text{span}_\alpha(H_\alpha) = H$. □

PROOF OF THEOREM 2.2. Excluding the trivial case, let $T \notin C_{00}$. By Theorem 1.9, d_{T_0} and d_{T_1} are finite, and $T_0 \in C_0$. Let θ be the minimal annihilator of T_0. Let us assume that T_0 is not hyperreflexive. It follows from the theory of the Jordan model for C_0-contractions (see for example [2, Chapter 3]) that there exists a T-invariant subspace H_* such that the restriction $T|_{H_*}$ is quasisimilar to M_θ, and an operator $Y: K_\theta \to H$ such that $YM_\theta = TY$, clos range $Y = H_*$. Because the operators T_0 and $T|_{H_*}$ have the same minimal annihilator θ, it follows from [2, Theorem 4.1.25] that $T|_{H_0}$ is not hyperreflexive. Let

$$N \in \text{AlgLat}'(T|_{H_*}) \setminus \{T|_{H_*}\}',$$
$$A: H \to H: \quad h \mapsto Ah \stackrel{\text{def}}{=} NYP_\theta Xh, \quad h \in H,$$

where $X: H \to H^2$, $XT = SX$, clos range $X = H^2$ (such an X exists by Corollary 1.5).

Let us show that $A \in \text{AlgLat}' T$. If $Xh = 0$, then $Ah = 0$. Let us assume that $Xh \neq 0$, and let ω be the minimal common inner divisor of the functions Xh and θ. Then

$$Y(\omega H^2 \ominus \theta H^2) \subset \text{span}_{n \geq 0} YP_\theta S^n Xh \subset E_h \stackrel{\text{def}}{=} \text{clos}\{Bh : B \in \{T\}'\},$$

because $YP_\theta S^n X \in \{T\}'$ for all $n \geq 0$. Hence,

$$Ah = NYP_\theta Xh \in NY(\omega H^2 \ominus \theta H^2) \subset N(H_* \cap E_h).$$

The subspace $H_* \cap E_h$ is an invariant subspace of $T|_{H_*}$. Since the multiplicity of $T|_{H_*}$ equals 1, this subspace is hyperinvariant (see for example [2, Corollary 3.2.14]), and therefore $N(H_* \cap E_h) \subset H_* \cap E_h$. Thus, $Ah \in H_* \cap E_h \subset E_h$ and $A \in \text{AlgLat}' T$.

Since clos range $(YP_\theta X) = H_*$ and $N \notin \{T|_{H_*}\}'$, there exists an element $h \in H$ for which $TNYP_\theta Xh \neq NTYP_\theta Xh$. Taking into account the fact that $TYP_\theta X = YP_\theta XT$, we get $TAh \neq ATh$. Hence, $A \notin \{T\}'$ and T is not hyperreflexive.

Conversely, let us assume that T_0 is hyperreflexive. Let H_0 denote the canonical subspace $\{h : T^n h \to 0\}$ in which T_0 acts and set $H_1 \stackrel{\text{def}}{=} \text{clos range}\,\theta(T)$. The subspaces

$$H_0' \stackrel{\text{def}}{=} H \ominus H_0 \quad \text{and} \quad H_1' \stackrel{\text{def}}{=} H \ominus H_1 = \ker(\theta(T))^*$$

are hyperinvariant under T^*. If h is orthogonal to both of them, then $h = 0$ by Lemma 1.11. Thus, span$(H_0', H_1') = H$. Obviously, $T_1' \stackrel{\text{def}}{=} (T^*|_{H_1'})^* \in C_0$ and the minimal annihilator of T_1' is θ. Operators T_0 and T_1' are of class C_0 and have the same minimal annihilator θ. Since T_0 is hyperreflexive, it follows from Theorem 4.1.25 of [2] that T_1' is hyperreflexive too. By Lemmas 1.6 and 1.11, $d_{T_0'} < \infty$ and $T_0' \in C_{10}$; it is hyperreflexive by Corollary 1.4. The adjoint operators $(T_0')^*$ and $(T_1')^*$ are also hyperreflexive, and applying Corollary 2.4 to the operator T^*, we see that T^* is a hyperreflexive operator. Hence, T is hyperreflexive. □

In the proof of Theorem 2.2 we essentially used the fact that there exists a nonzero operator X such that $SX = XT$. Probably, the conditions on T in the theorem can be weakened.

CONJECTURE 2.5. *Let T be a contraction of class C_{*0} with canonical triangulation*

$$T = \begin{pmatrix} T_0 & * \\ 0 & T_1 \end{pmatrix} \quad \text{of type} \quad \begin{pmatrix} C_{00} & * \\ 0 & C_{10} \end{pmatrix}.$$

Suppose that there exists an operator $X \colon H \to H^2$ such that $XT = SX$. Then T is hyperreflexive if and only if T_0 is.

In general, there are contractions of class C_{10} for which there exist no nonzero operators X such that $SX = XT$. So, by Theorem 2 of [10], there exist invertible contractions T of class C_{10}; then $SX = XT$ easily implies $X = 0$. This fact leads to the following more general conjecture.

CONJECTURE 2.6. *The statement of Theorem 2.2 is true for all contractions of class C_{*0}.*

§3. Proof of the Main Theorem

Let there be at least two factors among $\Theta_{01}, \Theta_{11}, \Theta_{10}$ that are not unitary constants. Then at least one of Θ_{01}, Θ_{10} is not a unitary constant. Let it be Θ_{10}. Then by Corollary 1.5 there exists a nonzero operator $X \colon H \to H^2$ such that $SX = XT$. Since at least one of Θ_{01}, Θ_{11} is not a unitary constant, Θ is not inner and $T \notin C_{*0}$. It follows from Theorem 2.1 that T is not hyperreflexive.

In the case in which Θ_{10} is a unitary constant but Θ_{01} is not, the point spectrum of T^* does not fill $\mathbb{D}$ and by Lemma 1.6, $d_{T^*} < \infty$. Similarly, in this case there exists a nonzero operator X such that $SX = XT^*$ (cf. [11, Lemma 3]). Hence, T is not hyperreflexive by Theorem 2.1 applied to T^*.

Now assume that that one of $\Theta_{01}, \Theta_{11}, \Theta_{10}$ is not a unitary constant. Denote by T_0 the contraction with characteristic function Θ_{00} that appears in the canonical triangulation of T. Let us show that in this case T is hyperreflexive if and only if T_0 is. If Θ_{10} is not a unitary constant, then $T \in C_{*0}$ and the statement for this case is contained in Theorem 2.2. If Θ_{10} is a unitary constant but Θ_{01} is not, then by Lemma 1.6 T^* is of class C_{*0}, $d_{T^*} < \infty$, and the required statement follows from Theorem 2.2 (applied to T^*). If Θ_{11} is not a unitary constant, then T is a weak contraction, and Theorem 2.4 of [3] leads to the statement for this case.

For C_0-contractions, a reduction to the case of scalar inner characteristic function can be found in [2, Theorem 4.1.25]. For contractions with scalar inner characteristic function, the properties of hyperreflexivity and reflexivity coincide, and a criterion for them is given by Theorem 3.1 of [3].

Let us remark that if all the zeros of θ are simple and for the associated measure μ of the singular factor of θ, we have $\mu E = 0$ for every Beurling–Carleson set $E \subset \mathbb{T}$, then Theorem 3.1 of [3] asserts that idempotents of the algebra $H^\infty/\theta H^\infty$ form a complete family in this algebra in the $*$-weak topology. If the class of a function $\varphi \in H^\infty$ in $H^\infty/\theta H^\infty$ is an idempotent of this algebra, then the operator $\varphi(T)$ is a projection. If $A \in \operatorname{AlgLat}' T$, then A commutes with all projections of the form $\varphi(T)$, and hence with all operators of the form $\varphi(T)$, $\varphi \in H^\infty$. In particular, A commutes with T, i.e., $A \in \{T\}'$. Thus, we have a direct proof of the hyperreflexivity of general C_0-contractions whose minimal function satisfies the properties mentioned above. □

§4. Concluding remarks

REMARK 4.1. The formulation of the Main Theorem can be modified to cover contractions with a unitary direct summand.

Namely, for every contraction $T\colon H \to H$ there exists a representation of the form $T = U_s \oplus U_a \oplus T'$, where U_s is a singular unitary operator, U_a is an absolutely continuous unitary operator, and T' is completely nonunitary. *The nontriviality of the singular unitary direct summand does not affect the hyperreflexivity of T.* To prove this assertion, we shall show that the commutant $\{T\}' = \{U_s \oplus (U_a \oplus T')\}'$ splits, i.e.,

$$\{U_s \oplus (U_a \oplus T')\}' = \{U_s\}' \oplus \{U_a \oplus T'\}'.$$

Then

$$\operatorname{AlgLat}' T = \operatorname{AlgLat}'(U_s \oplus (U_a \oplus T')) = \operatorname{AlgLat}' U_s \oplus \operatorname{AlgLat}(U_a \oplus T')$$

by [12, Lemma 1.2], and thus T is hyperreflexive if and only if both U_s and $U_a \oplus T'$ are. But U_s is hyperreflexive as a unitary operator, and hence T and $U_a \oplus T'$ are hyperreflexive simultaneously. Now let us show that the commutant of T splits. Let $U_*\colon H_* \to H_*$ be the minimal unitary dilation of $U_a \oplus T'$ (which is an absolutely continuous unitary operator). Then $U_s \oplus U_*$ is a dilation of T. If $A \in \{T\}'$, then by the lifting theorem there exists an operator $\hat{A} \in \{U_s \oplus U_*\}'$ which is a dilation of A. The commutant of any unitary operator consists of operators which are decomposable with respect to the spectral measure; hence $\{U_s \oplus U_*\}' = \{U_s\}' \oplus \{U_*\}'$. Therefore,

$$\hat{A} \in \{U_s\}' \oplus \{U_*\}' \quad \text{and} \quad A \in \{U_s\}' \oplus \{U_a \oplus T'\}'.$$

This means that $\{T\}' = \{U_s\}' \oplus \{U_a \oplus T'\}'$.

For an absolutely continuous unitary operator, there exists a similar c.n.u. contraction of class C_{11}, (cf. [13, Lemma 2]). Hence, *the nontriviality of the absolutely continuous unitary part gives the same effect as the condition that Θ_{11} is not a unitary constant.*

REMARK 4.2. Our Main Theorem seems to be true for contractions T whose defect operator $D_T \stackrel{\text{def}}{=} (I - T^*T)^{1/2}$ belongs to the Hilbert–Schmidt class (such contractions will be called *almost isometric*).

The analogs of Theorem 1.3 and Lemma 1.8 are proved in [14] and [15] respectively. The proof of the analog of Lemma 1.6 can be obtained from the proof of Lemma 1.6 by replacing properties "to have finite defect index" and "to be of finite rank" by the properties "to be an almost isometric contraction" and "to be of trace class" respectively. The extension of Lemma 1.7 follows from Lemmas 2 and 3 of [16]. This allows us to prove the statement of Theorem 1.9 for almost isometric contractions in the same way. Hence, the statement of Theorem 2.2 is true for almost isometric contractions too. Only the extension of Theorem 1.10 is missing. It was conjectured in [**8,** Concluding Remarks to Lecture 1], but unfortunately we are not ready to present a proof here. If it were really true, then Lemma 1.11 and our Main Theorem could be extended with the same proofs.

References

1. H. Bercovici, C. Foiaş, and B. Sz.-Nagy, *Reflexive and hyperreflexive operators of class C_0*, Acta Sci. Math. (Szeged) **43** (1981), 5–13.
2. H. Bercovici, *Operator theory and arithmetic in H^∞*, Math. Surveys and Monographs, vol. 26, Amer. Math. Soc., Providence, RI, 1988.
3. V. V. Kapustin, *Reflexivity of operators: general methods and a criterion for almost isometric contractions*, Algebra i Analiz **4** (1992), no. 2, 141–160; English transl., St. Petersburg Math. J. **4** (1993), 319–335.
4. B. Sz.-Nagy and C. Foiaş, *Harmonic analysis of operators on Hilbert space*, North-Holland, Amsterdam, 1970.
5. B. Sz.-Nagy, *Diagonalization of matrices over H^∞*, Acta Sci. Math. (Szeged) **38** (1976), 223–238.
6. P. Y. Wu, *Approximate decompositions of certain contractions*, Acta Sci. Math. (Szeged) **44** (1982), 137–149.
7. _____, *Toward a characterization of reflexive contractions*, J. Operator Theory **13** (1985), 73–86.
8. N. K. Nikol′skiĭ, *Treatise on the shift operator*, Springer-Verlag, Berlin and Heidelberg, 1986.
9. P. Y. Wu, *Hyponormal operators quasisimilar to an isometry*, Trans. Amer. Math. Soc. **291** (1985), 229–239.
10. L. Kerchy, *On the spectra of contractions belonging to special classes*, J. Funct. Anal. **67** (1986), 153–166.
11. K. Takahashi, *Contractions with the bicommutant property*, Proc. Amer. Math. Soc. **93** (1985), 91–95.
12. J. B. Conway and P. Y. Wu, *The splitting of $\mathfrak{A}(T_1 \oplus T_2)$ and related questions*, Indiana Univ. Math. J. **26** (1977), 41–56.
13. L. Kerchy, *On invariant subspace lattices of C_{11}-contractions*, Acta Sci. Math. (Szeged) **43** (1981), 281–293.
14. M. Uchiyama, *Contractions and unilateral shifts*, Acta Sci. Math. (Szeged) **46** (1983), 345–356.
15. K. Takahashi and M. Uchiyama, *Every C_{00}-contraction with Hilbert–Schmidt defect operator is of class C_0*, J. Operator Theory **10** (1983), 331–335.
16. K. Takahashi, *C_1-contractions with Hilbert–Schmidt defect operators*, J. Operator Theory **12** (1984), 331–347.

Translated by THE AUTHOR

POMI, FONTANKA 27, ST. PETERSBURG 191011, RUSSIA
E-mail address: kapustin@lomi.spb.su

On the Definition of the 2-Category of 2-Knots

V. M. Kharlamov and V. G. Turaev

Dedicated to our mathematics teachers Yury Ionin and Kira Muranova

§1. Introduction

The aim of this paper is to define a 2-category of 2-knots in 4-dimensional Euclidean space. The categorical approach to knots and tangles in $\mathbb{R}^3$ introduced in [**Tu, Ye**] plays an important role in the construction of 3-dimensional topological quantum field theories (TQFT's) based on the theory of quantum groups (see [**RT1, RT2**]). The category of tangles consists of objects that are finite subsets of $\mathbb{R}$ considered up to isotopy in $\mathbb{R}$ and morphisms which are isotopy classes of tangles in $\mathbb{R}^2 \times [0, 1]$. Each tangle has several (≥ 0) bottom endpoints lying in $\mathbb{R} = \mathbb{R} \times 0 \times 0$ and several top endpoints lying in $\mathbb{R} = \mathbb{R} \times 0 \times 1$. Such a tangle is regarded as a morphism from the set of its bottom endpoints to the set of its top endpoints. For instance, links in $\mathbb{R}^3$ are just endomorphisms of the empty subset of $\mathbb{R}$. The composition of morphisms is defined by attaching one tangle on top of another one and compressing the result into $\mathbb{R}^2 \times [0, 1]$. This category of tangles admits a number of useful modifications. For instance, one may consider oriented tangles, framed tangles, colored tangles, etc. It is important that the category of tangles admits a purely algebraic description (in terms of generators and relations or in terms of universal properties). It is this fact which allows us to use the category of tangles in 3-dimensional TQFT's.

To extend these ideas to surfaces in $\mathbb{R}^4$, it is natural to involve the notion of 2-category. A 2-category is a category endowed with so-called 2-morphisms. More precisely, for any two (ordinary) morphisms $f \colon X \to Y, g \colon X \to Y$ we have a class of 2-morphisms "acting" from f into g. The 2-morphisms are subject to two composition operations $\circ$ and $\star$. The composition $\star$ transforms a pair of 2-morphisms $f \Rightarrow g$ and $g \Rightarrow h$ into a 2-morphism $f \Rightarrow h$. The composition $\circ$ transforms a pair of 2-morphisms $f \Rightarrow g$ and $f' \Rightarrow g'$ with $\text{source}(f') = \text{source}(g') = \text{target}(f) = \text{target}(g)$ into a 2-morphism $f'f \Rightarrow g'g$. For more details see §2.

The 2-categories seem to be adequate for an algebraic description of surfaces in $\mathbb{R}^4$. The idea is to use as 2-morphisms the isotopy classes of surfaces in $\mathbb{R}^2 \times [0, 1] \times [0, 1]$ interpolating between a tangle in $\mathbb{R}^2 \times [0, 1] \times 0$ and a tangle in $\mathbb{R}^2 \times [0, 1] \times 1$. The

---

1991 *Mathematics Subject Classification.* Primary 18D05, 57M99.

compositions ∘ and ⋆ of such 2-morphisms are obtained by attaching one such surface on top of another one along the third or the fourth coordinate respectively.

Two-categories of surfaces in $\mathbb{R}^4$ were first considered by J. E. Fisher [**F**]. Despite the apparent simplicity of the basic idea, this approach meets with an important difficulty. Namely, to define the composition ⋆ one must glue a surface in $\mathbb{R}^2 \times [0, 1] \times [0, 1]$ to a surface in $\mathbb{R}^2 \times [0, 1] \times [1, 2]$ along two isotopic tangles in $\mathbb{R}^2 \times [0, 1] \times 1$. To perform this gluing one has to fix an isotopy between the tangles in question. In general, different isotopies give rise to topologically different results, which leads to the absence of a well-defined composition ⋆. The problem is due to the fact that the space of tangles isotopic to a given one may have a nontrivial fundamental group. Note that a similar problem does not come up in the lower dimension because for any $n \geqslant 0$ the space of n-point subsets of $\mathbb{R}$ is contractible.

To overcome this problem, we must change the definition of 1-morphisms. We use as 1-morphisms the diagrams of tangles in $\mathbb{R} \times [0, 1]$ considered up to a certain equivalence relation. We look for a relation such that the space of diagrams equivalent to a given one is simply connected. There are different equivalence relations satisfying this condition. For instance, the identity relation (geometric coincidence of diagrams) obviously satisfies this condition. However, the resulting 2-categories are too large and cannot be described in algebraic terms. Thus, our aim is to find an equivalence relation satisfying the condition above but leading to a 2-category that is not excessively large, possibly admitting a purely algebraic description. Here we propose such an equivalence relation. For each diagram we consider the set A (resp. B) of points of local maximum (resp. minimum) of the projection $\mathbb{R} \times [0, 1] \to [0, 1]$ restricted to the diagram. Both these sets are ordered in accordance with the values of the projection. We consider the equivalence relation generated by ambient isotopies of diagrams in $\mathbb{R} \times [0, 1]$ preserving the order both in A and B. In other words, we consider those isotopies that never exchange the levels of two points of local maximum or two points of local minimum. Our main result asserts that the space of diagrams equivalent to the given one is simply connected. This leads to a 2-category of surfaces in $\mathbb{R}^4$ which seems to be suitable for an algebraic study. Similar to the setting of knots in $\mathbb{R}^3$, this 2-category admits various modifications involving oriented, framed, and colored tangles and surfaces.

The results of this paper and related results were reported by the first author at the 56th meeting of physicists and mathematicians in Strasbourg (RCP-25) in May 1993. This talk also included a discussion of the relationships between our work and the tetrahedron equations (see [**Za, M–S, Kh**], [**F, K–V, K–S**]). We do not discuss the tetrahedron equations here.

We thank D. Bennequin for a helpful discussion concerning the proof of Theorem 3.4.

§2. Two-categories

2.1. The notion of 2-category. The categories we are interested in are strictly associative and have strict units. That is why some authors call them *strict 2-categories*, see for example [**KV**]. We omit the adjective "strict" and call them 2-*categories*.

Let us recall the definition.

DEFINITION. A 2-category $\mathcal{A}$ is a collection of:
(a) three sets
$$\mathcal{A}_0, \mathcal{A}_1, \mathcal{A}_2$$
whose elements are called respectively *objects*, 1-*morphisms*, 2-*morphisms*;

(b) four maps
$$s_0, t_0 \colon \mathcal{A}_1 \to \mathcal{A}_0, \qquad s_1, t_1 \colon \mathcal{A}_2 \to \mathcal{A}_1;$$
s_0, s_1 and t_0, t_1 are called respectively *source* and *target maps*;

(c) two maps
$$I_0 \colon \mathcal{A}_0 \to \mathcal{A}_1, \qquad I_1 \colon \mathcal{A}_1 \to \mathcal{A}_2;$$
their images $I_0(A), A \in \mathcal{A}_0$ and $I_1(u), u \in \mathcal{A}_1$ are called *identity morphisms* and denoted respectively by 1_A and 1_u;

(d) a composition operation for 1-morphisms
$$(u, v) \mapsto u \circ v$$
defined when $t_0(v) = s_0(u)$;

(e) two composition operations for 2-morphisms
$$(F, G) \mapsto F \circ G, \qquad (F, G) \mapsto F \star G;$$
the first is defined when $t_0 s_1(G) = t_0 t_1(G) = s_0 s_1(F) = s_0 t_1(F)$, and the second, when $t_1(G) = s_1(F)$.

It is required that
(1) $t_0 s_1 = t_0 t_1$, $s_0 t_1 = s_0 s_1$;
(2) $s_0(1_A) = t_0(1_A) = A$, $s_1(1_u) = t_1(1_u) = u$;
(3) the composition $\circ$ of 1-morphisms is associative and the relations $1_B \circ u = u = u \circ 1_A$ are satisfied for any 1-morphism u with $s_0(u) = A$, $t_0(u) = B$ (i.e., in other words, $\mathcal{A}_0, \mathcal{A}_1, s_0, t_0, \circ$ form a category (called the *underlying category of* $\mathcal{A}$) with 2-sided identities $1_A, A \in \mathcal{A}_0$);
(4) the composition $\star$ of 2-morphisms is associative and the relations $1_v \star F = F = F \star 1_u$ are satisfied for any 2-morphism F with $s_1(F) = u$, $t_1(F) = v$ (i.e., $\mathcal{A}_1, \mathcal{A}_2, s_1, t_1, \star$ form a category with 2-sided identities $1_u, u \in \mathcal{A}_1$);
(5) the composition $\circ$ of 2-morphisms is associative and the relations $1_v \circ F = F = F \circ 1_u$ are verified for $u = 1_A, v = 1_B$ and any 2-morphism F with $s_0 t_1(F) = A$, $t_0 s_1(F) = B$ (i.e., $\mathcal{A}_0, \mathcal{A}_2, s_0 t_1, t_0 s_1, \circ$ form a category with 2-sided identities $1_u, u = 1_A, A \in \mathcal{A}_0$);
(6) for any 1-morphisms u, v such that $t_0(v) = s_0(u)$ we have $1_{u \circ v} = 1_u \circ 1_v$;
(7) for any 2-morphisms F, G such that $t_0 s_1(G) = s_0 t_1(F)$ we have
$$F \circ G = (F \circ 1_{t_1(G)}) \star (1_{s_1(F)} \circ G) = (1_{t_1(F)} \circ G) \star (F \circ 1_{s_1(G)}).$$

We use the symbols $u \colon A \to B$ and $A \xrightarrow{u} B$ to denote a 1-morphism $u \in \mathcal{A}_1$ with $A = s_0(u)$ and $B = t_0(u)$. A 2-morphism F with $s_1(F) = u$, $t_1(F) = v$ is denoted by the symbol $F \colon u \Rightarrow v$ and the pictures

$$u \left(\begin{array}{c} A \\ \Downarrow F \\ B \end{array} \right) v \qquad \text{and} \qquad \begin{array}{ccc} A & = & A \\ u \downarrow & \Downarrow F & \downarrow v \\ B & = & B \end{array}$$

where $s_0(u) = A$, $t_0(v) = B$. The figures

$$\left(\begin{array}{c} A \\ G \Downarrow \, \Downarrow F \\ B \end{array} \right) \qquad \text{and} \qquad \begin{array}{ccccc} A & = & A & = & A \\ \downarrow & \Downarrow G & \downarrow & \Downarrow F & \downarrow \\ B & = & B & = & A \end{array}$$

serve to depict $F \star G$ and the figures

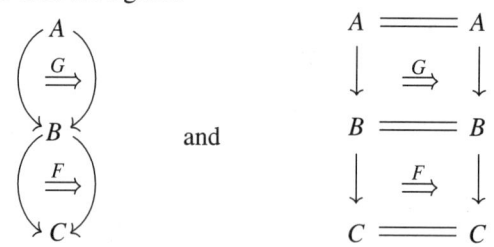

and

serve to depict $F \circ G$.

Note finally that in the situation indicated in the following pictures

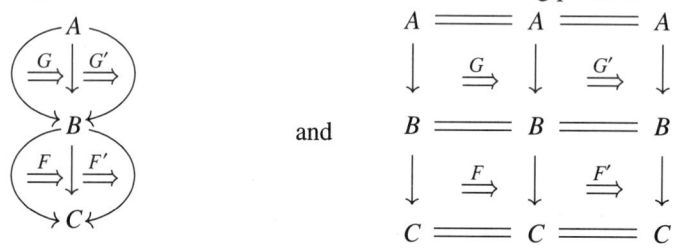

and

we have $(F' \star F) \circ (G' \star G) = (G \circ F) \star (G' \circ F')$, by the definition of 2-category. Thus we can give a well-defined meaning to compositions represented by this and similar plane cellular and polygonal decompositions.

2.2. Examples. We sketch several elementary examples. For more examples and a detailed discussion, see [**KV**].

(A) In the basic example coming from the theory of categories, $\mathcal{A}_0$, $\mathcal{A}_1$, $\mathcal{A}_2$ consist of all categories,[1] all functors between categories, and all natural transformations of functors respectively. Here, the composition $\star$ of two natural transformations $T: \varphi \to \psi$, $S: \psi \to \eta$ of functors φ, ψ and η from $\mathcal{L}$ to $\mathcal{M}$ ($\mathcal{L}, \mathcal{M} \in \mathcal{A}_0$) is their usual composition defined by the formula

$$(S \star T)_L = S_L \circ T_L,$$

where

$$\{T_L: \varphi(L) \to \psi(L) \mid L \in Ob\mathcal{L}\}, \qquad \{S_L: \psi(L) \to \eta(L) \mid L \in Ob\mathcal{L}\},$$

and

$$\{(S \star T)_L: \varphi(L) \to \eta(L) \mid L \in Ob\mathcal{L}\}$$

are families of 1-morphisms which present the functors T, S, and $S \star T$ respectively. To define the composition $\circ$ of two natural transformations

$$U = \{U_L: \varphi(L) \to \psi(L) \mid L \in Ob\mathcal{L}\}: \varphi \to \psi,$$
$$T = \{T_M: \varphi'(M) \to \psi'(M) \mid M \in Ob\mathcal{M}\}: \varphi' \to \psi',$$

where φ, ψ are functors from $\mathcal{L}$ to $\mathcal{M}$ and φ', ψ' are functors from $\mathcal{M}$ to $\mathcal{N}$ ($\mathcal{L}, \mathcal{M}, \mathcal{N} \in \mathcal{A}_0$), we set

$$(T \circ U)_L = \psi'(U_L) \circ T_{\varphi(L)}.$$

---

[1] The difficulty which is due to the fact that they do not constitute a set can be resolved in several well-known ways.

(B) Another example comes from homotopy theory. Considering points of a topological space X as objects, continuous paths in X as 1-morphisms, and homotopies of paths as 2-morphisms, we obtain a 2-category. Two types of composition of 2-morphisms correspond to two types of decomposition of a rectangle in two halves: these are the decompositions produced by vertical and horizontal pastings of rectangles. Indeed if, as usual, a path is defined to be a map $\xi \colon I \to X$ and a homotopy to be a map $H \colon I \times I \to X$, then the resulting 2-category will not be strict and will not satisfy the axioms enumerated in 2.1: thus, in particular, the composition of 1-morphisms will not be associative. To obtain a strict 2-category, it is sufficient to identify paths and, correspondingly, homotopies which differ solely by their parametrization.

(C) A simple example, that is closer to our interests, is provided by scanned surfaces in $\mathbb{R}^4$. This is a 2-category $\mathcal{A}$ such that $\mathcal{A}_0$ is formed by finite subsets of $\mathbb{R}^2$, $\mathcal{A}_1$ by one-dimensional proper submanifolds of $\mathbb{R}^2 \times \Delta$, Δ being a closed interval in $\mathbb{R}$, and $\mathcal{A}_2$ by two-dimensional proper submanifolds of $\mathbb{R}^2 \times \Delta \times \Delta'$, Δ' also being a closed interval. We define the composition of 1-morphisms $t \subset \mathbb{R}^2 \times \Delta$ and $t' \subset \mathbb{R}^2 \times \Delta'$ if the end point of Δ is the beginning point of Δ' and if in addition the target of t coincides with the source of t'. Taking similar precautions in the definition of compositions of 2-morphisms, we get a 2-category in the sense of 2.1.

§3. Spaces of tangle-diagrams

3.1. Systems of arcs and loops. In what follows we denote the closed strip $\mathbb{R}^1 \times [0, 1]$ by B.

We call a compact subset L of B a *regular system of arcs and loops* if it satisfies the following conditions: L is a 1-dimensional smooth submanifold of B except at a finite number of points, each exceptional point is a transversal double point and lies in the interior of B, each boundary point of L belongs to ∂B, and L is nowhere tangent to ∂B. Any regular system of arcs and loops can be presented as the image of a proper immersion of a disjoint union of intervals and circles.

For a system of arcs and loops $L \subset B$, we define the *height function* $h \colon L \to [0, 1]$ to be the restriction of the projection $B = \mathbb{R}^1 \times [0, 1] \to [0, 1]$.

A regular system of arcs and loops L is said to be *generic* if the height function $h \colon L \to [0, 1]$ is a Morse function whose critical values are pairwise distinct and none of the critical points is a double point of L. If h is a Morse function everywhere except at a finite number of points which are simple degenerate critical points (i.e., the third derivative of h (with respect to a local coordinate) is nonzero at these points, while the first and the second ones are equal to 0) and if, in addition, no two local maximum values or two local minimum values of h are equal, then the system L is said to be *weakly generic*. Note that, in particular, a weakly generic system may have a local maximum value equal to a local minimum value.

3.2. Tangle-diagrams. A 1-*diagram* (or a tangle-diagram) is a regular system of arcs and loops in B equipped with an overcrossing/undercrossing mark at each double point; the mark serves to distinguish the "upper" and the "lower" branches.[2] A 1-diagram is said to be *generic* if the underlying system of arcs and loops is generic. If the system of arcs and loops is weakly generic, then the diagram is called *weakly generic* too.

---

[2] This definition is motivated by the fact that any 1-diagram can be presented as the projection to B of a disjoint union of intervals and circles in $B \times \mathbb{R}$, cf. 3.4.

3.3. Spaces of diagrams. The set of 1-diagrams provided with the C^∞ topology is an infinite-dimensional topological space. This space has an infinite number of connected components: 1-diagrams belong to the same component if and only if they are isotopic.[3]

Let us denote the space of all 1-diagrams by D, the subspace of generic diagrams by D^0, and the subspace of weakly generic diagrams by D'. Note that isotopic generic (resp. weakly generic) 1-diagrams may belong to different components of D^0 (resp. D'). For example, the three diagrams shown in Figure 1 belong to three different components of D^0 and to two different components of D' (the second and the third diagrams belong to the same component of D').

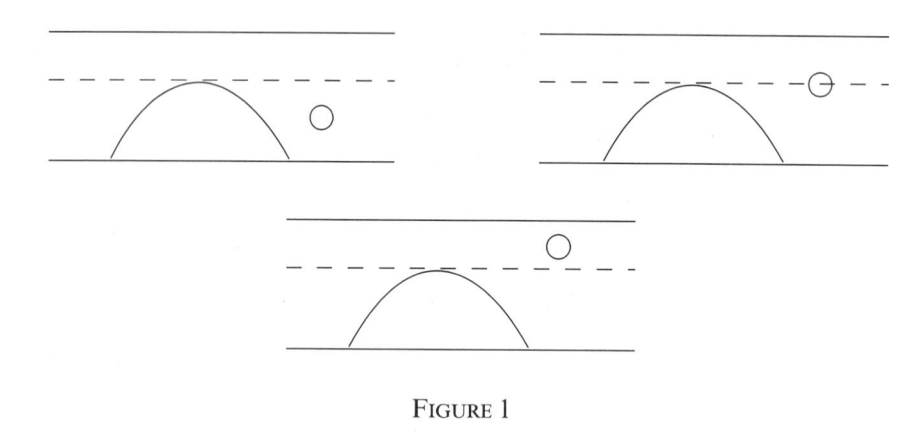

FIGURE 1

3.4. THEOREM. *Each component of D' is simply connected.*

This is the key result in the construction of a 2-category of 2-knots. A proof of Theorem 3.4 is given in §6.

Note that the space D^0 has the same property (see §5) and thus can replace D' in the construction of a 2-category of 2-knots. We prefer to use D' because it has less components. This simplifies the 2-category under construction. In contrast to D' and D^0, the space D has components that are not simply connected.

3.5. From diagrams to tangles. A *tangle* is a proper 1-dimensional compact smooth submanifold of $\mathbb{R}^2 \times [0,1]$. A tangle Γ is said to be *regular* if: (a) the standard projection $\varphi \colon \Gamma \to \mathbb{R}^1 \times [0,1]$ along the second axis of $\mathbb{R}^2$ is a proper immersion, (b) this immersion has only a finite number of multiple points each being a transversal double crossing and lying in the interior of $\mathbb{R}^1 \times [0,1]$. If, in addition, the composition of φ with the height function h (see 3.1) is a Morse function whose critical values are pairwise distinct and none of the critical points is a double point of the immersion, the tangle is called a *generic tangle*.

The image $\varphi(\Gamma)$ of a regular tangle is a regular system of arcs and loops 1-diagram and thus such a tangle gives rise, in a canonical way, to a 1-diagram (see 3.2). Note that due to the definition of a 1-diagram, a tangle represented by a 1-diagram is regular.

The image $\varphi(\Gamma)$ of a generic tangle is a generic system of arcs and loops. A tangle is generic if and only if its 1-diagram is generic.

---

[3]In other words, we consider the group of ambient C^∞ isotopies with the C^∞ topology and in the set of 1-diagrams we introduce the minimal topology such that the action of the group of ambient isotopies is continuous.

A *weakly generic tangle* is a regular tangle whose 1-diagram is weakly generic.

Let us provide the set of regular tangles with the C^∞ topology and denote this space by $\widetilde{D}$. The subspaces of generic and weakly generic tangles are denoted by $\widetilde{D}^0$ and $\widetilde{D}'$ respectively.

THEOREM. *Each component of $\widetilde{D}'$ is simply connected.*

PROOF. Regular tangles form a space which is fibered over the space of 1-diagrams: the fibration $p: \widetilde{D}' \to D'$ carries a tangle to its 1-diagram. Tangles with the same 1-diagram form a convex (infinite-dimensional) affine body. Thus the fibers of p are contractible. Therefore, the exactness of the homotopy sequence of p together with Theorem 3.4 gives the desired result.

3.6. REMARK. Since D^0 is simply connected, each component of $\widetilde{D}^0$ is also simply connected.

§4. Two-category of 2-knots

In this section we construct a 2-category of 2-knots in $\mathbb{R}^4$. We denote it by $\mathcal{F}$. To simplify the notation, set $K = \mathbb{R}^2 \times [0,1]$, $K_0 = \mathbb{R}^2 \times \{0\}$, and $K_1 = \mathbb{R}^2 \times \{1\}$.

4.1. Objects and 1-morphisms. Let us begin with a definition of the underlying 1-category.

Objects of $\mathcal{F}$ are finite subsets of $\mathbb{R}^1$ considered up to isotopies in $\mathbb{R}^1$. Since the only isotopy invariant is the number of points, the set of objects, $\mathcal{F}_0$, can be identified with $\mathbb{N}$.

A 1-*morphism* of $\mathcal{F}$ is a weakly generic 1-diagram considered up to isotopies in the class of weakly generic diagrams. In other words, a 1-morphism is a connected component of D', i.e., $\mathcal{F}_1 = \pi_0(D')$. For the definition of D' see 3.3.

The *source* $s_0(u)$ of a 1-morphism u represented by a 1-diagram d is the (finite) set $d \cap (\mathbb{R}^1 \times \{0\})$. Similarly, $t_0(u) = d \cap (\mathbb{R}^1 \times \{1\})$.

To any object A of $\mathcal{F}$ we assign the *identity morphism* 1_A represented by the diagram $d = A \times [0,1]$. To define the composition $u \circ v$ of 1-morphisms u, v such that $t_0(v) = s_0(u)$, represent them by 1-diagrams c, d respectively, put d into $\mathbb{R}^1 \times [0, 1/3]$ and c into $\mathbb{R}^1 \times [2/3, 1]$ and then insert in $\mathbb{R}^1 \times [1/3, 2/3]$ an isotopy in $\mathbb{R}^1$ between the target of d and the source of c (see Figure 2). The resulting diagram represents $u \circ v$.

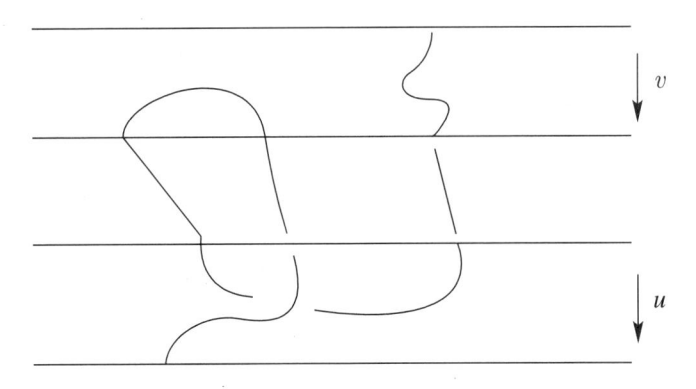

FIGURE 2. Composition of two 1-morphisms.

The composition is well defined because the configuration space of finite subsets of given cardinality in $\mathbb{R}$ is contractible.

Standard arguments show that the morphisms 1_A, $A \in \mathcal{F}_0$ are 2-sided identities with respect to the composition $\circ$ of 1-morphisms and the composition of 1-morphisms is associative. In particular, we conclude that $(\mathcal{F}_0, \mathcal{F}_1, s, t, \circ)$ *is a category*.

4.2. Two-morphisms. In the 2-category under construction, $\mathcal{F}$ and 2-morphisms are represented by surfaces in $K \times [0, 1] = \mathbb{R}^2 \times [0, 1]^2$ satisfying certain boundary conditions.

Consider a compact smooth proper two-dimensional submanifold Σ of the manifold $K \times [0, 1]$. The boundary of Σ is decomposed into 4 parts:

$$i(\Sigma) = \Sigma \cap (K_0 \times [0,1]), \qquad s(\Sigma) = \Sigma \cap (K \times \{0\}),$$
$$e(\Sigma) = \Sigma \cap (K_1 \times [0,1]), \qquad t(\Sigma) = \Sigma \cap (K \times \{1\}).$$

The surface Σ is said to be *regular* if
(1) $i(\Sigma)$, $e(\Sigma)$, $s(\Sigma)$, and $t(\Sigma)$ are proper 1-submanifolds of $K_0 \times [0,1], K_1 \times [0,1], K \times \{0\}$, and $K \times \{1\}$ respectively;
(2) $s(\Sigma)$ and $t(\Sigma)$ are weakly generic tangles in $K = K \times \{0\} = K \times \{1\}$;
(3) the projection $K_0 \times [0,1] = K_1 \times [0,1] = \mathbb{R}^2 \times [0,1] \to \mathbb{R}^1 \times [0,1]$ along the second coordinate of $\mathbb{R}^2$ restricted to $i(\Sigma)$ and $e(\Sigma)$ has no critical points and the image of $i(\Sigma)$, as well as that of $e(\Sigma)$, is the graph of an isotopy of a finite subset of $\mathbb{R}^1$.[4]

A *2-morphism* in $\mathcal{F}$ is represented by a regular surface; two surfaces Σ_1, Σ_2 represent the same 2-morphism if and only if they are isotopic in the class of regular surfaces.

For a 2-morphism F represented by a regular surface Σ, the corresponding 1-morphisms $s_1(F)$ and $t_1(F)$ are defined to be the 1-diagrams of the tangles $s(\Sigma)$ and $t(\Sigma)$. By assumption (2), these diagrams are well defined and weakly generic.

To define the *identity 2-morphism* 1_u, where u is a 1-morphism, we represent u by the 1-diagram of a weakly generic tangle $\gamma \subset K$ and take 1_u to be $\gamma \times [0, 1] \subset K \times [0, 1]$.

It is clear that s_1, t_1, s_0, t_0 and the indentity 2-morphisms satisfy relations (1) and (2) from the definition of a 2-category, see 2.1.

It remains to define the compositions of 2-morphisms.

First, let us define the *composition* $\circ$. Assume that F, G are 2-morphisms such that $t_0 s_1(G) = s_0 t_1(F)$. Then there exists a finite set $A \subset \mathbb{R}$ and regular surfaces Σ_1, Σ_2 representing 2-morphisms F and G and such that

$$e(\Sigma_2) = (A \times \{r\}) \times \{1\} \times [0,1], \qquad i(\Sigma_1) = (A \times \{r\}) \times \{0\} \times [0,1],$$

where r is an arbitrary constant.

Insert Σ_2 in $K_0 \times [0, 1/2] \times [0, 1]$ via the map

$$K_0 \times [0,1] \times [0,1] \to K_0 \times [0, 1/2] \times [0,1],$$
$$(x, y, z, t) \mapsto (x, y, z/2, t)$$

and Σ_1 in $K_0 \times [1/2, 1] \times [0, 1]$ via the map

$$K_0 \times [0,1] \times [0,1] \to K_0 \times [1/2, 1] \times [0,1],$$
$$(x, y, z, t) \mapsto (x, y, (1+z)/2, t).$$

---
[4]This is the case, for instance, if the images of $i(\Sigma)$ and $e(\Sigma)$ are properly embedded 1-submanifolds of $\mathbb{R}^1 \times [0,1]$ and the projection $\mathbb{R}^1 \times [0,1] \to [0,1]$ restricted to these submanifolds has no critical points.

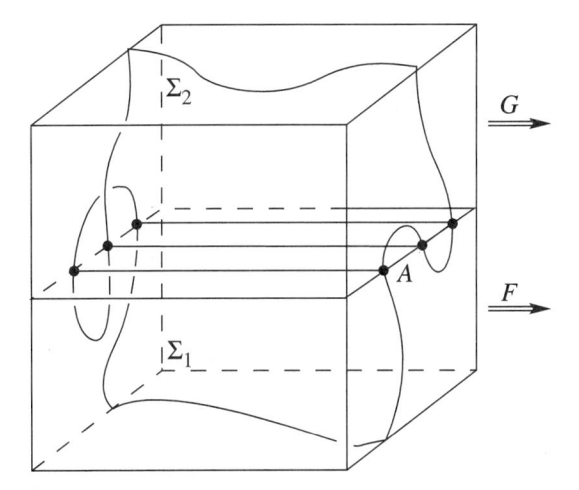

Their union Σ, smoothed if necessary along $\Sigma_1 \cap \Sigma_2 = \partial\Sigma_1 \cap \partial\Sigma_2$, represents the 2-morphism $F \circ G$. This 2-morphism is independent of the choice of r, A, and surfaces representing F and G.

To verify the independence of $F \circ G$ of the choice made in the definition of $\circ$, let us take other representations Σ'_1, Σ'_2 of F and G with

$$e(\Sigma'_2) = (A' \times \{r'\}) \times \{1\} \times [0, 1], \qquad i(\Sigma'_1) = (A' \times \{r'\}) \times \{0\} \times [0, 1].$$

Since Σ'_1, Σ'_2 are isotopic to Σ_1, Σ_2 in the class of regular surfaces, the composed surfaces Σ' and Σ are also isotopic in the class of regular surfaces. Moreover, reparametrizing the composed isotopy connecting Σ' and Σ we can obtain an isotopy (in the class of regular surfaces) joining Σ' to a regular surface Σ'' built from three pieces: the image of Σ_2 under the map

$$K_0 \times [0, 1] \times [0, 1] \to K_0 \times [0, 1/4] \times [0, 1],$$
$$(x, y, z, t) \mapsto (x, y, z/4, t),$$

the graph

$$A_{t,\tau} \times \{\tau\} \times \{t\}, \qquad \tau \in [1/4, 3/4], \, t \in [0, 1]$$

of a loop in the loop space $\mathrm{Map}(S^1, \mathrm{Emb}(A, \mathbb{R}))$ and the image of Σ_1 under the map

$$K_0 \times [0, 1] \times [0, 1] \to \mathbb{R}^2 \times [3/4, 1] \times [0, 1],$$
$$(x, y, z, t) \mapsto (x, y, (3+z)/4, t).$$

The configuration space $\mathrm{Emb}(A, \mathbb{R})$ is contractible and thus any loop in $\mathrm{Map}(S^1, \mathrm{Emb}(A, \mathbb{R}))$ is homotopic to the trivial loop. Hence the regular surface Σ'' is isotopic to Σ in the class of regular surfaces. This implies the required independence.

Now, let us define the *composition* $\star$. Assume that F, G are 2-morphisms such that $t_1(G) = s_1(F)$. Then there exists a weakly generic (and even generic) tangle γ and regular surfaces Σ_1, Σ_2 representing 2-morphisms F and G and such that $t(\Sigma_2) = s(\Sigma_1) = \gamma$. Put Σ_2 in $K \times [0, 1/2]$ by the map

$$K \times [0, 1] \to K \times [0, 1/2], \qquad (x, y, z, t) \mapsto (x, y, z, t/2)$$

and Σ_1 in $K \times [1/2, 1]$ by the map

$$K \times [0, 1] \to K \times [1/2, 1], \qquad (x, y, z, t) \mapsto (x, y, z, (1+t)/2).$$

Their union Σ, smoothed if necessary along their common boundary, represents a 2-morphism. This morphism is independent of the choice of auxiliary representatives and is, by definition, $F \star G$.

To verify its independence of the choice made in the definition of $\star$, let us take some other representatives Σ'_1, Σ'_2 of F and G with $t(\Sigma'_2) = s(\Sigma'_1) = \gamma'$. The same arguments as above show that, since Σ'_1, Σ'_2 are isotopic (in the class of regular surfaces) to Σ_1, Σ_2, the composed surface Σ' is isotopic (in the class of regular surfaces) to a regular surface Σ'' built from three parts: the image of Σ_2 under the map

$$K \times [0, 1] \to K \times [0, 1/4], \qquad (x, y, z, t) \mapsto (x, y, z, t/4),$$

the graph of a loop of $\widetilde{D}'$, and the image of Σ_1 under the map

$$K \times [0, 1] \to K \times [3/4, 1], \qquad (x, y, z, t) \mapsto (x, y, z, (3+t)/4).$$

By Theorem 3.5, each component of $\widetilde{D}'$ is simply connected, and thus any loop in this space is homotopic to the trivial loop. Hence Σ'' is isotopic to Σ in the class of regular surfaces. This implies the required independence.

4.3. THEOREM. *The collection* $(\mathcal{F}_0, \mathcal{F}_1, \mathcal{F}_2, s_0, t_0, s_1, t_1, \circ, \star)$ *is a 2-category.*

PROOF. This is a routine check of axioms (1)–(6). Some of the axioms, for example, axioms (1) and (5), are tautological. The others turn out to be trivial after some change of coordinates. We restrict ourselves to two typical verifications: the associativity of $\star$ and axiom (6).

To verify the relation $(F \star G) \star H = F \star (G \star H)$, let us represent 2-morphisms F, G, H which satisfy conditions $s_1(F) = t_1(G)$, $s_1(G) = t_1(H)$ by surfaces $\Sigma_1, \Sigma_2, \Sigma_3$ such that $t(\Sigma_2) = s(\Sigma_1)$ and $t(\Sigma_3) = s(\Sigma_2)$. Then the composed surfaces Σ and Σ' representing $(F \star G) \star H$ and $F \star (G \star H)$ differ by a homeomorphism

$$\Lambda: K \times [0, 1] \to K \times [0, 1] : (x, t) \mapsto (x, \lambda(t)),$$

where λ is a piecewise linear function transforming the partition

$$[0, 1] = [0, 1/4] \cup [1/4, 1/2] \cup [1/2, 1]$$

into the partition

$$[0, 1] = [0, 1/2] \cup [1/2, 3/4] \cup [3/4, 1].$$

Accompanying an isotopy of partitions by a piecewise linear transformation of the coordinate t, one sees that Σ and Σ' are isotopic in the class of regular surfaces and thus they define the same 2-morphism.

To check that

$$F \circ G = (1_{t_1(F)} \circ G) \star (F \circ 1_{s_1(G)}),$$

let us represent F and G by surfaces Σ_1 and Σ_2 such that

$$e(\Sigma_2) = (A \times \{r\}) \times \{1\} \times [0, 1], \qquad i(\Sigma_1) = (A \times \{r\}) \times \{0\} \times [0, 1],$$

where A is a finite set in $\mathbb{R}$ (see the definition of $\circ$ in 4.2). Following this definition, put Σ_1, Σ_2 in the two halves of $K \times [0, 1]$ to get the surface Σ representing $F \circ G$. Then moving

Σ_2 to the right and Σ_1 to the left along their common boundary $(A \times \{r\}) \times \{1/2\} \times [0, 1]$ we obtain a surface representing the same 2-morphism built from four blocks

$$\Sigma_{00} = \{(x, y, z/2, t/2) \mid (x, y, z, t) \in s(\Sigma_2) \times [0, 1]\},$$
$$\Sigma_{01} = \{(x, y, z/2, (1+t)/2) \mid (x, y, z, t) \in \Sigma_2\},$$
$$\Sigma_{10} = \{(x, y, (1+z)/2, t/2) \mid (x, y, z, t) \in \Sigma_1\},$$
$$\Sigma_{11} = \{(x, y, (1+z)/2, (1+t)/2) \mid (x, y, z, t) \in t(\Sigma_1) \times [0, 1]\}.$$

This surface represents the 2-morphism $(1_{t_1(F)} \circ G) \star (F \circ 1_{s_1(G)})$ and this gives the desired relation.

The equality

$$F \circ G = (F \circ 1_{t_1(G)}) \star (1_{s_1(F)} \circ G)$$

is proven similarly, we must move Σ_2 to left and Σ_1 to right.

§5. A simplified version of Theorem 3.4

5.1. THEOREM. *The space D^0 is simply connected.*

This theorem may be considered as a simplified version of Theorem 3.4. Its proof allows us to clarify certain ideas used in §6 to prove Theorem 3.4.

5.2. PROOF. Take a smooth loop in D^0. It is represented by a family of generic 1-diagrams $\{L_\alpha, \alpha \in S^1\}$ in B. The graph of this family is the image of a fiber-to-fiber map $F: l \times S^1 \to B \times S^1$, where $F|_{l \times \{\alpha\}}, \alpha \in S^1$ is a parametrization of L_α (here l is the corresponding disjoint sum of intervals and circles). This map is a proper immersion.

The composition F' of F with the projection

$$B \times S^1 = \mathbb{R} \times [0, 1] \times S^1 \to [0, 1] \times S^1$$

has, in general, critical points. They are extremum points (i.e., points of local maxima and minima) of the height functions $h_\alpha: L_\alpha \to [0, 1], \alpha \in S^1$ (see 3.1). According to the definition of generic diagrams, these critical points and the double points of the immersion (the latter correspond to the double points of L_α) form a finite number of disjoint circles in $l \times S^1$. Moreover, since the extremum points of a generic diagram lie on different levels of the height function, the union C_e of the circles formed by the critical points of F' is embedded by F' in $[0, 1] \times S^1$ and the image of each of these circles intersects each fiber of $[0, 1] \times S^1 \to S^1$ at one point, intersections being transversal. The union C_d of the circles formed by the double points is immersed by F' in $[0, 1] \times S^1$. Each circle of double points is embedded by F' in $[0, 1] \times S^1$ and its image intersects every fiber of the projection $[0, 1] \times S^1 \to S^1$ at one point. The latter assertion results, for example, from the observation that double points of a generic diagram can be encoded as follows. Take the left descending branch of a double point, it descends either to a minimum or to a bottom input of the diagram. We encode the double point by this minimum or bottom point and the number of double points met during the descent. This encoding pair is preserved when α is varied.

Since the curve $F'(C_e)$ is embedded in $[0, 1] \times S^1$ and intersects each fiber of $[0, 1] \times S^1 \to S^1$ at one point, it can be transformed by an isotopy into $[0, 1] \times S^1$ preserving the second coordinate in a family of horizontal (i.e., parallel to S^1) circles. This isotopy can be followed by an isotopy of F and thus by an isotopy of 1-diagrams.

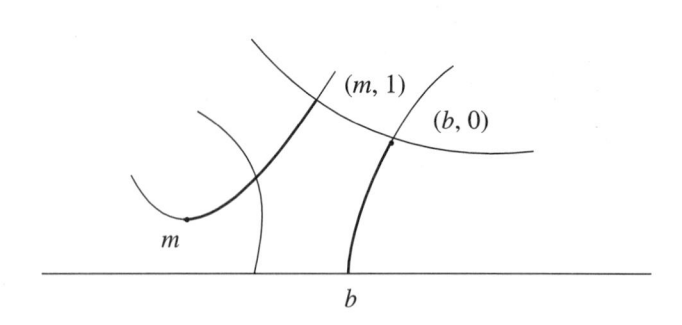

In this way we obtain a family of 1-diagrams whose extremum points preserve their levels.

At the next step we arrange that the double points stay on levels different from the levels of critical points. This is done by an inductive procedure that decreases the number of points in $F'(C_e) \cap F'(C_d)$ by 2.

This procedure goes as follows. Choose a component s of C_e and a pair of points in $F'(s) \cap F'(C_d)$ which both lie in the image of one and the same component s' of C_d. This pair of points can be chosen so that the arc θ' bounded by them in $F'(s')$ lies on the one side of $F'(s)$ and the arc θ bounded by them in $F'(s)$ contains no other pair of points belonging to the image of a component of C_d and bounding in this image an arc lying on the same side of $F'(s)$ as θ' (see Figure 3).

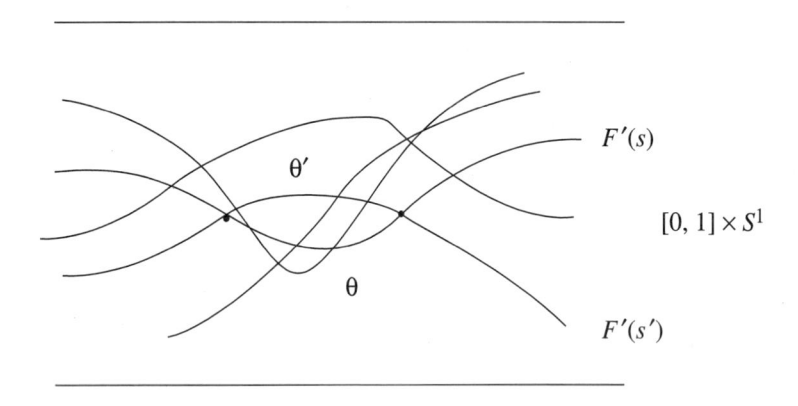

FIGURE 3

Then, for any double point d_α corresponding to a point of θ', the interior of the triangle δ_α in B bounded by the branches descending from this double point and by a horizontal segment of the same level as $F'(s_\alpha)$, $s_\alpha \in \theta$ (see Figure 4) does not contain any point of the diagram L_α.

Now, the double points d_α, $\alpha \in S^1$ can be pulled down simultaneously along the triangles δ_α to eliminate the two chosen points of $F'(s) \cap F'(s')$.

Iterating this procedure, we may ensure that $F'(C_e)$ is disjoint from $F'(C_d)$. A similar procedure allows us to eliminate intersections between images of different components of C_d.

The set $F'(C_e \cup C_d)$ is an embedded 1-submanifold of $[0, 1] \times S^1$ and each of its components transversally intersects every interval $[0, 1] \times \alpha$, $\alpha \in S^1$ at one point.

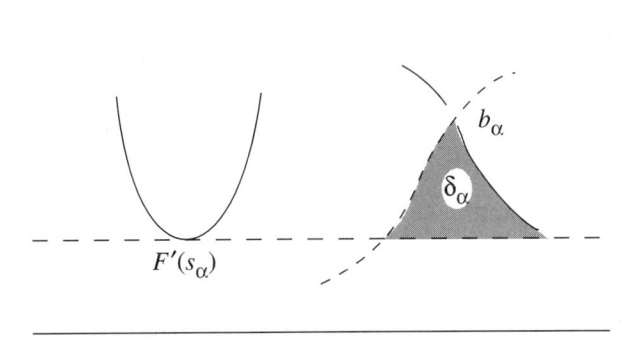

FIGURE 4

Therefore, $F'(C_e \cup C_d)$, like $F'(C_e)$, is isotopic in $[0, 1] \times S^1$ to a family of horizontal circles. This isotopy can be followed by an isotopy of F and thus by an isotopy of 1-diagrams. This means that we obtain a family of 1-diagrams whose extremum points and double points keep their levels. After that, again by an isotopy, the family of 1-diagrams can be made constant. This proves that the loop $\alpha \mapsto L_\alpha : S^1 \to D^0$ is contractible.

§6. Proof of Theorem 3.4

6.1. Configuration of pleats. In this subsection we consider a family of weakly generic 1-diagrams L_α, $\alpha \in S^1$, i.e., a loop in D'. We assume that this loop is generic and smooth.

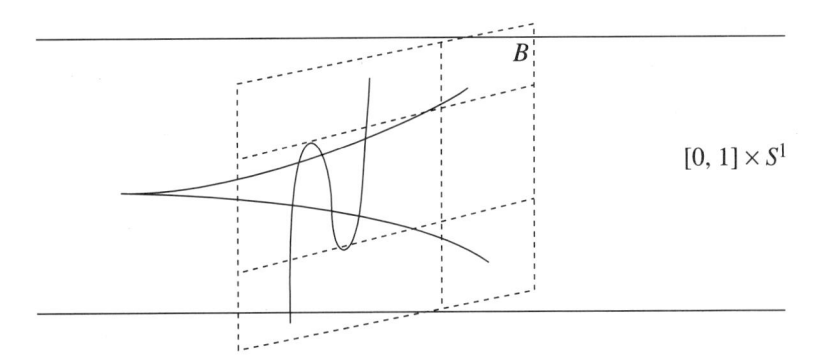

FIGURE 5

As in 5.2, we consider the graph of the family L_α, $\alpha \in S^1$ as the image of a fiber-to-fiber map $F : l \times S^1 \to B \times S^1$, l being the corresponding disjoint sum of intervals and circles.

The map F is a proper immersion and the critical point set $C_e \subset l \times S^1$ of the composition F' of F with the projection

$$B \times S^1 = \mathbb{R} \times [0, 1] \times S^1 \to [0, 1] \times S^1$$

is a union of a finite number of disjoint circles in $l \times S^1$. Since the loop is taken generic, the restriction of F' to C_e is a topological embedding locally. Moreover, the curve

$F'(C_e)$ is smooth and transversal to the fibers of $[0,1] \times S^1 \to S^1$ except at a finite number of points which are the *pleats* of the mapping $F': l \times S^1 \to [0,1] \times S^1$. Each pleat corresponds to a birth of a local maximum and a local minimum at a regular point of some L_α or to a death of such a pair (see Figure 5).

LEMMA. *Each component s of C_e which is not transversal to the fibers of* pr: $[0,1] \times S^1 \to S^1$ *contains exactly two pleats $b(s)$, $d(s)$ and bounds in $l \times S^1$ a disk $C(s)$. One pleat, $b(s)$, corresponds to the birth of a local maximum and a local minimum at a regular point of some L_b, $b \in S^1$, the other one, $d(s)$, corresponds to their death at a regular point of some L_d, $d \in S^1$. The disk $C(s)$ is formed by a family of connected arcs $\gamma_t \subset l \times \{be^{it}\}$, $t \in [0, T]$, $T \in \text{Arg}\, db^{-1}$ whose extremities δ_t and δ'_t underlie a local maximum and a local minimum of $L_{be^{it}}$.*

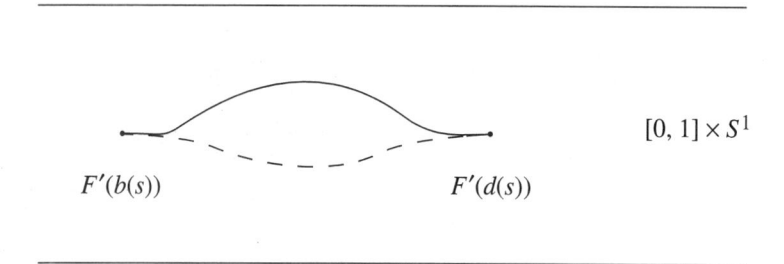

PROOF. Take a component s of C_e whose image $F'(s)$ is not transversal to the fibers of pr: $[0,1] \times S^1 \to S^1$. It contains at least one pleat corresponding either to a birth of a maximum and a minimum or to a death of such a pair. Suppose that there is a pleat $b \in s$ corresponding to a birth point; the other case is completely similar.

Let δ and δ' be the two arcs of s outgoing from b. They are different because one of them, say δ, is formed by points δ_t whose images Δ_t on the diagrams are points of local maximum and the other, δ', by points δ'_t whose images Δ'_t are points of local minimum. Let d and d' be the endpoints of δ and δ' different from b. If d and d' are distinct, then this means that the maximum points Δ_t are annihilated by some minimum points $m_t \in L_t$ different from Δ'_t and the minimum points Δ'_t are annihilated by some maximum points $M_t \in L_t$ different from Δ.

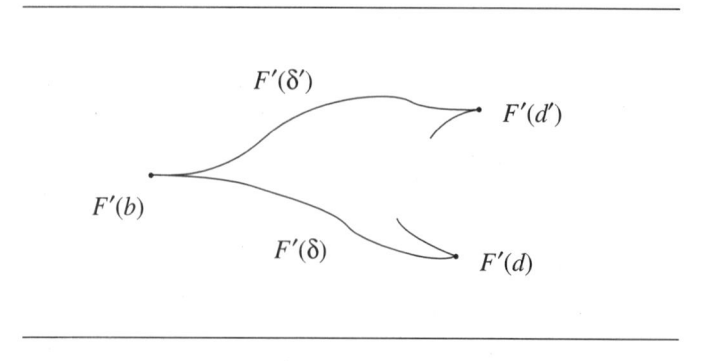

Suppose that the first event happens before the second one. Then $h_t(m_t)$ should be of an intermediate value with respect to $h_t(\Delta_t)$ and $h_t(\Delta'_t)$. At the moment of the birth of b such a minimum does not exist and the existing points of minimum cannot overgrow Δ_t. Minima descending from the levels higher than Δ'_t are not in elimination position. Each minimum m_t of an intermediate value with respect to $h_t(\Delta_t)$ and $h_t(\Delta'_t)$ which burns after b appears together with a maximum M'_t of an intermediate value also; this maximum M'_t, in its turn, cannot overgrow Δ'_t and thus does not allow m_t to reach Δ_t.

Thus we have proved that s contains exactly two pleats: a birth and a death.

The arcs γ_t which appear and live together with δ_t and δ'_t are mutually disjoint, since an extremum point belonging to the interior of γ_t should be burned after the birth of γ_t and thus should belong to another component of C_e (otherwise there are more than 2 pleats in the same component). That implies both the existence of the disk $C(s)$ and the last assertion of the lemma.

6.2. Elimination of a pleat. Consider the same situation as in 6.1 and preserve the notation introduced there.

LEMMA. *There is a loop in D' homotopic to the given one, L_α, $\alpha \in S^1$, and having two pleats less.*

PROOF. The proof is similar to that of Theorem 5.1. It contains several steps aimed to put the diagrams in a standard position.

Consider the pairs of pleats $(b(s), d(s))$ introduced in Lemma 6.1. Each pair has its proper period of life $T = T(s)$ and its proper dispersion

$$\max_{t \in [0, T(s)]} |h(\Delta_t(s)) - h(\Delta'_t(s))|,$$

where $\Delta_t(s)$ and Δ'_t are points of $L_{b e^{it}}$ corresponding to $\delta_t(s)$ and $\delta'_t(s)$, see Lemma 6.1. Let us take a pair with the minimal dispersion.

In the family of bands $B^t = [h(\Delta'_t(s)), h(\Delta_t(s))] \times \mathbb{R}$ consider the branch c_t of $L_{b e^{it}} \cap B^t$ containing the image $F'(\gamma_t)$ of the arc γ_t introduced in Lemma 6.1 (Figure 6).

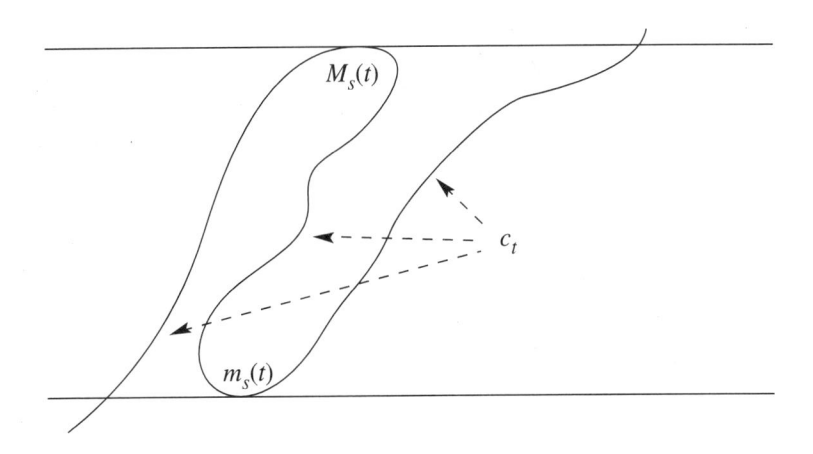

FIGURE 6

Because of the minimality of the dispersion, neither the central part of c_t nor its lateral sides contain points of local minimum or maximum. Thus by an ambient

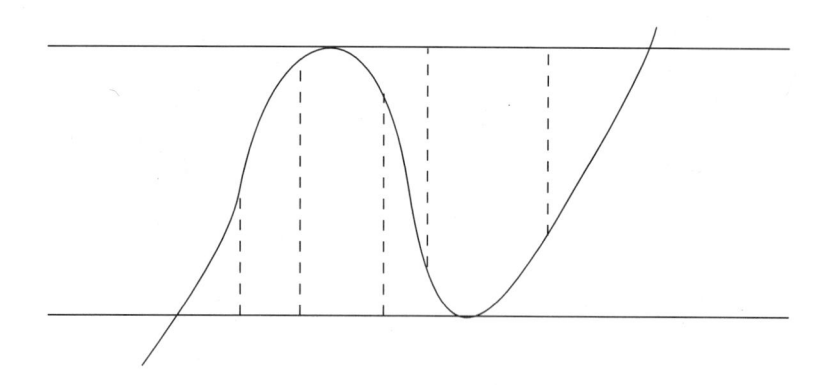

FIGURE 7

horizontal isotopy the family of diagrams can be transformed into a family such that for all $t \in [0, T]$ the curve c_t intersects each vertical interval in at most one point (Figure 7).

Then, by a trick similar to that used in the proof of Theorem 5.1 to eliminate pairs of intersection points, we pull out of the interior J^t of the part of B^t bounded by the lateral sides of c_t (Figure 8) all critical and double points which occasionally enter there. After that, by a horizontal ambient isotopy all branches occasionally intersecting c_t are made vertical inside J^t. Now, the vertical straightening of the branches c_t, $0 \leqslant t \leqslant T$ keeps the diagrams inside D' and eliminates the two pleats chosen.

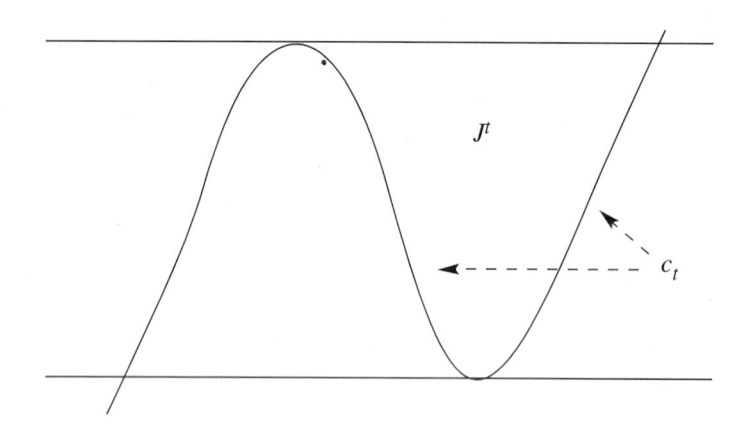

FIGURE 8

6.2. Proof of Theorem 3.4. Let us take a family of 1-diagrams L_α, $\alpha \in S^1$ representing a generic smooth loop in D' and preserve the notation introduced in 6.1 and 6.2.

According to Lemma 6.2, all pleats of the corresponding fiber-to-fiber map $F: l \times S^1 \to B \times S^1$ can be eliminated by a homotopy of the loop in D'. Thus we can suppose that the restriction of F' to $C_e \cup C_d$ is an immersion transversal to the fibers of pr: $[0, 1] \times S^1 \to S^1$.

By the definition of D', the image by F' of each component of C_e has no self-intersection. Intersection points between images of different components can be eliminated in pairs by a procedure similar to that used in 5.2.

After this elimination, we get a loop in D^0. According to 5.1, this loop can be contracted to a constant loop in D^0. Thus the initial loop is contractible in D'.

References

[F]   J. E. Fischer, Jr., *2-categories and 2-knots*, Duke Math. J. **75** (1994), 493–526.

[K–V]   M. M. Kapranov and V. A. Voevodsky, *Braided monoidal 2-categories, 2-vector spaces and Zamolodchikov tetrahedra equations*, Preprint (1991).

[K–S]   D. Kazhdan and Y. Soibelman, *Representations of the quantized function algebras, 2-categories and Zamolodchikov tetrahedra equation*, Selecta Math. **1** (1995), no. 1.

[Kh]   V. M. Kharlamov, *Movements of straight lines and the tetrahedron equations*, Publ. Dipart. Mat. Univ. Pisa (1992).

[M–S]   Yu. I. Manin and V. V. Schechtman, *Arrangements of hyperplanes, higher braid groups and higher Bruhat orders*, Adv. Stud. Pure Math., vol. 17, 1989, pp. 289–308.

[RT1]   N. Reshetikhin and V. Turaev, *Ribbon graphs and their invariants derived from quantum groups*, Comm. Math. Phys. **127** (1990), 1–26.

[RT2]   _____ , *Invariants of 3-manifolds via link polynomials and quantum groups*, Invent. Math. **103** (1991), 547–598.

[Tu]   V. Turaev, *Operator invariants of tangles and R-matrices*, Izv. Akad. Nauk SSSR Ser. Mat. **53** (1989); English transl., Math. USSR-Izv. **35** (1990), 411–444.

[Ye]   D. N. Yetter, *Markov algebras and braids*, Contemp. Math., vol. 78, Amer. Math. Soc., Providence, RI, 1988, pp. 705–730.

[Z]   A. B. Zamolodchikov, *Tetrahedron equations and integrable systems in three-dimensional space*, Zh. Èksper. Teoret. Fiz. **52** (1980), 325–336; English transl. in JETP **52** (1980).

Translated by THE AUTHORS

INSTITUT DE RECHERCHE MATHÉMATIQUE AVANCÉE, UNIVERSITÉ LOUIS PASTEUR, 7, RUE RENÉ DESCARTES, 67084 STRASBOURG, FRANCE

Application of Topology to Some Problems in Combinatorial Geometry

V. V. Makeev

Dedicated to #45 Boarding School on occasion of its 30th anniversary

The present paper is devoted to some problems in combinatorial geometry. In §1 we consider the following question: given a convex polygon A, is it true that for any planar convex figure F there exists an affine transformation f such that $f(A)$ is inscribed in (circumscribed about) F? We give the answer to this question provided that A is a centrally symmetric hexagon or a pentagon that satisfy some additional conditions. Similar problems in solid geometry are also considered. In the rest of the paper (§§2, 3), we deal with two more problems in combinatorial geometry: partitions of planar sets by a 5-ray configuration and a corollary to the Rattray theorem.

§1. Affine-inscribed and affine-circumscribed polygons

Throughout this paper by a *convex body* we mean a convex compact subset of $\mathbb{R}^n$ with the nonempty interior. A polytope $A \in \mathbb{R}^n$ is said to be *inscribed* in a convex body K if all the vertices of A belong to the boundary ∂K of K. A polytope A is said to be *circumscribed* about a convex body K if $K \subset A$ and K meets each face of A. A polytope A is said to be *affine-inscribed* in (*circumscribed* about) a convex body K if some affine image of A is inscribed in (circumscribed about) K.

Besicovich [4] and Dantzer proved that a regular hexagon is affine-inscribed in and affine-circumscribed about every planar figure (for further references see [7]).

This brings up the following natural problem: find all polytopes in $\mathbb{R}^n$ that are affine-inscribed in (affine-circumscribed about) every convex body of $\mathbb{R}^n$.

In the study of this problem we use the following method (for definiteness we consider the case of inscribed polytopes).

Let $A_1, \ldots, A_k$ be the vertices of a polytope $A \in \mathbb{R}^n$, let G be the group of orientation-preserving affine transformations of $\mathbb{R}^n$, and let H be the subgroup of G consisting of all transformations which take A into itself.

Consider the set

$$A_G = \{(g(A_1), \ldots, g(A_k)) \in \mathbb{R}^{nk} \mid g \in G\}.$$

1991 *Mathematics Subject Classification.* Primary 52A37, Secondary 57M35.

©1996, American Mathematical Society

Let $K \subset \mathbb{R}^n$ be a convex body with smooth boundary. We are looking for polytopes A such that A_G meets $(\partial K)^k \subset \mathbb{R}^{nk}$.

From standard considerations of general position, it follows that the manifold A_G is transversal to $(\partial K)^k$ for a dense set of bodies K, and, therefore, $A_G \cap (\partial K)^k$ is a smooth submanifold of codimension k in A_G. We call these bodies K *bodies in general position*. In particular, it follows that A cannot be affine-inscribed if $k > \dim A_G \geqslant \dim G = n^2 + n$. By definition, the set $A_G \cap (\partial K)^k$ is H-invariant, and we can consider the submanifold $(A_G \cap (\partial K)^k)/H$ of A_G/H (the latter is homeomorphic to G/H).

Since any two convex bodies in $\mathbb{R}^n$ with smooth boundaries are smoothly homotopic, and the homotopy $(\partial K_t)^k$, $0 \leqslant t \leqslant 1$, can be brought in general position with respect to A_G by a small perturbation, we see that for a dense set of compact sets in general position, each of the following homology classes is defined: the class of the manifold

$$\{g \in G \mid (g(A_1), \ldots, g(A_k)) \in (\partial K)^k\} \text{ in } H_*(G; \mathbb{Z})$$

and the class of $(A_G \cap (\partial K)^k)/H$ in $H_*(A_G/H; \mathbb{Z}_2) \cong H_*(G/H; \mathbb{Z}_2)$. Here we need, of course, the compactness of $A_G \cap \{(\partial K_t)^k \mid 0 \leqslant t \leqslant 1\}$ in A_G, which yields additional restrictions to K.

Let us especially note the case $k = \dim A_G$ in which, for some compact set K in general position, the number of elements of $(A_G \cap (\partial K)^k)/H$, i.e., the number of polytopes inscribed in K and affine-equivalent to A, is odd. In this case $(A_G \cap (\partial K)^k)/H$ yields a nontrivial element of $H_0(G/H; \mathbb{Z}_2) \cong \mathbb{Z}_2$. Then, for a dense set of convex bodies, $A_G \cap (\partial K)^k \neq \varnothing$.

THEOREM 1. *Let the points $A_1, \ldots, A_5$ of an ellipse be the vertices of a pentagon such that the sum of any two of its neighboring angles is greater than or equal to π. Let K be a strictly convex figure with C^1-smooth boundary and a fixed distinguished point A. Then some affine image of the pentagon is inscribed in K and has A as the image of A_1.*

Note that the restriction to the angles of the pentagon is of affine type, and we can reformulate it as follows: for any point O on the plane some of the segments $OA_1, \ldots, OA_5$ meet the pentagon. First let K be a disk and $A_1, \ldots, A_5$ lie on the boundary of K. Then the orientation-preserving affine transformations that inscribe the pentagon in K are rotations about the center of K. These rotations represent a generator of $H_1(G; \mathbb{Z}) \cong \mathbb{Z}$. Let K_t, $0 \leqslant t \leqslant 1$, be a smooth homotopy in the class of strictly convex bodies with smooth boundary. We shall prove that the set $A_G \cap \{(\partial K_t)^5 \mid 0 \leqslant t \leqslant 1\}$ of affine images $A_{1t}, \ldots, A_{5t}$ of $A_1, \ldots, A_5$ inscribed in the family K_t is compact. Consider the ellipses that pass through the affine images of $A_1, \ldots, A_5$ inscribed in K_t. The sizes of these ellipses are bounded from below and from above. If $A_{1t}, \ldots, A_{5t}$ are disposed on some K_t ($0 \leqslant t \leqslant 1$) clockwise and if the diameter of $A_{1t}A_{2t}\ldots A_{5t}$ tends to 0, then $\angle A_{2t}A_{1t}A_{5t} + \angle A_{4t}A_{5t}A_{1t}$ tends to 0 since ∂K_t is smooth. This proves that the diameters of the ellipses are bounded from below. From the standard compactness reasons, the sizes of the ellipses are bounded from above. Finally, the ratio of their semiaxes is bounded from below. Therefore, the major axes of the ellipses are uniformly bounded from below and from above. But if the small axes of some of ellipses tend to 0, then we can choose a subsequence of ellipses which tends to a line segment. We can also choose subsequences from $A_{1t}, \ldots, A_{5t}$ that tend to at least three points of this segment, which contradicts the strict convexity of K_t. Thus we have proved the compactness of the set of inscribed pentagons. Therefore, for a dense set of figures, the set $A_G \cap (\partial K)^5$ is homologous to the group of orientation-preserving rotations. We define a map $f : A_G \cap (\partial K)^5 \to \partial K$

by $f(g(A_1), \ldots, g(A_5)) = g(A_1)$. This map is a homeomorphism when K is a disk. Since the homology class of the image is defined for the figures in general position, the map f is surjective. Now, by passing to the limit, we obtain the statement of the theorem for all smooth strictly convex figures.

REMARKS. 1. Some theorems concerning inscribed affine-regular pentagons and octagons can be found in [5] and [6].

2. Theorem 1 can be proved for arbitrary points $A_1, \ldots, A_5$ of an ellipse, but in this case we need a higher smoothness class of ∂K. It is also more difficult to prove that the set of inscribed pentagons is compact. This will be done in the next paper.

Similarly one can prove the following statement.

THEOREM 2. *Let $A_1A_2A_3A_4A_5$ be a pentagon circumscribed about an ellipse. Let K be a strictly convex figure with C^1-smooth boundary ∂K and a fixed distinguished point $A \in \partial K$. Then there exists an affine transformation g such that $g(A_1 \ldots A_5)$ is circumscribed about K and the side $g(A_1A_2)$ is tangent to K at A.*

For a proof, one should consider a pentagon circumscribed about a circle and observe that affine transformations which circumscribe the pentagon about the circle are rotations about the center of this circle. The rest of the proof is similar to that of the previous theorem.

THEOREM 3. *An arbitrary centrally symmetric convex hexagon $A_1 \ldots A_6$ is affine-inscribed in every planar convex figure.*

Let K be a triangle. Then there are exactly three affine images of the configuration of the points $A_1, \ldots, A_6$ which lie on ∂K, namely, when one side from each pair of parallel sides of the hexagon belongs to some side of K (Figure 1). Thus $(A_G \cap (\partial K)^6)/H$ consists of three elements (or of one element if the hexagon is affine-regular). It is not hard to see that A_G is transversal to $(\partial K)^6$, and we have the same picture if we smooth the triangle. Therefore a nontrivial element of $H_0(G/H; \mathbb{Z}_2)$ is defined.

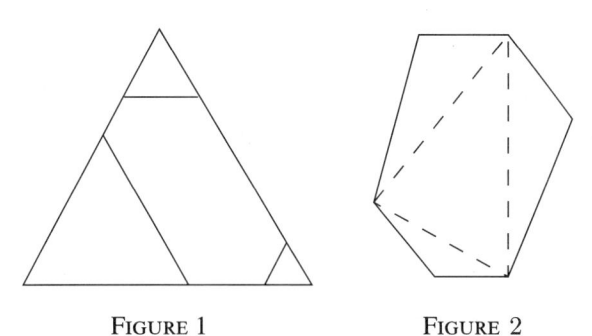

FIGURE 1   FIGURE 2

Thus the theorem is proved for a dense set of strictly convex figures with smooth boundaries. For arbitrary convex figures, the theorem can be proved by a standard passage to the limit.

REMARKS. 1. A centrally symmetric hexagon is likely to be circumscribed about every planar convex figure, since exactly three of its affine images are circumscribed about the triangle (or exactly one if the hexagon is regular). However, three vertices of these affine images coincide with the vertices of K (Figure 2) and A_G meets $(\partial K)^6$ at nonsmooth points. Therefore, this requires further study.

2. Is Theorem 3 valid for six arbitrary points of an ellipse? Do the theorems proved above remain valid if we replace the boundary of the convex figure by an arbitrary regular Jordan plane curve (or, more generally, by an arbitrary Jordan curve)?

3. The well-known results saying that a regular octagon is affine-inscribed in (affine-circumscribed about) every planar convex figure can also be proved by the above-mentioned method since only one of the affine images of the octagon is inscribed in (resp. circumscribed about) a square (Figure 3).

THEOREM 4. *Let $K \subset \mathbb{R}^3$ be a centrally symmetric strictly convex body. Then there exists a tetrahedron T such that the polytope $\{x - y \mid x, y \in T\}$ is inscribed in K and has a vertex at a preassigned point of ∂K.*

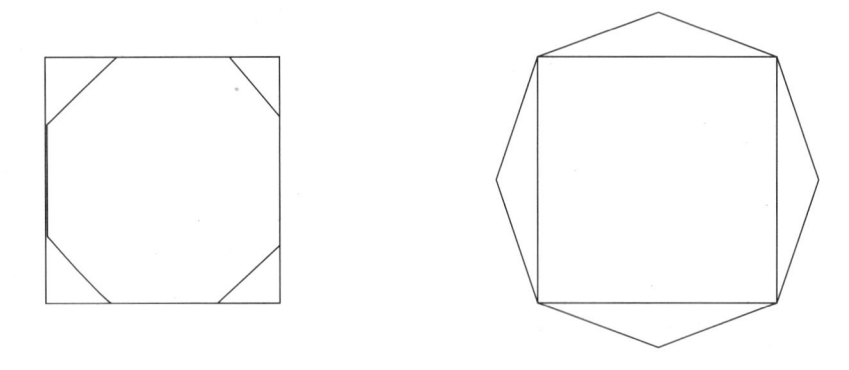

FIGURE 3

We call $\{x - y \mid x, y \in T\}$ the difference set of T. Let A be the difference set of a regular tetrahedron. Let A be inscribed in some ball $K \subset \mathbb{R}^3$. Then the set of orientation-preserving transformations of $\mathbb{R}^3$ which leave A inscribed in K is a group of orientation-preserving rotations about the center of K. This group represents a nontrivial element of $H_3(G; \mathbb{Z})$, i.e., a generator of a nonzero homology group of higher dimension. Thus $\{g \in G \mid gA \text{ is inscribed in } K\}$ represents a nontrivial element of $H_3(G; \mathbb{Z})$ for a dense set of centrally symmetric strictly convex bodies K with smooth boundary. But then the images of any vertex of A evidently cover ∂K. Now, for arbitrary centrally symmetric strictly convex bodies, the theorem can be proved by a standard passage to the limit. If the set of inscribed difference sets is not compact, it is possible to inscribe in some centrally symmetric body $K \subset \mathbb{R}^3$ the projection of the difference set of the tetrahedron on a plane so that all vertices of the difference set belong to ∂K. But this is impossible since, under projection on a plane or on a line, some vertex of the difference set does not go into a vertex of the projection. In the higher dimensional case, the latter statement does not hold, and the proof given above fails.

§2. Partitions of a plane figure

Every planar bounded measurable set can be divided into four parts of equal measure by two perpendicular lines. Are there any other configurations consisting of four rays with common vertex that have the same property? For example, is it true that if a configuration of four rays with common vertex divides some disk into four parts of equal measure, then every planar bounded measurable set can be divided into four parts of equal measure by means of this configuration?

The author does not know the answer to this question. However the following statement holds.

THEOREM 5. *For any planar bounded measurable set E there exist five rays with common vertex which divide the plane into five congruent sectors M_1, M_2, M_3, M_4, and M_5 such that $M_1 \cap E$, $M_2 \cap E$, $M_3 \cap E$, and $M_4 \cap E$ have the same area which is not less than the area of $M_5 \cap E$.*

PROOF. Let S^2 be the standard two-dimensional sphere with measure μ. We divide S^2 into five congruent sectors $M_1, \ldots, M_5$ by five half-planes through fixed antipodal points of S^2. Let $E \subset S^2$ be a measurable set. We prove that there exists a rotation $a \colon S^2 \to S^2$ such that $\mu(a(M_1) \cap E) = \cdots = \mu(a(M_4) \cap E) \geqslant \mu(a(M_5) \cap E)$, and Theorem 5 follows by the standard passage to the limit.

Consider the map $F \colon \mathrm{SO}(3) \to \mathbb{R}^5$ defined by the formula

$$F(a) = \left(\mu(a(M_1) \cap E) - \frac{\mu(E)}{5}, \ldots, \mu(a(M_5) \cap E) - \frac{\mu(E)}{5} \right).$$

The image $F(\mathrm{SO}(3))$ is contained in the four-dimensional subspace L of $\mathbb{R}^5$ given by the equation $x_1 + \cdots + x_5 = 0$. If $F(\mathrm{SO}(3))$ contains $0 \in \mathbb{R}^5$, the theorem is proved. Otherwise, consider the map

$$G = \frac{F}{\|F\|} \colon \mathrm{SO}(3) \to S^3 \subset L,$$

where S^3 is the unit sphere with the center at $0 \in L$.

The sectors $M_1, \ldots, M_5$ obviously determine a subgroup of $\mathrm{SO}(3)$ isomorphic to $\mathbb{Z}_5$. The group $\mathbb{Z}_5$ acts freely on $\mathrm{SO}(3)$ by right multiplications and on $S^3 \subset L$ by cyclic permutations of coordinates. By definition, G preserves the actions of $\mathbb{Z}_5$ and, therefore G is surjective (cf. [1, §1]). Thus we have a rotation, namely,

$$G^{-1}(1/\sqrt{20}, 1/\sqrt{20}, 1/\sqrt{20}, 1/\sqrt{20}, -2/\sqrt{5})$$

which satisfies the required conditions.

§3. Approximation of convex bodies by parallelepipedons

Lassak [3] proved that, for a planar convex figure, there exists two homothetic rectangles T_1 and T_2 with homothety ratio 2 such that $T_1 \subset K \subset T_2$. Is it true that, for each convex compact set $K \subset \mathbb{R}^n$, there exist two homothetic rectangular parallelepipedons T_1 and T_2, $T_1 \subset K \subset T_2$, with homothety ratio $1/n$?

THEOREM 6. *For any convex compact set $K \subset \mathbb{R}^n$ there exist similar rectangular parallelepipedons T_1 and T_2, $T_1 \subset K \subset T_2$, with similarity coefficient $1/n$.*

PROOF. By a theorem of Rattray [2], there exists a rectangular parallelepipedon T_2 which is tangent to K at $A_1^1, A_2^1, A_1^2, A_2^2, \ldots, A_1^n, A_2^n$ (where, for each i, A_1^i and A_2^i belong to parallel faces of T_2) such that segments $A_1^1 A_2^1, A_1^2 A_2^2, \ldots, A_1^n A_2^n$ are pairwise orthogonal. (This fact is proved in [2] only for centrally symmetric convex bodies. The general result follows from consideration of the difference set of the body.)

Let O be an arbitrary point of $\mathbb{R}^n$. Consider the 2^n vectors of the form
$$\frac{1}{n}\sum_{k=1}^{n} OA_{i(k)}^{k}, \quad \text{where } i(k) \in \{1, 2\}.$$
The endpoints of these vectors are vertices of the parallelepipedon T_1 which contains $\frac{1}{n}T_2$. Since K is convex, $T_1 \subset K$. $\square$

References

1. V. V. Makeev, *The Knaster problem and almost spherical sections*, Mat. Sb. **180** (1989), 424–431; English transl., Math. USSR-Sb. **66** (1990), 431–438.
2. B. Rattray, *An antipodal-point, orthogonal point theorem*, Ann. Math. **60** (1954), 502–512.
3. M. Lassak, *Approximation of convex bodies by rectangles*, Geom. Dedicata (1993), 111–117.
4. A. S. Besicovich, *Measure of a symmetry of convex bodies*, J. London Math. Soc. **23** (1945), 237–240.
5. W. Böhme, *Ein Satz über ebene konvexe Figuren*, Math.-Phys. Semesterber. **6** (1958), 153–156.
6. B. Grünbaum, *Affineregular polygons inscribed in plane convex sets*, Riveon Lematematika **13** (1959), 20–24.
7. _____, *Measures of symmetry for convex sets*, Convexity (V. Klee, ed.), Proc. Symp. Pure Math., vol. 7, Amer. Math. Soc., Providence, RI, 1963, pp. 233–270.

Translated by B. BEKKER

ST. PETERSBURG STATE UNIVERSITY, RUSSIA, MATH. & MECH. DEPT., BIBLIOTECHNAYA PL. 2, STARY PETERGOF, ST. PETERSBURG, 198904, RUSSIA

Homology and Cohomology of Hypersurfaces with Quadratic Singular Points in Generic Position

N. Yu. Netsvetaev

Dedicated to our teachers

ABSTRACT. We calculate the homology groups of hypersurfaces of sufficiently high degree in $\mathbb{C}P^{n+1}$, $n \geqslant 3$, with a fixed number and, maybe, a position of singular points. In the case of quadratic singularities, we use the results of the calculations to give a topological description (as specific as possible) of such a hypersurface by means of decomposing it into a connected sum. In this case the topological type of the hypersurface is determined by its dimension, degree, and the number of singular points.

§0. Nonsingular hypersurfaces

In what follows we use only the homology and cohomology with integer coefficients. We denote by $b_n(X)$ the Betti number rk $H_n(X)$.

Let $X \subset \mathbb{C}P^{n+1}$ be a nonsingular hypersurface of degree d. Its differential type is determined by n and d: $X \cong X_n(d)$. As a model hypersurface we can take the one given by the Fermat equation

$$X_n(d) = \{x_0^d + x_1^d + \cdots + x_{n+1}^d = 0\}.$$

The structure of $X_n(d)$ was studied in sufficient detail by W. Browder, R. S. Kulkarni, A. S. Libgober, J. W. Wood, and others (see [5, 6] and the references therein). Here we mention the basic facts only. It easily follows from the Lefschetz Hyperplane Section Theorem that if $i \neq n$, then $H_i X = H_i \mathbb{C}P^n$. Recall that

$$H_i \mathbb{C}P^n = \begin{cases} \mathbb{Z}, & \text{if } i \text{ is even}, i \leqslant 2n, \\ 0, & \text{otherwise}. \end{cases}$$

Let y be the generator of $H^2 X = \mathbb{Z}$. For n even let $h \in H_n X$ be the homology class dual to $y^{n/2}$. The class h is primitive (indivisible). Let $n \neq 2$. $X_n(d)$ admits a

---

1991 *Mathematics Subject Classification.* Primary 14P25; Secondary 14M07.

The research described in this publication was made possible in part by Grant No. R4C000 from the International Science Foundation.

©1996, American Mathematical Society

connected sum decomposition of the form

$$X_n(d;s) \cong M_n(d) \# a(S^n \times S^n),$$

where $b_n(M_n(d)) = 0$ or 2, for n odd, and $b_n(M_n(d)) - |\operatorname{sign} X_n(d)| \leqslant 5$, for n even. It can be shown that the manifold $M_n(d)$ is uniquely determined up to a diffeomorphism (cf. [7, 8]).

§1. Diffeomorphism and rigid isotopy. Generic hypersurfaces

Below we restrict our consideration mainly by hypersurfaces with quadratic singularities, though all that follows can be extended, with appropriate changes, to the case of arbitrary isolated singularities.

The following definition seems to be most suitable for the topological study of the hypersurfaces with isolated singularities.

DEFINITION. Two hypersurfaces with quadratic singularities are *diffeomorphic* if there is a homeomorphism f between them such that
  (1) f is a diffeomorphism outside the singular points and
  (2) in the vicinities of the corresponding singular points, there are suitable holomorphic coordinates $x_1, \ldots, x_{n+1}; y_1, \ldots, y_{n+1}$ such that the hypersurfaces are given by $\sum_{i=1}^{n+1} x_i^2 = 0$ and $\sum_{i=1}^{n+1} y_i^2 = 0$, and f is given by $x_i = y_i$, $i = 1, \ldots, n+1$.

In other words, f must be (locally) extendable to a diffeomorphism of some neighborhoods of the hypersurfaces, holomorphic in the vicinities of singular points. We easily see that a diffeomorphism is also a PL-homeomorphism with respect to suitable triangulations of the hypersurfaces.

A sufficient condition for diffeomorphism is yielded by rigid isotopy.

DEFINITION. By a *rigid isotopy* we mean a 1-parametric family of hypersurfaces of fixed type. (It is sufficient that the Milnor numbers of singular points do not vary during the isotopy.)

It would be interesting to find two hypersurfaces which are diffeomorphic but not rigidly isotopic.

In its turn, the simplest way to prove the rigid isotopy is to prove that the corresponding moduli space is irreducible. Thus, we see that all nonsingular hypersurfaces of the same dimension and degree are diffeomorphic. More generally, there are several cases where the rigid isotopy type (and so the topological and the diffeomorphism type) of a hypersurface is uniquely determined by its dimension, degree, and the number of singular points. For simplicity we restrict our consideration by quadratic singularities only.

NOTATION. Let $P_1, \ldots, P_s \in \mathbb{C}P^{n+1}$. We define $\phi(n; \{P_1, \ldots, P_s\})$ as the minimal possible degree of a hypersurface Y with quadratic singularities such that $\operatorname{Sing} Y = \{P_1, \ldots, P_s\}$. If the points $P_1, \ldots, P_s$ are in "general position", then the number $\phi(n; \{P_1, \ldots, P_s\})$ depends only on n and s, and we denote it by $\phi(n; s)$.

PROPOSITION. *The rigid isotopy type of a hypersurface of degree d with quadratic singular points $P_1, \ldots, P_s \in \mathbb{C}P^{n+1}$ is uniquely determined in each of the following cases*:
  (1) *the singular points P_i are fixed*;
  (2) *the singular points P_i are in generic position and the degree d is sufficiently high*: $d \geqslant \phi(n, s)$;

(3) *the singular points P_i are arbitrary and $d > 2s$.*

Case (1) was noted by Dimca [2]. Cases (2) and (3) can be proved by means of extending his arguments.

In the above cases (1)–(3) we shall speak about *generic* hypersurfaces.

§2. Estimates for the number of singular points

In dimensions greater than 2, the best estimates for the number of singular points are due to Varchenko [9].

DEFINITION. The *Arnold number* $A(m, d)$ is defined by

$$A(m,d) = \text{card}\left\{(a_1,\ldots,a_m) \in \mathbb{Z}^m : \frac{(m-2)d}{2} < \sum_{i=1}^{m} a_i \leq \frac{md}{2}\right\}.$$

PROPOSITION (Arnold's Conjecture). *Let $X \subset \mathbb{C}^{n+1}$ be a hypersurface of degree d and dimension n with s isolated singularities. Then $s < A(n+1, d)$.*

This estimate was shown to be exact for cubic hypersurfaces by Kalker [4] (note that $A(n+1, 3) = \binom{n+1}{[n/2]}$). S. V. Chmutov's hypersurfaces show that it is asymptotically precise up to a factor of $\sqrt{3}$, cf. [1]. For quartics a series of "good" examples was found by Goryunov; see [3], which contains an interesting discussion as well.

By certain combinatorial calculations one can deduce the following proposition.

PROPOSITION 2.1. *If n is even, then for $(n-4)(d-2) > 17$, $(n,d) \neq (6,12)$, we have the inequality $s < \min\{b_n^+(X_n(d)), b_n^-(X_n(d))\}$.*

§3. Homologically standard hypersurfaces

Let $X \subset \mathbb{C}P^{n+1}$ be a hypersurface with quadratic singularities. It easily follows from the Lefschetz Hyperplane Section Theorem that if $i \neq n, n+1$, then $H_i X = H_i \mathbb{C}P^n$. Let y be the generator of $H_2 X = \mathbb{Z}$. For n even let $h \in H_n X$ be the homology class dual to $y^{n/2}$.

DEFINITION. We say that the hypersurface X is *homologically standard*, if it satisfies the following conditions (1)–(3):
(1) $\text{Tors } H_n X = 0$;
(2) $H_{n+1} X = H_{n+1} \mathbb{C}P^n$;
(3) for n even, the class $h \in H_n X$ is indivisible.

REMARKS. 1. If X is homologically standard, then all its homology groups are uniquely determined by its dimension, degree, and the number of singular points: $H_i X = H_i \mathbb{C}P^n$ for $i \neq n$, and $H_n X$ is a free abelian group of rank $b_n X = b_n X(n, d) - s$.

2. Condition (2) is automatically fulfilled if n is even.
3. Condition (3) is fulfilled if $\deg X \notin 2\mathbb{Z}$.

We start with the following technical proposition, which was obtained in collaboration with Oleg Ivanov. Its proof will be published elsewhere.

PROPOSITION 3.1. *Let $X', Y \subset \mathbb{C}P^{n+1}$ be two hypersurfaces with isolated singularities, which intersect transversally. (In particular, $X' \cap \text{Sing } Y = Y \cap \text{Sing } X' = \emptyset$.) Assume that $\{W_t\}$, $|t| < \varepsilon$, is a perturbation of the union $W_0 = X' \cup Y$ such that for*

$t \neq 0$, W_t is a hypersurface having only isolated singularities, which are of the same type as those of X' and Y and lie at the same points. Then W_t is a homologically standard hypersurface.

REMARK. Actually, it is sufficient that the singularities of W_t lie "close" to those of X' and Y, i.e., W_t gives a rigid isotopy of W_0 outside a neighborhood of $X' \cap \operatorname{Sing} Y = Y \cap \operatorname{Sing} X' = \varnothing$.

PROPOSITION 3.2. *Let $X, Y \subset \mathbb{C}P^{n+1}$ be two hypersurfaces with quadratic singularities such that $\operatorname{Sing} X = \operatorname{Sing} Y$ and $\deg X > \deg Y$. Then the hypersurface X is homologically standard.*

PROOF. Let X' be a nonsingular hypersurface of degree $\deg X - \deg Y$ which transversally intersects Y. Consider the union $W_0 = Y \cup X'$. We can apply Proposition 3.2 to the pencil $W_t = tX + (1-t)W_0$, considering X as a perturbation of W_0. It follows that $X = W_1$ is homologically standard.

For the definition of numbers $\phi(n; \{P_1, \ldots, P_s\})$ and $\phi(n; s)$ used in the statement of the following theorem see §1.

THEOREM 3.3. *Let $X \subset \mathbb{C}P^{n+1}$ be a hypersurface with s quadratic singularities which satisfies one of the following conditions*:
  (1) $\deg X > \phi(n; \operatorname{Sing} X)$;
  (2) *singular points of X are in generic position and* $\deg X > \phi(n; s)$;
  (3) $\deg X > 2s$.
Then the hypersurface X is homologically standard.

We see that assertion (1) implies assertions (2) and (3). (Because in conditions of (2) we have $\phi(n; \operatorname{Sing} X) = \phi(n; s)$, and always $\phi(n; \operatorname{Sing} X) \leqslant 2s$.)

PROOF. As already mentioned, it is sufficient to prove assertion (1), which in turn is an easy corollary of Proposition 3.3.

Indeed, let Y be a hypersurface of degree $\phi(n; \operatorname{Sing} X)$ with quadratic singularities such that $\operatorname{Sing} Y = \operatorname{Sing} X$. By the assumption, $\deg X > \deg Y$, and so by Proposition 3.3, the hypersurface X is homologically standard.

§4. Corollaries of homological standardness

THEOREM 4.1. *The differential type of a homologically standard hypersurface X of dimension not less than 3 with quadratic singularities is uniquely determined by $\dim X = n$, $\deg X = d$, and $\operatorname{card}(\operatorname{Sing} X) = s$: $X \cong X_n(d; s)$.*

THEOREM 4.2. *Let X be a homologically standard hypersurface and let one of the following conditions* (1)–(3) *be fulfilled*:
  (1) n *is odd*;
  (2) n *is even*, $(n-4)(d-2) > 17$, *and* $(n, d) \neq (6, 12)$;
  (3) n *is even*, $n \geqslant 4$, *and* $s < \min\{b_+(X_n(d)), b_-(X_n(d))\}$.
Then we have a connected sum decomposition

$$(*) \qquad X \cong X_n(d; s) \cong M_n(d) \# (a-s)(S^n \times S^n) \# s(S^n \times S^n/\Delta),$$

where Δ is the diagonal and $M_n(d)$ and a depend on n and d only.

REMARK. Note that assertion (3) of Theorem 4.2 implies assertion (2). Also, it is obvious that the connected sum decomposition (∗) implies the inequality $s \leqslant \min\{b_+(X_n(d)), b_-(X_n(d))\}$.

In our proof of Theorems 4.1 and 4.2 we use the following approach. Using the estimates on the number of singular points discussed in §2, one can show that the property of a hypersurface to be homologically standard implies that the arrangement of the vanishing cycles in the homology group of a close nonsingular hypersurface is standard, which in turn allows us to describe the topological structure of the singular hypersurface.

§5. Topological structure of generic hypersurfaces

By combining results of the above sections we can give a rather explicit topological description of a generic hypersurface (see §1).

THEOREM 5.1. *Let $X \subset \mathbb{C}P^{n+1}$ be a hypersurface of degree d and dimension n with s quadratic singularities. Assume that one of the following conditions hold*:
(1) *the singular points P_i are fixed and the degree d is sufficiently high*: $d > \phi(n, \{P_1, \ldots, P_s\})$;
(2) *the singular points P_i are in generic position and the degree d is sufficiently high*: $d > \phi(n; s)$;
(3) *the singular points P_i are arbitrary and $d > 2s + 1$.*

Then the hypersurface X is homologically standard, hence its differential type is uniquely determined by $\dim X = n$, $\deg X = d$, *and* $\operatorname{card}(\operatorname{Sing} X) = s$: $X \cong X_n(d; s)$.

More precisely, we have a connected sum decomposition

$$X \cong X_n(d;s) \cong M_n(d) \# (a-s)(S^n \times S^n) \# s(S^n \times S^n / \Delta),$$

where Δ is the diagonal and $M_n(d)$ and a depend on n and d only.

References

1. S. V. Chmutov, *Examples of projective surfaces with many singularities*, J. Alg. Geom. **1** (1992), 191–196.
2. A. Dimca, *On the homology and cohomology of complete intersections with isolated singularities*, Compositio Math. **58** (1986), 321–339.
3. V. V. Goryunov, *Symmetric quartics with many nodes*, Singularities and Bifurcations (V. I. Arnold, ed.), Advances Soviet Math., vol. 21, Amer. Math. Soc., Providence, RI, 1994, pp. 147–161.
4. A. Kalker, *Cubic fourfolds with fifteen ordinary double points*, Ph. D. Thesis, Leiden, 1986.
5. R. S. Kulkarni and J. W. Wood, *Topology of nonsingular complex hypersurfaces*, Adv. Math. **35** (1980), 239–263.
6. A. S. Libgober and J. W. Wood, *Differentiable structures on complete intersections*. I, Topology **21** (1982), 469–482; Singularities, Proc. Symp. Pure Math., vol. **40**, Part 2, Amer. Math. Soc., Providence, RI, 1983, pp. 123–133.
7. N. Yu. Netsvetaev, *Diffeomorphism and stable diffeomorphism of simply connected manifolds*, Algebra i Analiz **2** (1990), no. 2, 112–120; English transl. in Leningrad Math. J. **2** (1991).
8. _____, *Diffeomorphism criteria for smooth manifolds and algebraic varieties*, Contemp. Math., vol. 431 (Part 3), Amer. Math. Soc., Providence, RI, 1992, pp. 453–459.
9. A. N. Varchenko, *On semicontinuity of the spectrum and an upper bound for the number of singular points of projective hypersurfaces*, Dokl. Akad. Nauk SSSR **270** (1983); English transl. in Soviet Math. Dokl. **27** (1983).

Translated by THE AUTHOR

ST. PETERSBURG STATE UNIVERSITY, RUSSIA MATH. & MECH. DEPT., BIBLIOTECHNAYA PL. 2, STARY PETERGOF, ST. PETERSBURG, 198904, RUSSIA
E-mail address: `nyun@lomi.spb.su`

Asymptotic Solutions to the Quantized Knizhnik–Zamolodchikov Equation and Bethe Vectors

V. O. Tarasov and A. N. Varchenko

Dedicated to Boarding School #45 on the occasion of its 30th anniversary

ABSTRACT. Asymptotic solutions to the quantized Knizhnik–Zamolodchikov equation associated with $\mathfrak{gl}_{N+1}$ are constructed. The leading term of an asymptotic solution is the Bethe vector, an eigenvector of the transfer-matrix of a quantum spin chain model. We show that the norm of the Bethe vector is equal to the product of the Hessian of a suitable function and an explicitly written rational function. This formula is an analog of the Gaudin–Korepin formula for the norm of the Bethe vector. It is shown that, generically, the Bethe vectors form a base for the $\mathfrak{gl}_2$ case.

Introduction

In this paper we consider the quantized Knizhnik–Zamolodchikov equation associated with the Lie algebra $\mathfrak{gl}_{N+1}$. We describe asymptotic solutions to this equation. They are naturally related to the Bethe vectors, which are eigenvectors of the transfer-matrix of a quantum spin chain model. This connection allows us to give a formula for norms of the Bethe vectors. It is an analog of the Gaudin–Korepin formula established in [G, K] for the $\mathfrak{gl}_2$ case and in [R] for the $\mathfrak{gl}_3$ case and some choice of modules.

Let $\mathfrak{g} = \mathfrak{gl}_{N+1}$. Let $V_1, \ldots, V_n$ be highest weight $\mathfrak{g}$-modules and let $V = V_1 \otimes \cdots \otimes V_n$. The *quantized Knizhnik–Zamolodchikov equation* (qKZ) on a V-valued function $\Psi(z_1, \ldots, z_n)$ is the system of difference equations

$$\Psi(z_1, \ldots, z_i + p, \ldots, z_n) = K_i(z_1, \ldots, z_n; p) \Psi(z_1, \ldots, z_n),$$
(0.1) $\quad K_i(z_1, \ldots, z_n; p) = R_{i,i-1}(z_i - z_{i-1} + p) \cdots R_{i1}(z_i - z_1 + p)$
$$\times L_i(\mu) R_{ni}^{-1}(z_n - z_i) \cdots R_{i+1,i}^{-1}(z_{i+1} - z_i),$$

$i = 1, \ldots, n$, where $p \in \mathbb{C}$ and $\mu \in \mathbb{C}^{N+1}$ are parameters, $R_{ij}(z) \in \mathrm{End}(V)$ is the Yangian R-matrix corresponding to a pair V_i, V_j and $L_i(\mu) \in \mathrm{End}(V)$ is a suitable

---

1991 *Mathematics Subject Classification.* Primary 81T70.

The first author is supported by Russian Foundation for Fundamental Research and the Academy of Finland. The second author is supported by NSF Grant DMS-9203929.

©1996, American Mathematical Society

operator, which act nontrivially only in the corresponding V_i. (For the original definition and applications of qKZ, cf. [**FR, S**]).

The remarkable property of qKZ in that system (0.1) is holonomic. In particular, this means that operators $K_i(z;0)$ pairwise commute. Their eigenvectors can be obtained by the algebraic Bethe ansatz. The idea of this construction is to find a special vector-valued function $w(t,z)$ of $z_1,\ldots,z_n$ and auxiliary variables $t_1,\ldots,t_\ell$ and determine its arguments t so that the value of this function will be an eigenvector. The equations which determine these special values of arguments are called the *Bethe ansatz equations*.

In this paper we study asymptotic solutions to qKZ, as $p \to 0$. We use integral representations for solutions to qKZ obtained in [**M, R, V**] for the $\mathfrak{gl}_2$ case and in [**TV**] for the general case. An integral representation has the form $\Phi(t,z;p)w(t,z)$. Here $\Phi(t,z;p)$ is a scalar function and $w(t,z)$ is an V-valued rational function, the same one as in the algebraic Bethe ansatz.

As $p \to 0$, the leading term of the asymptotics of $\Phi(t,z;p)$ is equal to $\exp(\tau(t,z)/p + a(p))\Xi(t,z)$, where $\tau(t,z)$, $a(p)$, $\Xi(t,z)$ are some functions. We showed in [**TV**] that the equations of critical points of the function $\tau(t,z)$ with respect to the variables t coincide with the Bethe equations.

Let $z^* \in \mathbb{C}^n$ and let $t^* \in \mathbb{C}^\ell$ be a critical point of the function $\tau(t,z^*)$ with respect to t. We showed in [**TV**] that the vector $w(t^*,z^*)$ is an eigenvector of the operator $K_i(z^*,0)$ with eigenvalue

$$\exp\left(\frac{\partial \tau}{\partial z_i}(t^*,z^*)\right) \quad \text{for } i = 1,\ldots,n.$$

Let t^* be a nondegenerate t-critical point of the function $\tau(t,z^*)$. In this paper we construct an asymptotic solution to qKZ in a neighborhood of z^* with a leading term of the form $\exp(\tau(t(z),z)/p)w(t(z),z)$, where $t(z)$ is the nondegenerate t-critical point of $\tau(t,z)$ analytically depending on z and such that $t(z^*) = t^*$.

This construction connects the Bethe vectors and the asymptotic solutions to qKZ.

We introduce the dual system of difference equations

$$(0.2) \qquad \widetilde{\Psi}(z_1,\ldots,z_i+p,\ldots,z_n) = K_i^{*-1}(z_1,\ldots,z_n;p)\widetilde{\Psi}(z_1,\ldots,z_n)$$

on the V-valued function $\widetilde{\Psi}(z)$ and establish a connection between asymptotic solutions to (0.1) and (0.2). We establish this connection using an integral representation $\widetilde{\Phi}(t,z;p)\widetilde{w}(t,z)$ for solutions to the dual qKZ.

As $p \to 0$, the leading term of the asymptotics of $\widetilde{\Phi}(t,z;p)$ is equal to $\exp(-\tau(t,z)/p - a(p))\Xi(t,z)$, where $\tau(t,z)$, $a(p)$, $\Xi(t,z)$ are the same functions as before. This means that the critical points are the same for the qKZ and for the dual qKZ.

Following the historical tradition, we call $\langle \widetilde{w}(t,z), w(t,z)\rangle$ the *norm of the Bethe vector* $w(t,z)$. Here $\langle\,,\,\rangle \colon V^* \otimes V \to \mathbb{C}$ is the canonical pairing.

We show that for every nondegenerate critical point (t^*,z^*) of the function $\tau(t,z)$ we have

$$(0.3) \qquad \langle \widetilde{w}(t^*,z^*), w(t^*,z^*)\rangle = \operatorname{const} \Xi^{-2}(t^*,z^*)H(t^*,z^*),$$

where const does not depend on μ and does not change under continuous deformations of the critical point (t^*,z^*). Here $H(t,z)$ is the Hessian of the function $\tau(t,z)$ as a function of t.

We prove that const $= (-1)^\ell$ for the $\mathfrak{gl}_2$ case and, therefore, give another proof of the Gaudin–Korepin formula [K] for the norm of the Bethe vector. It is plausible that our proof can be generalized to the $\mathfrak{gl}_{N+1}$ case.

To compute const in (0.3), we perform a detailed analysis of the Bethe ansatz equations for the $\mathfrak{gl}_2$ case. This allows us to prove that, generically, the Bethe vectors form a base in $V_1 \otimes \cdots \otimes V_n$ for the $\mathfrak{gl}_2$ case. This statement is usually called the *completeness* of Bethe vectors.

If $\mu = 0$, then the Bethe vectors are singular vectors in $V_1 \otimes \cdots \otimes V_n$. Using results from [RV, V2], we show that for $\mu = 0$ the Bethe vectors form a base in $\text{Sing}(V_1 \otimes \cdots \otimes V_n)$ for the $\mathfrak{gl}_2$ case.

All results except the last one, which is related to the case $\mu = 0$, admit natural deformations to the $U_q(\mathfrak{gl}_{N+1})$ case.

The paper is organized as follows. The first section contains some general facts about asymptotic solutions to difference equations. In §2 we recall basic facts about qKZ and construct asymptotic solutions to qKZ. The dual qKZ is considered in §3. In §4 we compute const in (0.3) for the $\mathfrak{gl}_2$ case. We discuss the case $\mu = 0$ in §5. We describe the deformations of the results to the $U_q(\mathfrak{gl}_{N+1})$ case in §6. Appendix 1 contains the necessary details about integral representations for solutions to qKZ and the dual qKZ. Appendix 2 presents a short remark concerning §4.

The first author greatly appreciates the hospitality of the Research Institute for Theoretical Physics in Helsinki. Part of this work was done when the authors visited the University of North Carolina and the University of Tokyo. The authors thank these institutions for their hospitality and R. Kashaev for his discussions.

§1. Asymptotic solutions to difference equations

Let M be a region in $\mathbb{C}^n$, $\mathcal{H}$ be the space of holomorphic functions in M, $z = (z_1, \ldots, z_n)$, and

$$\partial_i = \frac{\partial}{\partial z_i}, \quad Z_i = \exp(p\partial_i), \quad D_i = \frac{Z_i - 1}{p}$$

in the sense of formal power series, p being a formal variable.

Let V be a finite-dimensional vector space. Consider the operators

$$(1.1) \qquad K_i(z;p) = \sum_{s=0}^{\infty} K_{is}(z) p^s \in \mathcal{H}[[p]] \otimes \text{End}(V), \qquad i = 1, \ldots, n,$$

such that $K_{i0}^{-1}(z) \in \mathcal{H}$. The last assumption means that $K_i(z;p)$ are invertible in $\mathcal{H}[[p]] \otimes \text{End}(V)$.

Consider the system of formal difference equations

$$(1.2) \qquad Z_i \Psi(z;p) = K_i(z;p) \Psi(z;p), \qquad i = 1, \ldots, n.$$

In other words,

$$\Psi(z_1, \ldots, z_i + p, \ldots, z_n) = K_i(z;p) \Psi(z;p)$$

for all i.

Assume that the system is holonomic. This means that $K_i(z;p)$ obey the compatibility conditions

$$(1.3) \qquad Z_i K_j(z;p) \cdot K_i(z;p) = Z_j K_i(z;p) \cdot K_j(z;p)$$

for all i, j. In particular, $[K_{i0}(z), K_{j0}(z)] = 0$. We are interested in formal solutions $\Psi(z;p)$ to system (1.2) of the form

$$\Psi(z;p) = \exp(\tau(z)/p)\overset{\circ}{\Psi}(z;p), \qquad \tau(z) \in \mathcal{H}, \ \overset{\circ}{\Psi}(z;p) \in \mathcal{H}[[p]] \otimes V.$$

(1.4) DEFINITION. $\Psi(z;p)$ is an *asymptotic solution* to system (1.2) if

$$\exp(D_i\tau(z)) \cdot Z_i\overset{\circ}{\Psi}(z;p) = K_i(z;p)\overset{\circ}{\Psi}(z;p)$$

for all i.

These conditions are equivalent to (1.2).

(1.5) LEMMA. *Let $\Psi(z;p)$ be an asymptotic solution to system (1.2). Then*

$$K_{i0}(z)\Psi_0(z) = \exp(\partial_i\tau(z))\Psi_0(z).$$

The proof comes from the leading term of equality (1.4).

(1.6) THEOREM [**TV**, Theorem 5.1.5]. *Suppose $w(z) \in \mathcal{H}$ is a common eigenvector of $K_{i0}(z)$,*

$$K_{i0}(z)w(z) = E_i(z)w(z),$$

and, for any z, the space V is spanned by $w(z)$ and the subspaces $\mathrm{im}(K_{i0}(z) - E_i(z))$, $i = 1, \ldots, n$. Then
  i) $E_i(z) \in \mathcal{H}$;
  ii) *there is an asymptotic solution $\Psi(z;p)$ to system (1.2) such that $\Psi_0(z)$ is proportional to $w(z)$;*
  iii) *such a solution is unique modulo a scalar factor of the form*

$$\exp(\alpha/p)\beta(p), \qquad \alpha \in \mathbb{C}, \ \beta(p) \in \mathbb{C}[[p]].$$

REMARK. If the family $\{K_{i0}(z)\}_{i=1}^n$ acts semisimply in V for any z, then each common eigenvector of these operators obeys the conditions of Theorem (1.6).

Let V^* be the dual space, and $\langle\ ,\ \rangle : V^* \otimes V \to \mathbb{C}$ the canonical pairing. Consider the operators $K_i^*(z;p) \in \mathcal{H}[[p]] \otimes V^*$ and $\widetilde{K}_i(z;p) = (K_i^*(z;p))^{-1}$. They obey compatibility conditions

(1.7) $$Z_i\widetilde{K}_j(z;p) \cdot \widetilde{K}_i(z;p) = Z_j\widetilde{K}_i(z;p) \cdot \widetilde{K}_j(z;p)$$

similar to (1.3).

(1.8) DEFINITION. The holonomic system of formal difference equations

$$Z_i\widetilde{\Psi}(z;p) = \widetilde{K}_i(z;p)\widetilde{\Psi}(z;p), \qquad i = 1, \ldots, n,$$

is called the system *dual* to system (1.2).

Let $\Psi(z;p) = \exp(\tau(z)/p)\overset{\circ}{\Psi}(z;p)$ and $\widetilde{\Psi}(z;p)\exp(\widetilde{\tau}(z)/p)\overset{\bullet}{\Psi}(z;p)$ be asymptotic solutions to systems (1.2) and (1.8), respectively.

(1.9) LEMMA. *For all i we have*

$$Z_i\langle\overset{\bullet}{\Psi}(z;p), \overset{\circ}{\Psi}(z;p)\rangle = \exp(-D_i(\tau(z) + \widetilde{\tau}(z)))\langle\overset{\bullet}{\Psi}(z;p), \overset{\circ}{\Psi}(z;p)\rangle.$$

The proof immediately follows from the definition of asymptotic solutions.

(1.10) COROLLARY. i) $\langle \widetilde{\Psi}_0(z), \Psi_0(z) \rangle = \text{const}$;
ii) $\langle \widetilde{\Psi}_0(z), \Psi_0(z) \rangle = 0$, if there is j such that $\partial_j(\tau(z) + \tilde{\tau}(z)) \notin 2\pi i \mathbb{Z}$.

The proof is given by the zero and the first order terms in p of (1.9).

REMARK. The second part of Corollary (1.10) means that eigenvectors for operators $K_i(z)$ and $K_i^*(z)$ with different eigenvalues are orthogonal.

§2. Integral representations and asymptotic solutions to qKZ

Let $\mathfrak{g} = \mathfrak{gl}_{N+1}$ and $\{E_{ij}\}$ be the canonical generators of $\mathfrak{gl}_{N+1}$. The *Yangian* $Y = Y(\mathfrak{gl}_{N+1})$ is the unital associative algebra generated by elements T_{ij}^s, $i, j = 1, \ldots, N+1$, $s \in \mathbb{Z}_{>0}$, subject to the relations:

$$[T_{ij}^{s+1}, T_{kl}^t] - [T_{ij}^s, T_{kl}^{t+1}] = T_{kj}^s T_{il}^t - T_{kj}^t T_{il}^s \quad \text{for } s, t \geq 0,$$

where $T_{ij}^0 = \delta_{ij}$. Y is a Hopf algebra with the coproduct

$$\Delta \colon T_{ij}^s \mapsto \sum_{t=0}^{s} \sum_{k=1}^{N+1} T_{ik}^t T_{kj}^{s-t}.$$

There is a homomorphism of algebras $\varphi \colon Y \to U(\mathfrak{g})$, $T_{ij}^s \stackrel{\varphi}{\mapsto} E_{ji}\delta_{s1}$, which makes any $\mathfrak{g}$-module into a Y-module. For any $z \in \mathbb{C}$, the Yangian Y has a canonical automorphism θ_z:

$$\theta_z \colon T_{ij}^s \mapsto \sum_{t=1}^{s} \binom{s-1}{t-1} z^{s-t} T_{ij}^t.$$

Set $\varphi_z = \varphi \circ \theta_z$. For any two highest weight $\mathfrak{g}$-modules V_1 and V_2 with generating vectors v_1 and v_2, respectively, there is a unique R-matrix $R_{V_1 V_2}(z) \in \text{End}(V_1 \otimes V_2)$ such that for any $X \in Y$,

(2.1) $\quad R_{V_1 V_2}(z_1 - z_2)(\varphi_{z_1} \otimes \varphi_{z_2}) \circ \Delta(X) = (\varphi_{z_1} \otimes \varphi_{z_2}) \circ \Delta'(X) R_{V_1 V_2}(z_1 - z_2)$

in $\text{End}(V_1 \otimes V_2)$ and

(2.2) $\quad\quad\quad\quad\quad\quad\quad R_{V_1 V_2}(z) v_1 \otimes v_2 = v_1 \otimes v_2.$

Here $\Delta' = P \circ \Delta$ and P is a permutation of factors in $Y \otimes Y$. $R_{V_1 V_2}(z)$ preserves the weight decomposition of $V_1 \otimes V_2$ considered as a $\mathfrak{g}$-module; its restriction to any weight subspace of $V_1 \otimes V_2$ is a rational function in z.

Let V be a highest weight $\mathfrak{g}$-module. For any $\mu \in \mathbb{C}^{N+1}$ introduce $L(\mu) \in \text{End}(V)$:

$$L(\mu) = \exp\left(\sum_{i=1}^{N+1} \mu_i E_{ii}\right).$$

Let $V_1, \ldots, V_n$ be highest weight $\mathfrak{g}$-modules. If $R_{V_i V_j}(z) = \sum_d R_{(i)}^d(z) \otimes R_{(j)}^d(z)$, then we set

$$R_{ij}(z) = \sum_d 1 \otimes \cdots \otimes R_{(i)}^d(z) \otimes \cdots \otimes R_{(j)}^d(z) \otimes \cdots \otimes 1 \in \text{End}(V_1 \otimes \cdots \otimes V_n),$$

where $R^d_{(i)}(z)$ stands for the ith factor and $R^d_{(j)}(z)$, for the jth factor. Let us also set

$$L_i(\mu) = 1 \otimes \cdots \otimes L(\mu) \otimes \cdots \otimes 1 \in \mathrm{End}\,(V_1 \otimes \cdots \otimes V_n),$$

where $L(\mu)$ stands for the ith factor.

Let $p \in \mathbb{C}$ and $z = (z_1, \ldots, z_n)$. Denote by Z_i the p-shift operator:

$$Z_i \colon \Psi(z_1, \ldots, z_n) \mapsto \Psi(z_1, \ldots, z_i + p, \ldots, z_n).$$

(2.3) DEFINITION. The operators

$$\begin{aligned}K_i(z;p) = &\,R_{i,i-1}(z_i - z_{i-1} + p) \cdots R_{i1}(z_i - z_1 + p) \\ &\times L_i(\mu)\,R_{ni}^{-1}(z_n - z_i) \cdots R_{i+1,i}^{-1}(z_{i+1} - z_i),\end{aligned}$$

$i = 1, \ldots, n$, are called the *qKZ operators*.

The qKZ operators preserve the weight decomposition of the $\mathfrak{g}$-module $V_1 \otimes \cdots \otimes V_n$ and their restrictions to any weight subspace are rational functions in z.

(2.4) THEOREM [**FR**, Theorem (5.4)]. *The qKZ operators satisfy the compatibility conditions*

$$Z_i K_j(z;p) \cdot K_i(z;p) = Z_j K_i(z;p) \cdot K_j(z;p)$$

(cf. (1.3)).

(2.5) DEFINITION. The *quantized Knizhnik–Zamolodchikov equation* (qKZ) is the holonomic system of difference equations for a $V_1 \otimes \cdots \otimes V_n$-valued function $\Psi(z;p)$,

$$Z_i \Psi(z;p) = K_i(z;p)\,\Psi(z;p)$$

for $i = 1, \ldots, n$, [**FR**] (cf. (1.2)).

Fix $\lambda \in \mathbb{Z}_{\geq 0}^N$. Let $(\Lambda_1(1), \ldots, \Lambda_{N+1}(1)), \ldots, (\Lambda_1(n), \ldots, \Lambda_{N+1}(n))$ be the highest weights of the $\mathfrak{g}$-modules $V_1, \ldots, V_n$, respectively. Let $V_\lambda = (V_1 \otimes \cdots \otimes V_n)_\lambda$ be the weight subspace:
(2.6)
$$V_\lambda = \left\{ v \in V_1 \otimes \cdots \otimes V_n \;\middle|\; E_{ii} v = \left(\lambda_{i-1} - \lambda_i + \sum_{m=1}^n \Lambda_i(m)\right) v,\; i = 1, \ldots, N+1 \right\},$$

where $\lambda_0 = \lambda_{N+1} = 0$. Later on we will be interested in solutions to system (2.5) with values in V_λ.

Set $\ell = \sum_{i=1}^N \lambda_i$. Let

$$t = (t_{11}, \ldots, t_{1\lambda_1}, t_{21}, \ldots, t_{2\lambda_2}, \ldots, t_{N1}, \ldots, t_{N\lambda_N}) \in \mathbb{C}^\ell.$$

(2.7) DEFINITION. The function

$$\Phi(t,z;p) = \prod_{m=1}^{n} \prod_{i=1}^{N} \exp(z_m \mu_i \Lambda_i(m)/p) \prod_{i=1}^{N} \prod_{j=1}^{\lambda_i} \exp(t_{ij}(\mu_{i+1} - \mu_i)/p)$$

$$\times \prod_{m=1}^{n} \prod_{i=1}^{N} \prod_{j=1}^{\lambda_i} \frac{\Gamma((t_{ij} - z_m + \Lambda_i(m))/p)}{\Gamma((t_{ij} - z_m + \Lambda_{i+1}(m))/p)}$$

$$\times \prod_{i=1}^{N} \prod_{j=2}^{\lambda_i} \prod_{k=1}^{j-1} \frac{\Gamma((t_{ik} - t_{ij} - 1)/p)}{\Gamma((t_{ik} - t_{ij} + 1)/p)}$$

$$\times \prod_{i=1}^{N-1} \prod_{j=1}^{\lambda_i} \prod_{k=1}^{\lambda_{i+1}} \frac{\Gamma((t_{i+1,k} - t_{ij} + 1)/p)}{\Gamma((t_{i+1,k} - t_{ij})/p)}$$

is called the *phase function* of the weight subspace V_λ. Here Γ is the gamma-function.

Introduce the lexicographical ordering on the set of pairs (i,j): $(i,j) < (k,l)$ if $i < k$ or $i = k$ and $j < l$. Let $a, b, \ldots$ stay for $(i,j), (k,l), \ldots$. Let Q_a be the p-shift operator with respect to the variable t_a.

(2.8) DEFINITION.

$$\nabla_a \Phi(t,z) = \lim_{p \to 0} ((\Phi(t,z;p))^{-1} Q_a \Phi(t,z;p)),$$

$$\nabla^2_{ab} \Phi(t,z) = (\nabla_b \Phi(t,z))^{-1} \frac{\partial}{\partial t_a} \nabla_b \Phi(t,z),$$

$$H(t,z) = \det[\nabla^2_{ab} \Phi(t,z)]_{\ell \times \ell}.$$

Below we give a description of the space of admissible functions that will be used in the paper. Let $\mathcal{H}^0_\lambda$ be the space spanned by entries of the operators $K_i(z;p)$ restricted to V_λ, $i = 1, \ldots, n$. Let $\mathcal{H}_\lambda$ be the space spanned by products $g_1 \cdots g_s$, where $g_i \in \mathcal{H}^0_\lambda$ for all i and $s \in \mathbb{Z}_{\geq 0}$. Consider the following linear functions:

(2.9) $\quad t_{ij} - z_m + \Lambda_i(m) - p, \quad t_{ij} - z_m + \Lambda_{i+1}(m), \quad t_{ij} - t_{ik} - 1, \quad t_{i+1,l} - t_{ij},$

$m = 1, \ldots, n, i = 1, \ldots, N, j = 1, \ldots, \lambda_i, k = 1, \ldots, j-1, l = 1, \ldots, \lambda_{i+1}$. Let $\mathcal{F}_0$ be the space spanned by the products $g_1^{-1} \cdots g_s^{-1}$, $s \in \mathbb{Z}_{\geq 0}$, where each g_i is a linear function from the list (2.9) and $g_i \neq g_j$ for $i \neq j$. Set $\mathcal{F} = \mathbb{C}[t,z,p] \otimes \mathcal{F}_0$.

(2.10) DEFINITION. Let $\mathcal{Q}_\lambda$ be the space, spanned over $\mathbb{C}$ by the discrete differentials $Q_a(\Phi w) - \Phi w$, $a = 1, \ldots, \ell$, $w \in \mathcal{F} \otimes \mathcal{H}_\lambda$.

(2.11) DEFINITION. Let $w(t,z;p) \in \mathcal{F} \otimes V_\lambda$. The product $\Phi(t,z;p) w(t,z;p)$ is said to give an *integral representation* for solutions to system (2.5) if $Z_i(\Phi w) - K_i \Phi w \in \mathcal{Q}_\lambda \otimes V_\lambda$, $i = 1, \ldots, n$.

(2.12) THEOREM [**TV**, Theorem 1.5.2]. *There exists an integral representation for solutions to qKZ* (2.5) *associated with* $\mathfrak{gl}_{N+1}$.

The $\mathfrak{gl}_2$ case was considered in [**M, R2, V1**]. Explicit formulas for an integral representation are given in [**TV**] and Appendix 1.

REMARK. In [**TV**], we defined $\Phi(t, z; p)$ and $w(t, z; p)$ and proved that the differences $Z_i(\Phi w) - K_i \Phi w$ are discrete differentials. We did not specify the singularities of these differences, but the proof in [**TV**] clearly shows that these differences belong to $\mathcal{Q}_\lambda \otimes V_\lambda$.

(2.13) DEFINITION. A point (t, z) is called a *critical point* if $\nabla_a \Phi(t, z) = 1$ for $a = 1, \ldots, \ell$. A critical point (t, z) is called *nondegenerate* if $H(t, z) \neq 0$.

Let $M \subset \mathbb{C}^\ell$ be an open region such that all $K_i(z; 0)$ and $K_i^{-1}(z; 0)$ are regular in M. The qKZ operators $K_i(z; p)$ have power series expansions

$$K_i(z; p) = \sum_{s=0}^{\infty} K_{is}(z) p^s,$$

where $K_{is}(z)$ are also regular in M (cf. (1.1)). Now we are in a position related to §1, and we are interested in asymptotic solutions to system (2.5) as $p \to 0$.

REMARK. Actually, we must consider restrictions of qKZ operators to V_λ which are rational functions in z, p. In this case we can take M to be the complement to the singularities of $K_i(z; 0)$, $K_i^{-1}(z; 0)$.

Set $\chi(x) = x \log x - x$. Introduce $\tau(t, z)$ as follows:

(2.14)
$$\begin{aligned}
\tau(t, z) = &\sum_{m=1}^{n} \sum_{i=1}^{N} z_m \mu_i \Lambda_i(m) + \sum_{i=1}^{N} \sum_{j=1}^{\lambda_i} t_{ij}(\mu_{i+1} - \mu_i) \\
&+ \sum_{m=1}^{n} \sum_{i=1}^{N} \sum_{j=1}^{\lambda_i} (\chi(t_{ij} - z_m + \Lambda_i(m)) - \chi(t_{ij} - z_m + \Lambda_{i+1}(m))) \\
&+ \sum_{i=1}^{N-1} \sum_{j=1}^{\lambda_i} \sum_{k=1}^{\lambda_{i+1}} (\chi(t_{i+1,k} - t_{ij} + 1) - \chi(t_{i+1,k} - t_{ij})) \\
&+ \sum_{i=1}^{N} \sum_{j=2}^{\lambda_i} \sum_{k=1}^{j-1} (\chi(t_{ik} - t_{ij} - 1) - \chi(t_{ik} - t_{ij} + 1)).
\end{aligned}$$

(2.15) LEMMA. $\nabla_a \Phi(t, z) = \exp\left(\dfrac{\partial}{\partial t_a} \tau(t, z)\right)$.

(2.16) COROLLARY. $\nabla_{ab}^2 \Phi(t, z) = \nabla_{ba}^2 \Phi(t, z)$.

Let $p = \rho e^{i\vartheta}$, $\rho = |p|$. Choose the branch of $\log x$ such that

$$|\operatorname{Im} \log x - \vartheta| < \pi.$$

Let $\mathfrak{S}_\vartheta \subset \mathbb{C}^{\ell+n}$ be the cuts defining the corresponding branch of $\tau(t, z)$. Set

$$\Theta = \sum_{m=1}^{n} \sum_{i=1}^{N} \lambda_i (\Lambda_i(m) - \Lambda_{i+1}(m)) - \sum_{i=1}^{N} \lambda_i (\lambda_i - 1) + \sum_{i=1}^{N-1} \lambda_i \lambda_{i+1}.$$

Let $\mathcal{F}_\circ$ be the space of polynomials in t, z and the following rational functions:

$$(t_{ij} - z_m + \Lambda_i(m))^{-1}, \quad (t_{ij} - z_m + \Lambda_{i+1}(m))^{-1},$$
$$(t_{ij} - t_{ik} - 1)^{-1}, \quad (t_{i+1,l} - t_{ij})^{-1}, \quad (t_{i+1,l} - t_{ij} + 1)^{-1},$$
$$m = 1, \ldots, n, i = 1, \ldots, N, j, k = 1, \ldots, \lambda_i, l = 1, \ldots, \lambda_{i+1}.$$

(2.17) LEMMA. *Let $(t,z) \notin \mathfrak{S}_\vartheta$. As $\rho \to 0$, $\Phi(t,z;p)$ have the asymptotic expansion*

$$\Phi(t,z;p) \simeq \exp(-p^{-1}\log p \cdot \Theta + p^{-1}\tau(t,z))\Xi(t,z)\left(1 + \sum_{s=1}^{\infty} \phi_s(t,z)p^s\right),$$

where

$$\Xi(t,z) = \bigg(\prod_{m=1}^{n}\prod_{i=1}^{N}\prod_{j=1}^{\lambda_i} \frac{t_{ij} - z_m + \Lambda_i(m)}{t_{ij} - z_m + \Lambda_{i+1}(m)}$$

$$\times \prod_{i=1}^{N}\prod_{j=2}^{\lambda_i}\prod_{k=1}^{j-1} \frac{t_{ik} - t_{ij} - 1}{t_{ik} - t_{ij} + 1} \prod_{i=1}^{N-1}\prod_{j=1}^{\lambda_i}\prod_{k=1}^{\lambda_{i+1}} \frac{t_{i+1,k} - t_{ij} + 1}{t_{i+1,k} - t_{ij}}\bigg)^{-1/2}$$

and $\phi_s(t,z) \in \mathcal{F}_\circ$. Here $\log p = \log \rho + i\vartheta$.

The lemma follows from the Stirling formula.

Let $(t^*, z^*) \notin \mathfrak{S}_\vartheta$, $z^* \in M$ be a nondegenerate critical point. Consider the quadratic form

$$(2.18) \qquad S(x) = e^{-i\vartheta}\sum_{a=1}^{\ell}\sum_{b=1}^{\ell} x_a x_b \nabla^2_{ab}\Phi(t^*,z^*), \qquad x \in \mathbb{C}^\ell.$$

Let $W \subset \mathbb{C}^\ell$, $\dim_\mathbb{R} W = \ell$, be a real hyperplane such that the restriction of $S(x)$ to W is negative. Let $D \subset \mathbb{C}^\ell$ be a small disk:

$$(2.19) \qquad D = \{t \in \mathbb{C}^\ell \mid t - t^* \in W, |t - t^*| < \varepsilon\},$$

where ε is a small positive number. We have

$$(2.20) \qquad \max_{t \in \partial D} \operatorname{Re} S(t - t^*) < 0,$$

where ∂D is a boundary of D.

A point (t^*, z^*) is a nondegenerate critical point if and only if t^* is a nondegenerate solution to the system of equations

$$\nabla_a \Phi(t, z^*) = 1, \qquad a = 1, \ldots, \ell.$$

Hence, in a neighborhood of z^* we can define a holomorphic function $t(z)$ such that $(t(z), z)$ is a nondegenerate critical point and $t(z^*) = t^*$.

Later on we assume that p is small and $(t^*, z^*) \notin \mathfrak{S}_\vartheta$. Set

$$(2.21) \qquad I_a = \frac{1}{2\pi i}\frac{\partial}{\partial t_a}\tau(t^*,z^*) \quad \text{and} \quad I(t) = \exp\bigg(-2\pi i p^{-1}\sum_{a=1}^{\ell} I_a t_a\bigg).$$

It is clear that $\partial \tau(t(z),z)/\partial t_a = 2\pi i I_a$, and $I(t)$ is a p-periodic function with respect to all t_a. Set

$$(2.22) \qquad \Psi(z;p) = p^{-\ell/2}\exp(p^{-1}\log p \cdot \Theta)\int_D I(t)\Phi(t,z;p)w(t,z;p)\,d^\ell t,$$

where $w(t, z; p) \in \mathcal{F} \otimes V_\lambda$, and set

$$\widehat{\tau}(t, z) = \tau(t, z) - 2\pi i \sum_{a=1}^{\ell} I_a t_a.$$

(2.23) LEMMA. *As $p \to 0$, $\Psi(z; p)$ has the asymptotic expansion*

$$\Psi(z; p) \simeq (-2\pi)^{\ell/2} \exp(\widehat{\tau}(t(z), z)/p) \Xi(t(z), z) H^{-1/2}(t(z), z)$$
$$\times \left(w(t(z), z; 0) + \sum_{s=1}^{\infty} \psi_s(t(z), z) p^s \right),$$

where $\psi_s(t, z) \in \mathcal{F}_\circ \otimes V_\lambda$.

The lemma is a direct corollary of Lemma (2.17) and the method of steepest descend. (Cf., for example, §11 in [**AGV**].)

REMARK. Let $F(t, z; p) = \exp(\widehat{\tau}(t, z)/p) \Xi(t, z) f(t, z; p)$, where

$$\bar{\tau}(t, z) = e^{-i\vartheta} \widehat{\tau}(t, z) \quad \text{and} \quad f(t, z; p) \in \mathcal{F}.$$

It follows from (2.20) that for a small fixed ε and z close to z^* we have

$$\max_{t \in \partial D} \text{Re}(\bar{\tau}(t, z) - \bar{\tau}(t(z), z)) < 0.$$

Let $B \subset \mathbb{C}^\ell$ be a small ball centered at t^*. Let

$$B^- = \{ t \in B \mid \text{Re}(\bar{\tau}(t, z^*) - \bar{\tau}(t^*, z^*)) < 0 \}.$$

It is well known that $H_\ell(B, B^-, \mathbb{Z}) = \mathbb{Z}$ and, moreover, if D and D' are two cycles generating the same element in H_ℓ, then

$$\int_D F(t, z; p) d^\ell t \quad \text{and} \quad \int_{D'} F(t, z; p) d^\ell t$$

have the same asymptotic expansions. (Cf. §11 in [**AGV**].) Therefore, we can take D to be any cycle generating H_ℓ.

(2.24) THEOREM. *Let $\Phi(t, z; p) w(t, z; p)$ be an integral representation for solutions to qKZ (2.5). The asymptotic expansion of $\Psi(z; p)$ as $p \to 0$ gives an asymptotic solution to system (2.5) in the sense of (1.4).*

PROOF. For any a, $I_a \in \mathbb{Z}$ and $Q_a I(t) = I(t)$, because (t^*, z^*) is a critical point. Hence, we have $Z_i(I\Phi w) - K_i I\Phi w \in \mathcal{Q}_\lambda$. For any $\Omega(t, z; p) \in \mathcal{Q}_\lambda$, from Lemma (2.17) and the method of steepest descend we obtain

(2.25) $$\exp(p^{-1} \log p \cdot \Theta - p^{-1} \tau(t(z), z)) \int_D \Omega(t, z; p) d^\ell t = O(p^\infty)$$

as $p \to 0$. Now the theorem follows from Lemma (2.23) and (2.25). □

(2.26) COROLLARY.

$$K_i(z^*; 0) w(t^*, z^*; 0) = \exp\left(\frac{\partial \tau}{\partial z_i}(t^*, z^*) \right) w(t^*, z^*; 0), \qquad i = 1, \ldots, n.$$

PROOF. It follows from Theorem (2.24), Lemmas (1.5), (2.23) and the relation
$$\left.\frac{\partial \widehat{\tau}(t(z),z)}{\partial z_i}\right|_{z=z^*} = \left.\frac{\partial \tau(t,z)}{\partial z_i}\right|_{(t,z)=(t^*,z^*)}. \qquad \square$$

(2.27) DEFINITION. A critical point (t, z) is called an *offdiagonal* critical point if $t_{ij} \neq t_{ik}$ for $(i, j) \neq (i, k)$, and a *diagonal* critical point, otherwise.

(2.28) THEOREM. *Let (t^*, z^*) be a diagonal nondegenerate critical point, and let $\Psi(z; p)$ be defined by (2.22). Then $\exp(-\widehat{\tau}(t(z),z)/p)\Psi(z; p) = O(p^\infty)$ as $p \to 0$.*

PROOF. Suppose, for example, that $t_{11}^* = t_{12}^*$. Let $\bar{t}$ be obtained from t by the permutation of t_{11} and t_{12}. From the explicit formulas for $w(t, z; p)$ (cf. Appendix 1), we obtain

(2.29) $$w(\bar{t}, z; p) = \frac{t_{11} - t_{12} + 1}{t_{11} - t_{12} - 1} w(t, z; p).$$

From (2.7) and (2.21) it follows that
$$\frac{I(\bar{t})\Phi(\bar{t},z;p)}{I(t)\Phi(t,z;p)} = -\frac{t_{11} - t_{12} - 1}{t_{11} - t_{12} + 1} (1 + O(p^\infty)).$$

Hence, as $p \to 0$, the integrand in the right-hand side of (2.22) is an odd function with respect to the permutation of t_{11} and t_{12} in the asymptotic sense. On the other hand, the quadratic form $S(x)$ (cf. (2.18)) is preserved by the permutation of x_{11} and x_{12}. This means that
$$S(x) = \alpha(x_{11} - x_{12})^2 + \bar{S}(x),$$
where $\bar{S}(x)$ is a quadratic form in $x_{11} + x_{12}$ and all other x_{ij}'s. Now it is clear that we can find the required real hyperplane W, which is invariant under the permutation of x_{11} and x_{12}. Therefore, the disk D is also invariant, and the integral in the right-hand side of (2.22) must vanish in the asymptotic sense, as $p \to 0$. $\square$

§3. Dual qKZ and norms of Bethe vectors

Let V be a highest weight $\mathfrak{g}$-module with generating vector v. The restricted dual space V^* admits the natural structure of a right $\mathfrak{g}$-module. Let $v^* \in V^*$ be such that $\langle v^*, v \rangle = 1$ and $\langle v^*, v' \rangle = 0$ for any weight vector v', which is not proportional to v. If V is irreducible, V^* is an irreducible right lowest weight $\mathfrak{g}$-module with generating vector v^*.

Let V_1, V_2 be two highest weight $\mathfrak{g}$-module with generating vectors v_1, v_2, respectively. Let $R_{V_1 V_2}(z)$ be the corresponding R-matrix defined by (2.1), (2.2). Set $\widetilde{R}_{V_1 V_2}(z) = (R^*_{V_1 V_2}(z))^{-1}$. For any $X \in Y$ we have

(3.1) $$\widetilde{R}_{V_1 V_2}(z_1 - z_2)(\varphi_{z_1} \otimes \varphi_{z_2})\Delta(X) = (\varphi_{z_1} \otimes \varphi_{z_2})\Delta'(X)\widetilde{R}_{V_1 V_2}(z_1 - z_2)$$

in $\text{End}(V_1^* \otimes V_2^*)$ and

(3.2) $$\widetilde{R}_{V_1 V_2}(z) v_1^* \otimes v_2^* = v_1^* \otimes v_2^*.$$

Let us introduce the *dual qKZ operators* $\widetilde{K}_i(z; p) = (K_i^*(z; p))^{-1}$. They obey compatibility conditions

(3.3) $$Z_i \widetilde{K}_j(z;p) \cdot \widetilde{K}_i(z;p) = Z_j \widetilde{K}_i(z;p) \cdot \widetilde{K}_j(z;p)$$

similar to (2.4).

(3.4) DEFINITION. The *dual qKZ* is the holonomic system of difference equations for a $V_1^* \otimes \cdots \otimes V_n^*$-valued function $\widetilde{\Psi}(z;p)$:

$$Z_i \widetilde{\Psi}(z;p) = \widetilde{K}_i(z;p) \widetilde{\Psi}(z;p)$$

for $i = 1, \ldots, n$, (cf. (1.8), (2.5)).

Fix $\lambda \in \mathbb{Z}_{\geq 0}^N$. Let $(\Lambda_1(1), \ldots, \Lambda_{N+1}(1)), \ldots, (\Lambda_1(n), \ldots, \Lambda_{N+1}(n))$ be highest weights of $\mathfrak{g}$-modules $V_1, \ldots, V_n$, respectively.

Let $V_\lambda^* = (V_1^* \otimes \cdots \otimes V_n^*)_\lambda$ be the dual weight subspace:
(3.5)
$$V_\lambda^* = \left\{ v^* \in V_1^* \otimes \cdots \otimes V_n^* \;\middle|\; E_{ii} v^* = \left(\lambda_{i-1} - \lambda_i + \sum_{m=1}^n \Lambda_i(m) \right) v^*, \quad i = 1, \ldots, N+1 \right\},$$

where $\lambda_0 = \lambda_{N+1} = 0$. Later on we shall be interested in solutions to system (3.4) with values in V_λ^*.

Set $\ell = \sum_{i=1}^N \lambda_i$. Let

$$t = (t_{11}, \ldots, t_{1\lambda_1}, t_{21}, \ldots, t_{2\lambda_2}, \ldots, t_{N1}, \ldots, t_{N\lambda_N}) \in \mathbb{C}^\ell.$$

(3.6) DEFINITION. The function $\widetilde{\Phi}(t,z;p) = \Xi^2(t,z) \Phi^{-1}(t,z;p)$ is called the *phase function* of the weight subspace V_λ^*.

Let $\widetilde{\mathcal{Q}}_\lambda$ be the space spanned over $\mathbb{C}$ by the discrete differentials

$$Q_a(\widetilde{\Phi} w) - \widetilde{\Phi} w, \qquad a = 1, \ldots, \ell, \; w \in \mathcal{F} \otimes \mathcal{H}_\lambda.$$

(3.7) DEFINITION. Let $\widetilde{w}(t,z;p) \in \mathcal{F} \otimes V_\lambda^*$. The product $\widetilde{\Phi}(t,z;p) \widetilde{w}(t,z;p)$ is said to give an *integral representation* for solutions to system (3.4) if $Z_i(\widetilde{\Phi}\widetilde{w}) - \widetilde{K}_i \widetilde{\Phi}\widetilde{w} \in \widetilde{\mathcal{Q}}_\lambda \otimes V_\lambda^*$, $i = 1, \ldots, n$.

Integral representations for solutions to dual *qKZ* (3.4) can be obtained similarly to the case of *qKZ* (2.5). Explicit formulas are given in Appendix 1.

Let $p = \rho e^{i\vartheta}$, $\rho = |p|$ and $(t,z) \notin \mathfrak{S}_\vartheta$. As $\rho \to 0$, the function $\widetilde{\Phi}(t,z;p)$ has the asymptotic expansion

(3.8) $$\widetilde{\Phi}(t,z;p) \simeq \exp(p^{-1} \log p \cdot \Theta - p^{-1} \tau(t,z)) \Xi(t,z) \left(1 + \sum_{s=1}^\infty \widetilde{\phi}_s(t,z) p^s \right),$$

where $\widetilde{\phi}_s(t,z) \in \mathcal{F}_\circ$. Here $\log p = \log \rho + i\vartheta$. (Cf. (2.17).)

Let $M \subset \mathbb{C}^\ell$ be an open region such that all $K_i(z;0)$ and $K_i^{-1}(z;0)$ are regular in M. Let $(t^*, z^*) \notin \mathfrak{S}_\vartheta$, $z^* \in M$ be a nondegenerate critical point. Let $\widetilde{D} \subset \mathbb{C}^\ell$ be a small disk:

(3.9) $$\widetilde{D} = \{ t \in \mathbb{C}^\ell \mid t - t^* \in iW, \; |t - t^*| < \varepsilon \},$$

where ε is a small positive number and W is defined in (2.19). Set

(3.10) $$\widetilde{\Psi}(z;p) = p^{-\ell/2} \exp(-p^{-1} \log p \cdot \Theta) \int_{\widetilde{D}} I(t) \widetilde{\Phi}(t,z;p) \widetilde{w}(t,z;p) \, d^\ell t,$$

where $\widetilde{w}(t, z; p) \in \mathcal{F} \otimes V_\lambda^*$ (cf. (2.22)). As $p \to 0$, the function $\widetilde{\Psi}(z; p)$ has the asymptotic expansion

(3.11)
$$\widetilde{\Psi}(z; p) \simeq (2\pi)^{\ell/2} \exp(-\widehat{\tau}(t(z), z)/p) \Xi(t(z), z) H^{-1/2}(t(z), z)$$
$$\times \left(\widetilde{w}(t(z), z; 0) + \sum_{s=1}^{\infty} \widetilde{\psi}_s(t(z), z) p^s \right),$$

where $\widetilde{\psi}_s(t, z) \in \mathcal{F}_\circ \otimes V_\lambda^*$ (cf. (2.23)).

(3.12) THEOREM. *Let $\widetilde{\Phi}(t, z; p) \widetilde{w}(t, z; p)$ be an integral representation for solutions to the dual qKZ (3.4). Then the asymptotic expansion of $\widetilde{\Psi}(z; p)$ as $p \to 0$ gives an asymptotic solution to system (3.4) in the sense of (1.4).*

The proof is completely similar to the proof of Theorem (2.24).

(3.13) COROLLARY.

$$K_i^*(z^*; 0) \widetilde{w}(t^*, z^*; 0) = \exp\left(\frac{\partial \tau}{\partial z_i}(t^*, z^*) \right) \widetilde{w}(t^*, z^*; 0), \qquad i = 1, \ldots, n.$$

Let us consider $\mu \in \mathbb{C}^{N+1}$ as an additional set of variables. Let (t^*, z^*, μ^*) be an offdiagonal nondegenerate critical point (with respect to t). Let $t(z, \mu)$ be a holomorphic function such that $(t(z, \mu), z, \mu)$ is a nondegenerate critical point and $t(z^*, \mu^*) = t^*$. Recall that $w(t, z, \mu; p)$ and $\widetilde{w}(t, z, \mu; p)$ in the integral representations do not depend on μ and p at all. Furthermore, $H(t, z, \mu)$ and $\Xi(t, z, \mu)$ do not depend on μ as well.

(3.14) THEOREM. *Let*

$$\mathfrak{H}(z, \mu) = \Xi^2(t(z, \mu), z) H^{-1}(t(z, \mu), z) \langle \widetilde{w}(t(z, \mu), z), w(t(z, \mu), z) \rangle.$$

Then

$$\frac{\partial}{\partial z_i} \mathfrak{H}(z, \mu) = 0, \qquad i = 1, \ldots, n,$$
$$\frac{\partial}{\partial \mu_j} \mathfrak{H}(z, \mu) = 0, \qquad j = 1, \ldots, N+1.$$

PROOF. The first equality immediately follows from Theorems (2.24), (3.12), Lemmas (2.23), (3.11) and Corollary (1.10). To prove the second one, we note that the asymptotic expansions of $\Psi(z; p)$ and $\widetilde{\Psi}(z; p)$ give asymptotic solutions to systems (2.5) and (3.4), respectively, not only for μ^*, but for any μ close to μ^* as well. Computing the asymptotic expansion of

$$\frac{\partial}{\partial \mu_i} \langle \widetilde{\Psi}(z, \mu; p), \Psi(z, \mu; p) \rangle \quad \text{as } p \to 0,$$

we see that the leading term of this asymptotic expansion vanishes. On the other hand, this leading term is given by the left-hand side of the second equality. □

(3.15) COROLLARY. *For any offdiagonal nondegenerate critical point* (t, z),

$$\langle \widetilde{w}(t, z), w(t, z) \rangle = \text{const}\, \Xi^{-2}(t, z)\, H(t, z),$$

where const *does not depend on* μ *and does not change under continuous deformations of the critical point* (t, z).

(3.16) CONJECTURE. *For any offdiagonal critical point* (t, z), *we have*

$$\langle \widetilde{w}(t, z), w(t, z) \rangle = (-1)^\ell\, \Xi^{-2}(t, z)\, H(t, z).$$

(3.17) CONJECTURE. *Let* (t, z) *and* $(\tilde{t}, z)$ *be offdiagonal critical points such that*

$$\{t_{ij} \mid j = 1, \ldots, \lambda_i\} \neq \{\tilde{t}_{ij} \mid j = 1, \ldots, \lambda_i\}$$

for some i. Then

$$\langle \widetilde{w}(\tilde{t}, z), w(t, z) \rangle = 0.$$

We prove these conjectures for the $\mathfrak{gl}_2$ case in the next section using Corollary (3.15). Namely, we compute suitable limits of the right- and left-hand sides of (3.15) and check that const $= (-1)^\ell$ for that limit. Probably this proof can be generalized to other Lie algebras.

A combinatorial proof of Conjecture (3.16) for the $\mathfrak{gl}_2$ case was given in [**K**], and for the $\mathfrak{gl}_3$ case (with a special choice of $\mathfrak{g}$-modules) in [**R**]. An analog of Theorem (3.14) for the differential Knizhnik–Zamolodchikov equation was proved in [**RV**]. Analogs of Conjectures (3.16) and (3.17) were proved in [**V2**] and [**RV**], respectively, for the $\mathfrak{sl}_2$ case.

REMARK. For historical reasons, $\langle \widetilde{w}(t, z), w(t, z) \rangle$ is called the *norm* of the Bethe vector $w(t, z)$.

§4. Proof of Conjectures (3.16) and (3.17) in the $\mathfrak{gl}_2$ case

In this section we deduce Conjecture (3.16) for the $\mathfrak{gl}_2$ case from Corollary (3.15).

For the rest of this section we assume that $N = 1$. Without loss of generality we may assume that $\Lambda_1(m) = 0$, $m = 1, \ldots, n$, and $\mu_2 = 0$. Set $y_m = z_m - \Lambda_2(m)$, $m = 1, \ldots, n$, and $\kappa = \exp(\mu_1)$. We assume that all y_m, z_m are generic, unless the converse is indicated explicitly.

The original system of equations for critical points is

$$(4.1) \quad \kappa^{-1} \prod_{m=1}^{n} \frac{t_a - z_m}{t_a - y_m} \prod_{\substack{b=1 \\ b \neq a}}^{\ell} \frac{t_a - t_b - 1}{t_a - t_b + 1} = 1, \qquad a = 1, \ldots, \ell.$$

We replace it by the system of algebraic equations

$$(4.2) \quad \prod_{m=1}^{n}(t_a - z_m) \prod_{\substack{b=1 \\ b \neq a}}^{\ell}(t_a - t_b - 1) = \kappa \prod_{m=1}^{n}(t_a - y_m) \prod_{\substack{b=1 \\ b \neq a}}^{\ell}(t_a - t_b + 1),$$

$a = 1, \ldots, \ell$. Both systems (4.1) and (4.2) are preserved by the natural action of the symmetric group $\mathbf{S}_\ell$ on the variables $t_1, \ldots, t_\ell$.

Define $\mathfrak{Z} \subset \mathbb{C}^\ell$ by the equation

(4.3)
$$\prod_{a=1}^{\ell} \left(\prod_{m=1}^{n} (t_a - z_m)(t_a - y_m) \prod_{\substack{b=1 \\ b \neq a}}^{\ell} (t_a - t_b - 1) \right) = 0.$$

(4.4) LEMMA. *Systems* (4.1) *and* (4.2) *are equivalent for* $\kappa \neq 0$.

PROOF. We must show that system (4.2) has no solutions belonging to $\mathfrak{Z}$ if $\kappa \neq 0$. This can be done by a direct analysis of this system. As an example, consider the case $\ell = 2$. Take a solution $t \in \mathfrak{Z}$. Suppose $t_1 = z_m$. Then from the first equation, $t_2 = t_1 + 1$ and the second equation cannot be satisfied. If $t_1 = t_2 + 1$, then from the first equation, $t_1 = y_m$ for some m, and the second equation cannot be satisfied. Similarly, we can start from $t_1 = y_m$ or $t_1 = t_2 - 1$. All the other cases can be obtained by the action of the symmetric group $\mathfrak{S}_2$. □

(4.5) LEMMA. *All the solutions to system* (4.2) *remain finite for any* $\kappa \neq \exp(2\pi i r/s)$, $s = 1, \ldots, \ell$, $r = 0, \ldots, s$.

PROOF. Suppose that there is a solution $t(\kappa)$ to system (4.2) which tends to infinity as $\kappa \to \kappa_0$ and κ_0 is not a root of unity. We can assume that $t_a(\kappa) \to \infty$ for $a \leq f$ and $t_a(\kappa)$ remain finite for $a > f$. Fix a $u \in \mathbb{C}$ such that $u \neq t_a(\kappa_0)$ for $a = f+1, \ldots, \ell$. Set $x_a = (t_a - u)^{-1}$, $a = 1, \ldots, \ell$. The system

(4.6)
$$\prod_{m=1}^{n} (1 - x_a(z_m - u)) \prod_{\substack{b=1 \\ b \neq a}}^{\ell} (x_a - x_b + x_a x_b)$$
$$= \kappa \prod_{m=1}^{n} (1 - x_a(y_m - u)) \prod_{\substack{b=1 \\ b \neq a}}^{\ell} (x_a - x_b - x_a x_b), \quad a = 1, \ldots, \ell,$$

is equivalent to system (4.2) in the region $t_a \neq 0$, $x_a \neq 0$, $a = 1, \ldots, \ell$.

Let $x(\kappa)$ be the solution to system (4.6), corresponding to $t(k)$ under the transformation described above. We have $x_a(\kappa_0) = 0$ for $a \leq f$, $x_a(\kappa_0) \neq 0$ for $a > f$ and $x_a(\kappa) \neq 0$ for $\kappa \neq \kappa_0$, $a = 1, \ldots, \ell$. Taking the product of the first f equations of system (4.6), we obtain

$$\prod_{a=1}^{f} \prod_{\substack{b=1 \\ b \neq a}}^{f} (x_a - x_b + x_a x_b) \prod_{a=1}^{f} \left(\prod_{m=1}^{n} (1 - x_a(z_m - u)) \prod_{b=f+1}^{\ell} (x_a - x_b + x_a x_b) \right)$$

$$= \kappa^f \prod_{a=1}^{f} \prod_{\substack{b=1 \\ b \neq a}}^{f} (x_a - x_b - x_a x_b)$$

$$\times \prod_{a=1}^{f} \left(\prod_{m=1}^{n} (1 - x_a(y_m - u)) \prod_{b=f+1}^{\ell} (x_a - x_b - x_a x_b) \right).$$

It is easy to see that the first products in the left- and right-hand sides above coincide. Moreover, for κ close to κ_0, they must be zero. Otherwise, we could cancel these products and come to a contradiction in the limit as $\kappa \to \kappa_0$. Therefore,

$$\prod_{a=1}^{f} \prod_{\substack{b=1 \\ b \neq a}}^{f} (x_a(\kappa) - x_b(\kappa) + x_a(\kappa) x_b(\kappa)) = 0,$$

and the corresponding solution $t(\kappa)$ to the original system (4.2) belongs to $\mathfrak{Z}$, which is impossible (cf. the proof of Lemma (4.4)). Hence, there are no required solutions $x(\kappa)$ to system (4.6). And the original system (4.2) has no solutions which tend to infinity as $\kappa \to \kappa_0$. □

For now, set $\kappa = 0$ and consider the corresponding system

$$(4.7) \qquad \prod_{m=1}^{n} (t_a - z_m) \prod_{\substack{b=1 \\ b \neq a}}^{\ell} (t_a - t_b - 1) = 0, \qquad a = 1, \ldots, \ell.$$

The set of solutions to this system modulo, the action of the symmetric group $\mathbf{S}_\ell$, is in one-to-one correspondence with

$$\left\{ \eta \in \mathbb{Z}_{\geq 0}^{n\ell} \,\Big|\, \sum_{i=1}^{n} \sum_{j=1}^{\ell} \eta_{ij} = \ell, \, \eta_{ij} = 0 \implies \eta_{ij'} = 0 \text{ for } j' > j \right\}.$$

A solution is fixed by the conditions

$$\#\{a \mid t_a = z_i + j - 1\} = \eta_{ij}, \qquad i = 1, \ldots, n, \; j = 1, \ldots, \ell.$$

A solution is called an *offdiagonal* solution if $t_a \neq t_b$, for $a \neq b$, and a *diagonal* solution, otherwise.

(4.8) LEMMA. *The multiplicity of any offdiagonal solution to system (4.7) is equal to* 1.

The lemma immediately follows from the following lemma.

(4.9) LEMMA. *Let $Q_1, \ldots, Q_l$ be homogeneous polynomials in variables $x_1, \ldots, x_l$. Assume that $x_a = 0$, $a = 1, \ldots, l$, is an isolated solution to the system*

$$Q_a(x) = 0, \qquad a = 1, \ldots, l.$$

Then the multiplicity of this solution is equal to $\prod_{a=1}^{l} \deg Q_a$.

(4.10) LEMMA. *Let $t(\kappa)$ be a solution to system (4.2) which is a deformation of a diagonal solution $t(0)$ to system (4.7). Then $t(k)$ is a diagonal critical point.*

PROOF. Let $j_0 = \min\{j \mid \eta_{ij} > 1 \text{ for some } i\}$. Let i_0 be such that $\eta_{i_0 j_0} > 1$. Let, for example, $t_1 = t_\ell = z_{i_0} + j_0 - 1$. Let $t^- \in \mathbb{C}^{\ell-1}$ be obtained from $t \in \mathbb{C}^\ell$ by removing the last coordinate.

Consider the new system

$$\prod_{m=1}^{n}(t_1 - z_m)\prod_{b=2}^{\ell-1}(t_1 - t_b - 1) = -\kappa \prod_{m=1}^{n}(t_1 - y_m)\prod_{b=2}^{\ell-1}(t_1 - t_b + 1),$$

(4.11)
$$(t_a - t_1 - 1)\prod_{m=1}^{n}(t_a - z_m)\prod_{\substack{b=1\\b\neq a}}^{\ell-1}(t_a - t_b - 1)$$

$$= \kappa(t_a - t_1 + 1)\prod_{m=1}^{n}(t_a - y_m)\prod_{\substack{b=1\\b\neq a}}^{\ell-1}(t_a - t_b + 1),$$

$a = 2, \ldots, \ell-1$. It is obtained from system (4.2) by the substitution $t_\ell = t_1$. Incidentally, the first and the last equations of (4.2) coincide after this substitution. Any solution to system (4.11) gives rise to a diagonal solution to system (4.2) if we set $t_\ell = t_1$. It follows from Lemma (4.10) that the solution $t(0)$ to system (4.7) has the same multiplicity as the solution $t^-(0)$ to system (4.11) at $\kappa = 0$. This means that any deformation of the diagonal solution $t(0)$ can be obtained from some deformation of the solution $t^-(0)$ and, therefore, is diagonal. □

(4.12) LEMMA. *For generic κ there are precisely $\binom{n+\ell-1}{n-1}$ offdiagonal critical points modulo the action of the symmetric group $\mathbf{S}_\ell$. All of them are nondegenerate.*

PROOF. It follows from Lemmas (4.5), (4.8), and (4.9) that for generic κ the number of offdiagonal solutions to system (4.2) is the same as to system (4.7), and all of them are nondegenerate. It is a simple combinatorial exercise to count offdiagonal solutions to system (4.7). □

(4.13) LEMMA. *Offdiagonal solutions to system (4.2) remain finite for any $\kappa \neq 1$.*

PROOF. Define the rational function $r(u; t, \kappa)$ as follows:

(4.14) $\quad r(u; t, \kappa) = \prod_{m=1}^{n}(u - z_m)\prod_{a=1}^{\ell}\frac{u - t_a - 1}{u - t_a} + \kappa \prod_{m=1}^{n}(u - y_m)\prod_{a=1}^{\ell}\frac{u - t_a + 1}{u - t_a}.$

It depends continuously on t, κ in the sense that for almost all $u \in \mathbb{C}$ we have $r(u; t, \kappa) \to r(u; t_0, \kappa_0)$ as $(t, \kappa) \to (t_0, \kappa_0)$.

Let $t(\kappa)$ be an offdiagonal solution to system (4.2). Then $r(u; t(\kappa), \kappa)$ is a polynomial in u. Its coefficients are continuous functions of κ. From (4.14) we obtain

(4.15) $\quad r(u; t(\kappa), \kappa) = (1+\kappa)u^\ell - \left((1-\kappa)\ell + \sum_{m=1}^{n}(z_m + \kappa y_m)\right)u^{\ell-1} + \cdots.$

Suppose that the solution $t(\kappa)$ tends to infinity as $\kappa \to \kappa_0 \neq 1$. We can assume that $t_a(\kappa) \to \infty$ for $a > f$ and the functions $t_a(\kappa)$ remain finite for $a \leq f$. Then as $\kappa \to \kappa_0$, $r(u; t(\kappa), \kappa)$ tends to

$$\prod_{m=1}^{n}(u - z_m)\prod_{a=1}^{f}\frac{u - t_a(\kappa_0) - 1}{u - t_a(\kappa_0)} + \kappa_0 \prod_{m=1}^{n}(u - y_m)\prod_{a=1}^{f}\frac{u - t_a(\kappa_0) + 1}{u - t_a(\kappa_0)}.$$

Hence,

$$r(u; t(\kappa), \kappa) \to (1+\kappa_0)u^\ell - \left((1-\kappa_0)f + \sum_{m=1}^{n}(z_m + \kappa_0 y_m)\right)u^{\ell-1} + \cdots,$$

in contradiction with (4.15). □

PROOF OF CONJECTURE (3.16). First consider only generic κ. In this case, all offdiagonal critical points are nondegenerate and each of them can be obtained by a continuous deformation from a certain offdiagonal solution to system (4.7). According to Theorem (3.14), it suffices to check the conjecture only in the limit as $\kappa \to 0$. As we know, the function $\Xi^2(t,z) H^{-1}(t,z) \langle \tilde{w}(t,z), w(t,z) \rangle$ is preserved by the action of the symmetric group $\mathbf{S}_\ell$. So, we can take the most convenient solution for a given $\mathbf{S}_\ell$-orbit. In the sequel we do not indicate dependence on z explicitly.

Denote by $e = E_{12}$ and $f = E_{21}$ the offdiagonal generators of $\mathfrak{gl}_2$ and introduce the canonical monomial bases in $V_1 \otimes \cdots \otimes V_n$ and $V_1^* \otimes \cdots \otimes V_n^*$:

(4.16) $\qquad F^v = f^{v_1} v_1 \otimes \cdots \otimes f^{v_n} v_n, \qquad E^v = e^{v_1} v_1^* \otimes \cdots \otimes e^{v_n} v_n^*.$

They are dual to each other, up to a normalization:

(4.17) $\qquad \langle E^v, F^{v'} \rangle = \delta_{vv'} \prod_{m=1}^{n} \prod_{j=1}^{v_m} j(y_m - z_m - j + 1)$

(cf. Appendix 2).

Set $\mathcal{Z}_\ell = \{v \in \mathbb{Z}_{\geq 0}^n \mid \sum_{i=1}^n v_i = \ell\}$. The sets $\{F^v\}_{v \in \mathcal{Z}_\ell}$ and $\{E^v\}_{v \in \mathcal{Z}_\ell}$ form bases in the weight subspaces V_ℓ and V_ℓ^*, respectively. On $\mathcal{L}_\ell$ define a lexicographical ordering: $v < v'$ if $v_1 < v_1'$, or $v_1 = v_1'$, $v_2 < v_2'$, etc. It induces an ordering on the monomial bases.

Fix $v \in \mathcal{Z}_\ell$. Let t^* be the following offdiagonal solution to system (4.7):

(4.18) $\qquad t_a^* = z_i + a - \ell_{i-1} - 1 \quad \text{for } \ell_{i-1} < a \leq \ell_i,$

where $\ell_i = \sum_{j=1}^i v_j$, $\ell_0 = 0$, $\ell_n = \ell$. It is related to the previous description of solutions as:

$$\eta_{ij} = \begin{cases} 1 & \text{for } j \leq v_i, \\ 0 & \text{for } j > v_i. \end{cases}$$

Suppose for a moment that $b \ll a$ if $b \leq \ell_m < a$ for some m. From explicit formulas for the vectors $w(t)$ and $\tilde{w}(t)$ in the integral representations for solutions to qKZ, we see

$$w(t) \prod_{a=2}^{\ell} \prod_{b<a} \frac{t_a - t_b - 1}{t_a - t_b}$$

$$= F^v \prod_{a=2}^{\ell} \prod_{b \ll a} \frac{t_a - t_b - 1}{t_a - t_b} \prod_{m=1}^{n} \prod_{a > \ell_m} (t_a - z_m) \prod_{m=1}^{n} \prod_{a > \ell_{m-1}} (t_a - y_m)^{-1}$$

(4.19)
$$+ \sum_{v' > v} F^{v'} \theta_{vv'}(t) + o(1),$$

$$\widetilde{w}(t) \prod_{a=2}^{\ell} \prod_{b<a} \frac{t_a - t_b - 1}{t_a - t_b}$$

$$= E^\nu \prod_{a=2}^{\ell} \prod_{b \ll a} \frac{t_a - t_b + 1}{t_a - t_b} \prod_{m=1}^{n} \prod_{a \leqslant \ell_{m-1}} (t_a - z_m) \prod_{m=1}^{n} \prod_{a \leqslant \ell_m} (t_a - y_m)^{-1}$$

(4.20)
$$+ \sum_{\nu' < \nu} E^{\nu'} \theta_{\nu\nu'}(t) + o(1),$$

where $\theta_{\nu\nu'}(t) = O(1)$, as $t \to t^*$ (cf. Appendix 1). Hence only the first terms of the right-hand sides above contribute to the leading part of the pairing $\langle \widetilde{w}(t), w(t) \rangle$. The function $\Xi^2(t)$ reads as follows

$$\Xi^2(t) = \prod_{a=2}^{\ell} \prod_{b<a} \frac{t_a - t_b - 1}{t_a - t_b + 1} \prod_{m=1}^{n} \prod_{a=1}^{\ell} \frac{t_a - y_m}{t_a - z_m}$$

(cf. (2.17)). For any m, set

$$u_{m0} = z_m, \qquad u_{mi} = t_{\ell(m-1)+i-1} - t^*_{\ell(m-1)+i-1}, \qquad i = 1, \ldots, \nu_m.$$

Taking into account (4.17), we get

(4.21) $$\Xi^2(t) \langle \widetilde{w}(t), w(t) \rangle = (-1)^\ell \prod_{m=1}^{n} \prod_{i=0}^{\nu_m - 1} (u_{m,i+1} - u_{mi})^{-1} (1 + o(1)).$$

As $t \to t^*$, the leading term of the matrix of second derivatives

$$\frac{\partial^2 \tau(t)}{\partial t_a \partial t_b} = \nabla_{ab} \Phi(t)$$

is block-diagonal, consisting of n blocks (cf. (2.14)). The mth block is the following three-diagonal $\nu_m \times \nu_m$ matrix:

$$\begin{pmatrix} \frac{1}{u_{m1}-u_{m0}} + \frac{1}{u_{m2}-u_{m1}} & \frac{1}{u_{m1}-u_{m2}} & & \\ \frac{1}{u_{m1}-u_{m2}} & \frac{1}{u_{m2}-u_{m1}} + \frac{1}{u_{m3}-u_{m2}} & \ddots & \\ & \ddots & \ddots & \frac{1}{u_{m,\nu_m-1}-u_{m,\nu_m}} \\ & & \frac{1}{u_{m,\nu_m-1}-u_{m,\nu_m}} & \frac{1}{u_{m,\nu_m}-u_{m,\nu_m-1}} \end{pmatrix}.$$

Adding to each row all the subsequent rows, we can obtain a lower-triangular matrix and compute its determinant. Finally,

(4.22) $$H(t) = \prod_{m=1}^{n} \prod_{i=0}^{\nu_m-1} (u_{m,i+1} - u_{mi})^{-1} (1 + o(1)),$$

as $t \to t^*$. Comparing (4.21) and (4.22), we come to

$$\lim_{t \to t^*} (\Xi^2(t) H^{-1}(t) \langle \widetilde{w}(t), w(t) \rangle) = (-1)^\ell.$$

As we know, for any $\mathbf{S}_\ell$-orbit of offdiagonal solutions to system (4.7), we can find a $\nu \in \mathcal{Z}_\ell$ such that this orbit contains solution (4.18). So, Conjecture (3.16) is proved for generic κ, y, z.

For the general (not generic) case, let t be an offdiagonal critical point. Making, if necessary, a small deformation of $\mu, z, \Lambda(1), \ldots, \Lambda(n)$, we come to the generic case, considered above, and find an offdiagonal nondegenerate critical point $\tilde{t}$ close to t. For the generic case, Conjecture (3.16) is already proved, and both the left-hand side and the right-hand side of the equality are continuous functions. Therefore, Conjecture (3.16) holds for the general case as well. □

PROOF OF CONJECTURE (3.17). First, consider the generic case. Both t and $\tilde{t}$ are nondegenerate critical points. Therefore, due to Corollaries (2.26) and (3.13), $w(t,z)$ and $\tilde{w}(\tilde{t},z)$ are eigenvectors of the operators $K_1(z;0)$ and $K_1^*(z;0)$ with eigenvalues

$$\exp\left(\frac{\partial \tau}{\partial z_1}(t,z)\right) \quad \text{and} \quad \exp\left(\frac{\partial \tau}{\partial z_1}(\tilde{t},z)\right),$$

respectively. Here t and $\tilde{t}$ are obtained by continuous deformations from offdiagonal solutions to system (4.7) belonging to different orbits of the symmetric group $\mathbf{S}_\ell$. Using the explicit formulas

$$\exp\left(\frac{\partial \tau}{\partial z_1}(t,z)\right) = \prod_{a=1}^{\ell} \frac{t_a - z_1 + \Lambda_2(1)}{t_a - z_1},$$

$$\exp\left(\frac{\partial \tau}{\partial z_1}(\tilde{t},z)\right) = \prod_{a=1}^{\ell} \frac{\tilde{t}_a - z_1 + \Lambda_2(1)}{\tilde{t}_a - z_1}$$

we see that generically

$$\exp\left(\frac{\partial \tau}{\partial z_1}(t,z)\right) \neq \exp\left(\frac{\partial \tau}{\partial z_1}(\tilde{t},z)\right),$$

since this is the case as $\kappa \to 0$. But the eigenvectors of operators $K_1(z;0)$ and $K_1^*(z;0)$ with different eigenvalues must be orthogonal, $\langle \tilde{w}(\tilde{t},z), w(t,z) \rangle = 0$. To complete the proof in the general (not generic) case, we use the same deformation arguments as those at the end of the proof of Conjecture (3.16) above. □

Let $\mathfrak{C}(z,\mu)$ be a set of all different offdiagonal critical points modulo the action of the symmetric group $\mathbf{S}_\ell$. The vectors $w(t,z)$ and $\tilde{w}(t,z)$ are preserved by the action of $\mathbf{S}_\ell$ modulo multiplication by a scalar factor.

(4.23) THEOREM. *Let $z, \mu, \Lambda(1), \ldots, \Lambda(n)$ be generic. Then $\{w(t,z)\}_{t\in\mathfrak{C}(z,\mu)}$ and $\{\tilde{w}(t,z)\}_{t\in\mathfrak{C}(z,\mu)}$ are bases in V_ℓ and V_ℓ^*, respectively. They are dual to each other up to normalization.*

PROOF. The first statement follows from (4.19) and (4.20). The second one coincides with Conjecture (3.17), which is proved above. □

REMARK. The statement of Theorem (4.23) is often called the *completeness* of Bethe vectors.

§5. qKZ and bases of singular vectors

In this section we always assume that $\mu = 0$. We shall prove analogues of Theorem (4.23) for this special case.

Let $V_1 \otimes \cdots \otimes V_n$ be a tensor product of highest weight $\mathfrak{g}$-modules. Let V_λ be the weight subspace (2.6). Set

$$\text{Sing } V = \{v \in V_1 \otimes \cdots \otimes V_n \mid E_{i,i+1}v = 0, \ i = 1, \ldots, N\},$$
$$\text{Sing } V_\lambda = V_\lambda \cap \text{Sing } V.$$

Let $\Phi(t, z; p)w(t, z)$ be an integral representation for the solutions to qKZ (2.5). Let (t^*, z^*) be a nondegenerate critical point. Let $\Psi(z; p)$ be defined by (2.22).

(5.1) LEMMA. $\exp(-\widehat{\tau}(t(z), z)/p) E_{i,i+1} \Psi(z; p) = O(p^\infty)$, $i = 1, \ldots, N$, as $p \to 0$.

PROOF. Let $t^{ij} \in \mathbb{C}^{\ell-1}$ be obtained from $t \in \mathbb{C}^\ell$ by removing the coordinate t_{ij}. Let $\lambda^i = (\lambda_1, \ldots, \lambda_i - 1, \ldots, \lambda_N)$ and $w_{ij}(t, z) = w(t^{ij}, z) \in V_{\lambda^i}$. Set

$$\Phi_{ij}(t, z; p) = \Phi(t, z; p) \prod_{k=1}^{j-1} \frac{t_{ik} - t_{ij} - 1}{t_{ik} - t_{ij} + 1} \prod_{k=1}^{\lambda_{i+1}} \frac{t_{i+1,k} - t_{ij} + 1}{t_{i+1,k} - t_{ij}}.$$

Let Q_{ij} be the p-shift operator with respect to the variable t_{ij}. We have

$$(5.2) \qquad E_{i,i+1} \Phi w = \sum_{j=1}^{\lambda_i} (Q_{ij}(\Phi_{ij} w_{ij}) - \Phi_{ij} w_{ij})$$

(cf. Appendix 1). Now the statement follows from the method of steepest descend. □

(5.3) LEMMA. *Let (t, z) be an offdiagonal critical point. Then $w(t, z) \in \text{Sing } V_\lambda$.*

PROOF. The statement directly follows from (5.2) and the following equations for critical points:

$$\nabla_{ij} \Phi(t, z) = 1, \qquad i = 1, \ldots, N, \ j = 1, \ldots, \lambda_i.$$

If (t, z) is a nondegenerate critical point, the statement also follows from Lemmas (2.23) and (5.1). □

For the $\mathfrak{gl}_2$ case, this lemma was proved in [**FT2**]. For the $\mathfrak{gl}_{N+1}$ case it was formulated in [**KiR, KR**]; the proof is given in [**Kid**].

Let us consider the $\mathfrak{gl}_2$ case in more detail. The equations for critical points are

$$(5.4) \qquad \prod_{m=1}^{n} \frac{t_a - z_m + \Lambda_1(m)}{t_a - z_m + \Lambda_2(m)} \prod_{\substack{b=1 \\ b \neq a}}^{\ell} \frac{t_a - t_b - 1}{t_a - t_b + 1} = 1, \qquad a = 1, \ldots, \ell.$$

Let $s \in \mathbb{C}$. Carry out the change of variables $x = sz \in \mathbb{C}^n$, $u = st \in \mathbb{C}^\ell$. In the new variables u system (5.4) reads as follows:

$$(5.5) \qquad \prod_{m=1}^{n} \frac{u_a - x_m + s\Lambda_1(m)}{u_a - x_m + s\Lambda_2(m)} \prod_{\substack{b=1 \\ b \neq a}}^{\ell} \frac{u_a - u_b - s}{u_a - u_b + s} = 1, \qquad a = 1, \ldots, \ell.$$

As $s \to 0$, system (5.5) turns into

$$(5.6) \qquad \sum_{m=1}^{n} \frac{\Lambda_1(m) - \Lambda_2(m)}{u_a - x_m} - \sum_{\substack{b=1 \\ b \neq a}}^{\ell} \frac{2}{u_a - u_b} = 0, \qquad a = 1, \ldots, \ell.$$

Both systems (5.5) and (5.6) are preserved by the natural action of the symmetric group $\mathbf{S}_\ell$ on variables $u_1, \ldots, u_\ell$.

(5.7) LEMMA [V2]. *Let $\Lambda_1(m) - \Lambda_2(m) < 0$, $m = 1, \ldots, n$. Then for generic x all solutions to system (5.6) are nondegenerate. There are $\dim \operatorname{Sing} V_\lambda$ different solutions modulo the action of the symmetric group $\mathbf{S}_\ell$.*

Let $\mathfrak{C}(z)$ be the set of all offdiagonal critical points modulo the action of the symmetric group $\mathbf{S}_\ell$, see (2.13) and (2.27).

(5.8) THEOREM. *Let $\Lambda_1(m) - \Lambda_2(m) < 0$, $m = 1, \ldots, n$. Then for generic z all offdiagonal critical points are nondegenerate. Moreover, $\#\mathfrak{C}(z) = \dim \operatorname{Sing} V_\lambda$ and $\{w(t,z)\}_{t \in \mathfrak{C}(z)}$ is a base in $\operatorname{Sing} V_\lambda$.*

PROOF. Let $\mathfrak{C}_\circ(z) \subset \mathfrak{C}(z)$ be the subset of nondegenerate critical points. It follows from Lemma (5.7) that $\#\mathfrak{C}_\circ(z) \geqslant \dim \operatorname{Sing} V_\lambda$ for generic z. On the other hand, from Lemma (5.3) and Conjectures (3.16) and (3.17), we obtain $\#\mathfrak{C}_\circ(z) \leqslant \dim \operatorname{Sing} V_\lambda$. Therefore, $\#\mathfrak{C}_\circ(z) = \dim \operatorname{Sing} V_\lambda$. Moreover, $\{w(t,z)\}_{t \in \mathfrak{C}_\circ(z)}$ and $\{\widetilde{w}(t,z)\}_{t \in \mathfrak{C}_\circ(z)}$ are bases in $\operatorname{Sing} V_\lambda$ and $(\operatorname{Sing} V_\lambda)^*$, respectively. Let (t, z) be a critical point such that $(t, z) \in \mathfrak{C}(z) \setminus \mathfrak{C}_\circ(z)$. Then $w(t,z) \neq 0$ (cf. Appendix 1), $w(t,z) \in \operatorname{Sing} V_\lambda$ and $\langle \widetilde{v}, w(t,z) \rangle = 0$ for any $\widetilde{v} \in (\operatorname{Sing} V_\lambda)^*$. This is impossible. Hence, $\mathfrak{C}(z) = \mathfrak{C}_\circ(z)$. □

(5.9) COROLLARY. *For generic $z, \Lambda(1), \ldots, \Lambda(n)$ all offdiagonal critical points are nondegenerate. Moreover, $\#\mathfrak{C}(z) = \dim \operatorname{Sing} V_\lambda$ and $\{w(t,z)\}_{t \in \mathfrak{C}(z)}$ is a base in $\operatorname{Sing} V_\lambda$.*

The proof is absolutely similar to the proof of Theorem (5.8), if in it we use Theorem (5.8) instead of Lemma (5.7).

Assume now that $\Lambda_1(m) - \Lambda_2(m) \in \mathbb{Z}_{\geqslant 0}$, $m = 1, \ldots, n$. Let $V_1, \ldots, V_n$ be the irreducible $\mathfrak{g}$-modules with highest weights $\Lambda(1), \ldots, \Lambda(n)$, respectively.

(5.10) LEMMA [RV]. *For generic x there are $\dim \operatorname{Sing} V_\lambda$ different nondegenerate solutions to system (5.6), modulo the action of the symmetric group $\mathbf{S}_\ell$.*

An offdiagonal critical point (t, z) is called a *nontrivial* critical point if $w(t,z) \neq 0$, and a *trivial* critical point, otherwise. Let $\mathfrak{C}(z)$ be a set of all different nontrivial critical points modulo the action of the symmetric group $\mathbf{S}_\ell$.

(5.11) THEOREM. *For any z all trivial critical points are degenerate. For generic z all nontrivial critical points are nondegenerate. Moreover, $\#\mathfrak{C}(z) = \dim \operatorname{Sing} V_\lambda$ and $\{w(t,z)\}_{t \in \mathfrak{C}(z)}$ is a base in $\operatorname{Sing} V_\lambda$.*

PROOF. Let $\mathfrak{C}_\circ(z)$ be a set of all different offdiagonal nondegenerate critical points modulo the action of the symmetric group $\mathbf{S}_\ell$. Trivial critical points are degenerate by Conjecture (3.16). Therefore, $\mathfrak{C}_\circ(z) \subset \mathfrak{C}(z)$. If follows from Lemma (5.10) that $\#\mathfrak{C}_\circ(z) \geqslant \dim \operatorname{Sing} V_\lambda$ for generic z. On the other hand, from Lemma (5.3) and Conjectures (3.16) and (3.17), we obtain $\#\mathfrak{C}_\circ(z) \leqslant \dim \operatorname{Sing} V_\lambda$. Therefore, $\#\mathfrak{C}_\circ(z) = \dim \operatorname{Sing} V_\lambda$. Moreover, $\{w(t,z)\}_{t \in \mathfrak{C}_\circ(z)}$ and $\{\widetilde{w}(t,z)\}_{t \in \mathfrak{C}_\circ(z)}$ are bases in $\operatorname{Sing} V_\lambda$ and $(\operatorname{Sing} V_\lambda)^*$, respectively. Let (t, z) be a critical point such that $(t, z) \notin \mathfrak{C}_\circ(z)$. Then

$w(t,z) \in \text{Sing } V_\lambda$ and $\langle \tilde{v}, w(t,z) \rangle = 0$ for any $\tilde{v} \in (\text{Sing } V_\lambda)^*$. Hence, $w(t,z) = 0$ and $\mathfrak{C}(z) = \mathfrak{C}_\circ(z)$. □

Analogs of Theorems (5.8) and (5.11) for the differential Knizhnik–Zamolodchikov equation were proved in [**V2**] and [**RV**], respectively.

§6. Asymptotic solutions to qKZ, the $U_q(\mathfrak{gl}_{N+1})$ case

In this section we describe the q-deformations of the results given in §§2–4. All proofs are completely similar. The notation used in this section differs slightly from that used above. The reader should take care to avoid confusion.

Let $\mathfrak{g} = \mathfrak{gl}_{N+1}$, $q \in \mathbb{C}$, and q not be a root of unity. Set

$$[k]_q = \frac{q^k - q^{-k}}{q - q^{-1}}, \qquad [k]_q! = \prod_{m=1}^k [m]_q.$$

Then $U_q(\widehat{\mathfrak{g}})$ is the unital associative algebra, generated by the elements

$$k_0^{\pm 1}, \ldots, k_{N+1}^{\pm 1}, \; e_0, \ldots, e_N, \; f_0, \ldots, f_N,$$

subject to the relations

(6.1) $k_0 k_{N+1}^{-1}$ is a central element, $[k_i, k_j] = 0$,

$$k_i k_i^{-1} = k_i^{-1} k_i = 1, \quad k_{N+1} k_{N+1}^{-1} = k_{N+1}^{-1} k_{N+1} = 1,$$

$$k_i e_i k_i^{-1} = q e_i, \quad k_{i+1} e_i k_{i+1}^{-1} = q^{-1} e_i, \quad k_i f_i k_i^{-1} = q^{-1} f_i, \quad k_{i+1} f_i k_{i+1}^{-1} = q f_i,$$

$$k_i e_j k_i^{-1} = e_j, \quad k_i f_j k_i^{-1} = f_j, \qquad i \neq j, j+1 \pmod{N+1},$$

$$[e_i, e_j] = 0, \quad [f_i, f_j] = 0, \qquad i \neq j \pm 1 \pmod{N+1},$$

(a) $\quad e_i^2 e_j - [2]_q e_i e_j e_i + e_j e_i^2 = 0, \qquad i = j \pm 1 \pmod{N+1},$

(b) $\quad f_i^2 f_j - [2]_q f_i f_j f_i + f_j f_i^2 = 0, \qquad i = j \pm 1 \pmod{N+1},$

$$[e_i, f_j] = \delta_{ij} \frac{k_i k_{i+1}^{-1} - k_{i+1} k_i^{-1}}{q - q^{-1}},$$

$i, j = 0, \ldots, N$. For $N = 1$, relations (a) and (b) should be replaced by

$$e_i^3 e_j - [3]_q e_i^2 e_j e_i + [3]_q e_i e_j e_i^2 - e_j e_i^3 = 0, \qquad i = j \pm 1,$$

$$f_i^3 f_j - [3]_q f_i^2 f_j f_i + [3]_q f_i f_j f_i^2 - f_j f_i^3 = 0, \qquad i = j \pm 1,$$

respectively. $U_q(\widehat{\mathfrak{g}})$ is a Hopf algebra with the coproduct $\Delta: U_q(\widehat{\mathfrak{g}}) \to U_q(\widehat{\mathfrak{g}}) \otimes U_q(\widehat{\mathfrak{g}})$:

$$k_i \mapsto k_i \otimes k_i, \qquad\qquad i = 0, \ldots, N+1,$$

$$e_i \mapsto e_i \otimes 1 + k_i k_{i+1}^{-1} \otimes e_i, \qquad i = 0, \ldots, N,$$

$$f_i \mapsto f_i \otimes k_{i+1} k_i^{-1} + 1 \otimes f_i, \qquad i = 0, \ldots, N,$$

which is opposite to the coproduct used in [**FR**].

Further, $U_q(\mathfrak{g})$ is a Hopf subalgebra in $U_q(\widehat{\mathfrak{g}})$, generated by the elements $k_1^{\pm 1}, \ldots,$ $k_{N+1}^{\pm 1}, e_1, \ldots, e_N, f_1, \ldots, f_N$. There is a homomorphism of algebras $\varphi: U_q(\widehat{\mathfrak{g}}) \to$

$U_q(\mathfrak{g})$:

$$k_i \mapsto k_i, \quad e_i \mapsto e_i, \quad f_i \mapsto f_i, \quad i = 1, \ldots, N,$$
$$k_0 \mapsto k_{N+1}, \quad k_{N+1} \mapsto k_{N+1},$$
$$e_0 \mapsto (-1)^{N-1}[f_1, \ldots, f_N]_q \prod_{i=1}^{N+1} k_i, \quad f_0 \mapsto [e_1, \ldots, e_N]_q \prod_{i=1}^{N+1} k_i^{-1},$$

where $[x_1, \ldots, x_N]_q = [\ldots [x_1, x_2]_q, \ldots, x_N]_q$, $[x_1, x_2]_q = q^{-1} x_1 x_2 - q x_2 x_1$.

This homomorphism makes any $U_q(\mathfrak{g})$-module into an $U_q(\widehat{\mathfrak{g}})$-module. For any $z \in \mathbb{C}$, $z \neq 0$, $U_q(\widehat{\mathfrak{g}})$ has the automorphism θ_z:

$$k_i \mapsto k_i, \quad i = 1, \ldots, N+1,$$
$$e_i \mapsto z^{-1} e_i, \quad f_i \mapsto z f_i, \quad i = 1, \ldots, N,$$
$$e_0 \mapsto z^{N-1} e_0, \quad f_0 \mapsto z^{1-N} f_0.$$

Set $\varphi_z = \varphi \circ \theta_z$. For any two highest weight $U_q(\mathfrak{g})$-modules V_1 and V_2 with generating vectors v_1 and v_2, respectively, there is a unique R-matrix $R_{V_1 V_2}(z) \in \operatorname{End}(V_1 \otimes V_2)$ such that for any $X \in U_q(\widehat{\mathfrak{g}})$,

$$(6.2) \qquad R_{V_1 V_2}(z_1/z_2)(\varphi_{z_1} \otimes \varphi_{z_2}) \circ \Delta(X) = (\varphi_{z_1} \otimes \varphi_{z_2}) \circ \Delta'(X) R_{V_1 V_2}(z_1/z_2)$$

in $\operatorname{End}(V_1 \otimes V_2)$ and

$$(6.3) \qquad R_{V_1 V_2}(z) v_1 \otimes v_2 = v_1 \otimes v_2.$$

Here $\Delta' = P \circ \Delta$ and P are a permutation of factors in $U_q(\widehat{\mathfrak{g}}) \otimes U_q(\widehat{\mathfrak{g}})$. Clearly $R_{V_1 V_2}(z)$ preserves the weight decomposition of $V_1 \otimes V_2$ considered as an $U_q(\mathfrak{g})$-module; its restriction to any weight subspace of $V_1 \otimes V_2$ is a rational function in z.

Let V be a highest weight $U_q(\mathfrak{g})$-module. For any $\mu \in \mathbb{C}^{N+1}$, introduce $L(\mu) \in \operatorname{End}(V)$:

$$L(\mu) = \prod_{i=1}^{N+1} k_i^{\mu_i}.$$

It is well defined for any highest weight $U_q(\mathfrak{g})$-module.

Let $p \in \mathbb{C}$, $p \neq 0$, and $z = (z_1, \ldots, z_n)$. Denote by Z_i the p-shift operator:

$$Z_i : \Psi(z_1, \ldots, z_n) \mapsto \Psi(z_1, \ldots, p z_i, \ldots, z_n).$$

(6.4) DEFINITION. The operators

$$K_i(z; p) = R_{i,i-1}(p z_i / z_{i-1}) \cdots R_{i1}(p z_i / z_1) L_i(\mu) R_{ni}^{-1}(z_n / z_i) \cdots R_{i+1,i}^{-1}(z_{i+1}/z_i),$$

$i = 1, \ldots, n$, are called qKZ operators.

The qKZ operators preserve the weight decomposition of the $U_q(\mathfrak{g})$-module $V_1 \otimes \cdots \otimes V_n$ and their restrictions to any weight subspace are rational functions in z.

(6.5) THEOREM [**FR**, Theorem (5.4)]. *The qKZ operators obey the compatibility conditions*

$$Z_i K_j(z; p) \cdot K_i(z; p) = Z_j K_i(z; p) \cdot K_j(z; p).$$

(6.6) DEFINITION. The *quantized Knizhnik–Zamolodchikov equation* (*qKZ*) is the holonomic system of difference equations for a $V_1 \otimes \cdots \otimes V_n$-valued function $\Psi(z;p)$,

$$Z_i \Psi(z;p) = K_i(z;p) \Psi(z;p)$$

for $i = 1, \ldots, n$, [**FR**].

Fix $\lambda \in \mathbb{Z}_{\geq 0}^N$. Let $(\Lambda_1(1), \ldots, \Lambda_{N+1}(1)), \ldots, (\Lambda_1(n), \ldots, \Lambda_{N+1}(n))$ be highest weights of $U_q(\mathfrak{g})$-modules $V_1, \ldots, V_n$, respectively. Let $V_\lambda = (V_1 \otimes \cdots \otimes V_n)_\lambda$ be the weight subspace:

(6.7) $\quad V_\lambda = \{ v \in V_1 \otimes \cdots \otimes V_n \mid k_i v = q^{\lambda_{i-1} - \lambda_i + \sum_{m=1}^n \Lambda_i(m)} v, \ i = 1, \ldots, N+1 \}$,

where $\lambda_0 = \lambda_{N+1} = 0$. In the sequel we shall be interested in solutions to system (6.6) with values in V_λ.

Further on we assume that $p \in (0,1)$ and $q = p^{-\nu}$, $\nu \in \mathbb{C}$. Set

$$(u,p)_\infty = \prod_{j=0}^\infty (1 - p^j u).$$

Set $\ell = \sum_{i=1}^N \lambda_i$. Let

$$t = (t_{11}, \ldots, t_{1\lambda_1}, t_{21}, \ldots, t_{2\lambda_2}, \ldots, t_{N1}, \ldots, t_{N\lambda_N}) \in \mathbb{C}^\ell.$$

(6.8) DEFINITION. The function

$$\Phi(t,z;p) = \prod_{m=1}^n \prod_{i=1}^N z_m^{-\nu \mu_i \Lambda_i(m)} \prod_{i=1}^N \prod_{j=1}^{\lambda_i} t_{ij}^{\nu(\mu_i - \mu_{i+1})}$$

$$\times \prod_{m=1}^n \prod_{i=1}^N \prod_{j=1}^{\lambda_i} \frac{(q^{2\Lambda_{i+1}(m)} t_{ij}/z_m, p)_\infty}{(q^{2\Lambda_i(m)} t_{ij}/z_m, p)_\infty} \left(\frac{t_{ij}}{z_m} \right)^{\nu(\Lambda_i(m) - \Lambda_{i+1}(m))}$$

$$\times \prod_{i=1}^N \prod_{j=2}^{\lambda_i} \prod_{k=1}^{j-1} \frac{(q^2 t_{ik}/t_{ij}, p)_\infty}{(q^{-2} t_{ik}/t_{ij}, p)_\infty} \left(\frac{t_{ij}}{t_{ik}} \right)^{2\nu}$$

$$\times \prod_{i=1}^{N-1} \prod_{j=1}^{\lambda_i} \prod_{k=1}^{\lambda_{i+1}} \frac{(t_{i+1,k}/t_{ij}, p)_\infty}{(q^2 t_{i+1,k}/t_{ij}, p)_\infty} \left(\frac{t_{i+1,k}}{t_{ij}} \right)^\nu$$

is called the *phase function* of the weight subspace V_λ.

Introduce the lexicographical ordering on the set of pairs (i,j): $(i,j) < (k,l)$ if $i < k$ or $i = k$ and $j < l$. Let $a, b, \ldots$ stay for $(i,j), (k,l), \ldots$. Let Q_a be the *p*-shift operator with respect to the variable t_a.

(6.9) DEFINITION. Set

$$\nabla_a \Phi(t,z) = \lim_{p \to 1} ((\Phi(t,z;p))^{-1} Q_a \Phi(t,z;p)),$$

$$\nabla_{ab}^2 \Phi(t,z) = (\nabla_b \Phi(t,z))^{-1} t_a \frac{\partial}{\partial t_a} \nabla_b \Phi(t,z),$$

$$H(t,z) = \det[\nabla_{ab}^2 \Phi(t,z)]_{\ell \times \ell}.$$

Let $\mathcal{H}_\lambda^0$ be the space spanned by entries of operators $K_i(z;p)$ restricted to V_λ, $i = 1, \ldots, n$. Let $\mathcal{H}_\lambda$ be the space spanned by products $g_1 \cdots g_s$, where $g_i \in \mathcal{H}_\lambda^0$ for all i and $s \in \mathbb{Z}_{\geq 0}$. Consider the following linear functions:

$$(6.10) \qquad q^{2\Lambda_i(m)} t_{ij} - pz_m, \quad q^{\Lambda_{i+1}(m)} t_{ij} - z_m, \quad t_{ij} - q^2 t_{ik}, \quad t_{i+1,l} - t_{ij},$$

where $m = 1, \ldots, n$, $i = 1, \ldots, N$, $j = 1, \ldots, \lambda_i$, $k = 1, \ldots, j-1$, and $l = 1, \ldots, \lambda_{i+1}$. Let $\mathcal{F}_0$ be the space spanned by the products $g_1^{-1} \cdots g_s^{-1}$, $s \in \mathbb{Z}_{\geq 0}$, where each g_i is a linear function from the list (6.10) and $g_i \neq g_j$ for $i \neq j$. Set $\mathcal{F} = \mathbb{C}[t, z, p, p^{-1}] \otimes \mathcal{F}_0$.

(6.11) DEFINITION. Let $\mathcal{Q}_\lambda$ be the space spanned by the discrete differentials $Q_a(\Phi w) - \Phi w$, $a = 1, \ldots, \ell$, $w \in \mathcal{F} \otimes \mathcal{H}_\lambda$.

(6.12) DEFINITION. Let $w(t, z; p) \in \mathcal{F} \otimes V_\lambda$. The product $\Phi(t, z; p) w(t, z; p)$ is said to give an *integral representation* for solutions to system (6.6) if $Z_i(\Phi w) - K_i \Phi w \in \mathcal{Q}_\lambda \otimes V_\lambda$, $i = 1, \ldots, n$.

(6.13) THEOREM [**TV**, Theorem (1.5.2)]. *There exists an integral representation for solutions to qKZ (6.6) associated with $U_q(\mathfrak{gl}_{N+1})$.*

The $\mathfrak{gl}_2$ case was considered in [**M, R2, V**]. Explicit formulas for an integral representation are given in [**TV**] and in Appendix 1.

REMARK. In [**TV**], we defined $\Phi(t, z; p)$ and $w(t, z; p)$ and proved that the differences $Z_i(\Phi w) - K_i \Phi w$ are discrete differentials. We did not specify the singularities of these differences, but the proof in [**TV**] clearly shows that these differences belong to $\mathcal{Q}_\lambda \otimes V_\lambda$.

(6.14) DEFINITION. A point (t, z) is called a *critical point* if $\nabla_a \Phi(t, z) = 1$ for all a. A critical point (t, z) is called *nondegenerate* if $H(t, z) \neq 0$.

Set $p = e^{-\delta}$, $\delta > 0$, $v = \upsilon/\delta$, $\upsilon \in \mathbb{C}$. Then $q = e^\upsilon$.

Let $M \subset \mathbb{C}^\ell$ be an open region such that all $K_i(z; 1)$ and $K_i^{-1}(z; 1)$ are regular in M. The qKZ operators $K_i(z; p)$ have power series expansions

$$(6.15) \qquad K_i(z; p) = \sum_{s=0}^\infty K_{is}(z) \delta^s,$$

where $K_{is}(z)$ is also regular in M. Now we are in a position related to §1, and we are interested in asymptotic solutions to system (6.6) as $p \to 1$. The variables z, p in §1 correspond to variables $\log z, \delta$ in this section.

REMARK. Actually, we must consider restrictions of qKZ operators to V_λ that are rational functions in z, p. In this case we can take M to be the complement to the singularities of $K_i(z; 1)$, $K_i^{-1}(z; 1)$.

The dilogarithm function $\mathrm{Li}_2(u)$ is defined by

$$\mathrm{Li}_2(x) = -\int_0^u \log(1-t) \frac{dt}{t}, \qquad u \in (0, 1),$$

and can be analytically continued to $\mathbb{C} \setminus [1, \infty)$. Further, we always take the following branch of the logarithm

$$(6.16) \qquad \mathrm{Im} \log x \in (0, 2\pi).$$

Set $\chi(x, y) = -\text{Li}_2(xq^{2y}) - vy\log x$. Introduce $\tau(t, z)$ as follows:

(6.17)
$$\tau(t, z) = \sum_{m=1}^{n} \sum_{i=1}^{N} v\mu_i \Lambda_i(m) \log z_m + \sum_{i=1}^{N} \sum_{j=1}^{\lambda_i} v(\mu_{i+1} - \mu_i) \log t_{ij}$$
$$+ \sum_{m=1}^{n} \sum_{i=1}^{N} \sum_{j=1}^{\lambda_i} (\chi(t_{ij}/z_m, \Lambda_i(m)) - \chi(t_{ij}/z_m, \Lambda_{i+1}(m)))$$
$$+ \sum_{i=1}^{N-1} \sum_{j=1}^{\lambda_i} \sum_{k=1}^{\lambda_{i+1}} (\chi(t_{i+1,k}/t_{ij}, 1) - \chi(t_{i+1,k}/t_{ij}))$$
$$+ \sum_{i=1}^{N} \sum_{j=2}^{\lambda_i} \sum_{k=1}^{j-1} (\chi(t_{ik}/t_{ij}, -1) - \chi(t_{ik}/t_{ij}, 1)).$$

(6.18) LEMMA. $\nabla_a \Phi(t, z) = \exp\left(t_a \frac{\partial}{\partial t_a} \tau(t, z)\right).$

(6.19) COROLLARY. $\nabla^2_{ab} \Phi(t, z) = \nabla^2_{ba} \Phi(t, z).$

Let $\mathfrak{S} \subset \mathbb{C}^{\ell+n}$ be the cuts defining the branch of $\tau(t, z)$. Let $\mathcal{F}_\circ$ be a space of polynomials in t, z and the following rational functions:

$$(q^{2\Lambda_i(m)} t_{ij} - z_m)^{-1}, \quad (q^{2\Lambda_{i+1}(m)} t_{ij} - z_m)^{-1}, \quad (t_{ij} - q^2 t_{ik})^{-1},$$
$$(t_{i+1,l} - t_{ij})^{-1}, \quad (q^2 t_{i+1,l} - t_{ij})^{-1},$$
$$m = 1, \ldots, n, \ i = 1, \ldots, N, \ j, k = 1, \ldots, \lambda_i, \ l = 1, \ldots, \lambda_{i+1}.$$

Set $\Theta = \sum_{i=1}^{N} \lambda_i(\lambda_i - 1)/2$.

(6.20) LEMMA. *Let* $(t, z) \notin \mathfrak{S}$. *As* $p \to 1$, $\Phi(t, z; p)$ *has the asymptotic expansion*

$$\Phi(t, z; p) \simeq \exp(-\tau(t, z)/\delta) \Xi(t, z) q^{\Theta} \left(1 + \sum_{s=1}^{\infty} \phi_s(t, z)\delta^s\right),$$

where

$$\Xi(t, z) = \left(\prod_{m=1}^{n} \prod_{i=1}^{N} \prod_{j=1}^{\lambda_i} \frac{q^{2\Lambda_{i+1}(m)} t_{ij} - z_m}{q^{2\Lambda_i(m)} t_{ij} - z_m}\right.$$
$$\left.\times \prod_{i=1}^{N} \prod_{j=2}^{\lambda_i} \prod_{k=1}^{j-1} \frac{q^2 t_{ik} - t_{ij}}{t_{ik} - q^2 t_{ij}} \prod_{i=1}^{N-1} \prod_{j=1}^{\lambda_i} \prod_{k=1}^{\lambda_{i+1}} \frac{t_{i+1,k} - t_{ij}}{q^2 t_{i+1,k} - t_{ij}}\right)^{1/2}$$

and $\phi_s(t, z) \in \mathcal{F}_\circ$.

The lemma follows from the asymptotic expansion for $(u, p)_\infty$.

(6.21) LEMMA. *As* $p \to 1$, $(u, p)_\infty$ *has the following asymptotic expansion in* $\mathbb{C} \setminus [1, \infty)$:

$$(u, p)_\infty \simeq i(u - 1)^{1/2} \exp(-\text{Li}_2(u)/\delta) \left(1 + \sum f_s(u)\delta^s\right),$$

where $f_s(u) \in \mathbb{C}[u, (u-1)^{-1}]$.

Let $(t^*, z^*) \notin \mathfrak{S}$, $z^* \in M$ be a nondegenerate critical point. Consider the quadratic form
$$S(x) = \sum_{a=1}^{\ell} \sum_{b=1}^{\ell} x_a x_b \nabla^2_{ab} \Phi(t^*, z^*), \qquad x \in \mathbb{C}^{\ell},$$
a real hyperplane $W \subset \mathbb{C}^{\ell}$, $\dim_{\mathbb{R}} W = \ell$ such that the restriction of $S(x)$ to W is positive, and also consider a small disk
$$D = \{t \in \mathbb{C}^{\ell} \mid t = e^u t^*, u \in W, |u| < \varepsilon\},$$
where ε is a small positive number. Let $t(z)$ be a holomorphic function such that $(t(z), z)$ is a nondegenerate critical point and $t(z^*) = t^*$. Later on we assume that p is close to 1. Set

(6.22) $$I_a = \frac{t_a}{2\pi i} \frac{\partial}{\partial t_a} \tau(t^*, z^*) \quad \text{and} \quad I(t) = \prod_{a=1}^{\ell} t_a^{2\pi i I_a/\delta}.$$

It is clear that
$$t_a \frac{\partial}{\partial t_a} \tau(t(z), z) = 2\pi i I_a,$$
and $I(t)$ is a multiplicatively p-periodic function with respect to all t_a.

Set
$$D^{\ell} t = \frac{dt_1}{t_1} \wedge \cdots \wedge \frac{dt_{\ell}}{t_{\ell}}.$$

Set

(6.23) $$\Psi(z; p) = \delta^{-\ell/2} q^{-\Theta} \int_D I(t) \Phi(t, z; p) w(t, z; p) D^{\ell} t,$$

where $w(t, z; p) \in \mathcal{F} \otimes V_{\lambda}$, and set
$$\widehat{\tau}(t, z) = \tau(t, z) - 2\pi i \sum_{a=1}^{\ell} I_a \log t_a.$$

(6.24) LEMMA. *As $p \to 1$, $\Psi(z; p)$ has the asymptotic expansion*
$$\Psi(z; p) \simeq (2\pi)^{\ell/2} \exp(-\widehat{\tau}(t(z), z)/\delta) \, \Xi(t(z), z) H^{-1/2}(t(z), z)$$
$$\times \left(w(t(z), z; 1) + \sum_{s=1}^{\infty} \psi_s(t(z), z) \delta^s \right),$$
where $\psi_s(t, z) \in \mathcal{F}_{\circ} \otimes V_{\lambda}$.

(6.25) THEOREM. *Let $\Phi(t, z; p) w(t, z; p)$ be an integral representation for solutions to qKZ (6.6). The asymptotic expansion of $\Psi(z; p)$ as $p \to 1$ gives an asymptotic solution to system (6.6) in the sense of (1.4).*

(6.26) COROLLARY.
$$K_i(z^*; 1) w(t^*, z^*; 1) = \exp\left(z_i \frac{\partial \tau}{\partial z_i}(t^*, z^*) \right) w(t^*, z^*; 1), \qquad i = 1, \ldots, n.$$

(6.27) THEOREM. *Let (t^*, z^*) be a diagonal nondegenerate critical point, and let $\Psi(z; p)$ be defined by* (6.23). *Then* $\Psi(z; p) \exp(\widehat{\tau}(t(z), z)/\delta) = O(p^\infty)$ *as* $p \to 1$.

For any highest weight $U_q(\mathfrak{g})$-module V, the restricted dual space V^* admits the natural structure of a right $U_q(\mathfrak{g})$-module. Introduce the *dual qKZ operators* $\widetilde{K}_i(z; p) = (K_i^*(z; p))^{-1}$.

(6.28) DEFINITION. The *dual qKZ* is the holonomic system of difference equations for the $V_1^* \otimes \cdots \otimes V_n^*$-valued function $\widetilde{\Psi}(z; p)$:

$$Z_i \widetilde{\Psi}(z; p) = \widetilde{K}_i(z; p) \widetilde{\Psi}(z; p)$$

for $i = 1, \ldots, n$.

Let $V_\lambda^* = (V_1^* \otimes \cdots \otimes V_n^*)_\lambda$ be the dual weight subspace:

(6.29) $\quad V_\lambda^* = \{v^* \in V_1^* \otimes \cdots \otimes V_n^* \mid k_i v^* = q^{\lambda_{i-1} - \lambda_i + \sum_{m=1}^n \Lambda_i(m)} v^*, \, i = 1, \ldots, N+1\}$,

where $\lambda_0 = \lambda_{N+1} = 0$. In the sequel we shall be interested in solutions to system (6.28) with values in V_λ^*.

(6.30) DEFINITION. The function $\widetilde{\Phi}(t, z; p) = \Xi^2(t, z) \Phi^{-1}(t, z; p)$ is called the *phase function* of the weight subspace V_λ^*.

Let $\widetilde{\mathcal{Q}}_\lambda$ be the space spanned by the discrete differentials $Q_a(\widetilde{\Phi} w) - \widetilde{\Phi} w$, $a = 1, \ldots, \ell, w \in \mathcal{F}$.

(6.31) DEFINITION. Let $\widetilde{w}(t, z; p) \in \mathcal{F} \otimes V_\lambda^*$. The product $\widetilde{\Phi}(t, z; p) \widetilde{w}(t, z; p)$ is said to give an *integral representation* for solutions to system (6.28) if $Z_i(\widetilde{\Phi}\widetilde{w}) - \widetilde{K}_i \widetilde{\Phi} \widetilde{w} \in \widetilde{\mathcal{Q}}_\lambda \otimes V_\lambda^*$, $i = 1, \ldots, n$.

Integral representations for solutions to dual *qKZ* (6.28) can be obtained as in the case of *qKZ* (6.6). Explicit formulas are given in Appendix 1.

Let $(t, z) \notin \mathfrak{S}_\vartheta$. As $p \to 1$, the function $\widetilde{\Phi}(t, z; p)$ has the asymptotic expansion

(6.32) $\quad \widetilde{\Phi}(t, z; p) \simeq \exp(\tau(t, z)/\delta) \Xi(t, z) q^{-\Theta} \left(1 + \sum_{s=1}^\infty \widetilde{\phi}_s(t, z) \delta^s\right)$,

where $\widetilde{\phi}_s(t, z) \in \mathcal{F}_\circ$. Let $(t^*, z^*) \notin \mathfrak{S}_\vartheta$, $z^* \in M$ be a nondegenerate critical point. Let $\widetilde{D} \subset \mathbb{C}^\ell$ be the small disk

$$\widetilde{D} = \{t \in \mathbb{C}^\ell \mid t = e^{iu} t^*, u \in W, |u| < \varepsilon\}$$

where ε is a small positive number. Set

(6.33) $\quad \widetilde{\Psi}(z; p) = \delta^{-\ell/2} q^\Theta \int_{\widetilde{D}} I(t) \widetilde{\Phi}(t, z; p) \widetilde{w}(t, z; p) D^\ell t$,

where $\widetilde{w}(t, z; p) \in \mathcal{F} \otimes V_\lambda^*$. As $p \to 1$, the function $\widetilde{\Psi}(z; p)$ has the asymptotic expansion

(6.34)
$$\widetilde{\Psi}(z; p) \simeq (-2\pi)^{\ell/2} \exp(\widehat{\tau}(t(z), z)/\delta) \Xi(t(z), z) H^{-1/2}(t(z), z)$$
$$\times \left(\widetilde{w}(t(z), z; 1) + \sum_{s=1}^\infty \widetilde{\psi}_s(t(z), z) \delta^s\right),$$

where $\widetilde{\psi}_s(t, z) \in \mathcal{F}_\circ \otimes V_\lambda^*$.

(6.35) THEOREM. *Let $\widetilde{\Phi}(t,z;p)\widetilde{w}(t,z;p)$ be an integral representation for solutions to dual qKZ (6.28). Then the asymptotic expansion of $\widetilde{\Psi}(z;p)$ as $p \to 1$ gives an asymptotic solution to system (6.28) in the sense of (1.4).*

(6.36) COROLLARY.

$$K_i^*(z^*;1)\widetilde{w}(t^*,z^*;1) = \exp\left(z_i \frac{\partial \tau}{\partial z_i}(t^*,z^*)\right)\widetilde{w}(t^*,z^*;1), \qquad i = 1,\ldots,n.$$

Let us consider $\mu \in \mathbb{C}^{N+1}$ as an additional set of variables. Let (t^*, z^*, μ^*) be an offdiagonal nondegenerate critical point (with respect to t). Let $t(z, \mu)$ be a holomorphic function such that $(t(z,\mu), z, \mu)$ is a nondegenerate critical point and $t(z^*, \mu^*) = t^*$. Recall that $w(t, z, \mu; p)$ and $\widetilde{w}(t, z, \mu; p)$ in the integral representations do not depend on μ and p at all. Furthermore, $H(t, z, \mu)$ and $\Xi(t, z, \mu)$ do not depend on μ as well.

(6.37) THEOREM. *Let*

$$\mathfrak{H}(z,\mu) = \Xi^2(t(z,\mu),z)H^{-1}(t(z,\mu),z)\langle\widetilde{w}(t(z,\mu),z), w(t(z,\mu),z)\rangle.$$

Then

$$\frac{\partial}{\partial z_i}\mathfrak{H}(z,\mu) = 0, \qquad i = 1,\ldots,n,$$

$$\frac{\partial}{\partial \mu_j}\mathfrak{H}(z,\mu) = 0, \qquad j = 1,\ldots,N+1.$$

(6.38) COROLLARY. *For any offdiagonal nondegenerate critical point (t, z),*

$$\langle\widetilde{w}(t,z), w(t,z)\rangle = \mathrm{const}\,\Xi^{-2}(t,z)H(t,z),$$

where const *does not depend on μ and does not change under continuous deformations of the critical point (t, z).*

(6.39) CONJECTURE. *For any offdiagonal critical point (t, z), we have*

$$\langle\widetilde{w}(t,z), w(t,z)\rangle = (-1)^\ell(q-q^{-1})^{-\ell}\Xi^{-2}(t,z)H(t,z).$$

(6.40) CONJECTURE. *Let (t, z) and $(\tilde{t}, z)$ be offdiagonal critical points such that*

$$\{t_{ij} \mid j = 1,\ldots,\lambda_i\} \neq \{\tilde{t}_{ij} \mid j = 1,\ldots,\lambda_i\}$$

for some i. Then

$$\langle\widetilde{w}(\tilde{t},z), w(t,z)\rangle = 0.$$

These conjectures for the $U_q(\mathfrak{gl}_2)$ case can be proved using Corollary (6.38). A combinatorial proof of Conjecture (6.39) for the $U_q(\mathfrak{gl}_2)$ case was given in [**K**].

PROOF OF CONJECTURES (6.39) AND (6.40) IN THE $U_q(\mathfrak{gl}_2)$ CASE. This proof is completely similar to that of Conjectures (3.16) and (3.17) for the $\mathfrak{gl}_2$ case given in §4. We mention here only its key points.

We assume that $N = 1$. Without loss of generality we can suppose that $\Lambda_1(m) = 0$, $m = 1,\ldots,n$, and $\mu_2 = 0$. Set

$$y_m = q^{-2\Lambda_2(m)}z_m, \quad m = 1,\ldots,n, \quad \text{and} \quad \kappa = q^{\mu_1+\sum_{m=1}^n \Lambda_2(m)}.$$

We assume that all y_m, z_m are generic.

The original system of equations for critical points is

(6.41) $$\kappa^{-1} \prod_{m=1}^{n} \frac{t_a - z_m}{t_a - y_m} \prod_{\substack{b=1 \\ b \neq a}}^{\ell} \frac{t_a - q^2 t_b}{q^2 t_a - t_b} = 1, \qquad a = 1, \ldots, \ell.$$

We replace it by the system of algebraic equations

(6.42) $$\prod_{m=1}^{n} (t_a - z_m) \prod_{\substack{b=1 \\ b \neq a}}^{\ell} (t_a - q^2 t_b) = \kappa \prod_{m=1}^{n} (t_a - y_m) \prod_{\substack{b=1 \\ b \neq a}}^{\ell} (q^2 t_a - t_b), \qquad a = 1, \ldots, \ell.$$

Both systems (6.41) and (6.42) are preserved by the natural action of the symmetric group $\mathbf{S}_\ell$ on the variables $t_1, \ldots, t_\ell$. Denote by $\mathfrak{D} \subset \mathbb{C}^\ell$ the complement to the union of the coordinate hyperplanes $t_a = 0$, $a = 1, \ldots, \ell$. System (6.42) will be considered only in $\mathfrak{D}$.

(6.43) LEMMA. *Systems* (6.41) *and* (6.42) *are equivalent for* $\kappa \neq 0$.

(6.44) LEMMA. *All solutions to system* (6.42) *remain finite for any* $\kappa \neq q^{2(s-\ell)} e^{2\pi i r/s}$, $s = 1, \ldots, \ell$, $r = 0, \ldots, s$.

(6.45) LEMMA. *All solutions to system* (6.42) *remain in* $\mathfrak{D}$ *for any*

$$\kappa \neq q^{2(\ell-s)} e^{2\pi i r/s} \prod_{m=1}^{n} y_m z_m^{-1}, \qquad s = 1, \ldots, \ell, \ r = 0, \ldots, s.$$

(6.46) LEMMA. *The multiplicity of any offdiagonal solution to system* (6.42) *at* $\kappa = 0$ *is equal to* 1.

(6.47) LEMMA. *Let* $t(\kappa)$ *be a solution to system* (6.42), *which is a deformation of a diagonal solution* $t(0)$ *to this system at* $\kappa = 0$. *Then* $t(k)$ *is a diagonal critical point.*

(6.48) LEMMA. *For generic* κ, *there are* $\binom{n+\ell-1}{n-1}$ *offdiagonal critical points modulo the action of the symmetric group* $\mathbf{S}_\ell$. *All of them are nondegenerate.*

(6.49) LEMMA. *Offdiagonal solutions to system* (6.42) *remain finite for any* $\kappa \neq q^{2(s-\ell)}$. *Offdiagonal solutions to system* (6.42) *remain in* $\mathfrak{D}$ *for any* $\kappa \neq q^{2(\ell-s)} \prod_{m=1}^{n} y_m z_m^{-1}$.

The last formulas to be mentioned are related to the canonical monomial bases in $V_1 \otimes \cdots \otimes V_n$ and $V_1^* \otimes \cdots \otimes V_n^*$:

(6.50) $$F^v = f^{v_1} v_1 \otimes \cdots \otimes f^{v_n} v_n, \qquad E^v = e^{v_1} v_1^* \otimes \cdots \otimes e^{v_n} v_n^*,$$

where $e = e_1$, $f = f_1$ (cf. (4.16)). They are dual to each other up to normalization:

(6.51) $$\langle E^v, F^{v'} \rangle = \delta_{vv'} \prod_{m=1}^{n} \prod_{j=1}^{v_m} [j]_q [\Lambda_1(m) - \Lambda_2(m) - j + 1]_q. \qquad \square$$

Let $\mathfrak{C}(z, \mu)$ be a set of all different offdiagonal critical points modulo the action of the symmetric group $\mathbf{S}_\ell$. Vectors $w(t, z)$ and $\widetilde{w}(t, z)$ are preserved by the action of $\mathbf{S}_\ell$ modulo multiplication by a scalar factor.

(6.52) THEOREM. *Let z, μ, $\Lambda(1), \ldots, \Lambda(n)$ be generic. Then $\{w(t,z)\}_{t \in \mathfrak{C}(z,\mu)}$ and $\{\widetilde{w}(t,z)\}_{t \in \mathfrak{C}(z,\mu)}$ are bases in V_ℓ and V_ℓ^*, respectively. They are dual to each other up to normalization.*

All proofs are the same as in §§2–4.

Appendix 1

Here we recall the definition and main properties of the vectors $w(t,z)$, which appear in the integral representations for solutions to qKZ (cf. [TV] for details and references). The notation used here can differ from that used in [TV]. We describe integral representations for solutions to the dual qKZ as well. We also give explicit formulas for the action of some generators of $\mathfrak{gl}_{N+1}$ on the vectors $w(t,z)$.

Let $\mathfrak{g} = \mathfrak{gl}_{N+1}$, $Y = Y(\mathfrak{gl}_{N+1})$. Let $V_1, \ldots, V_n$ be $\mathfrak{g}$-modules with highest weights $\Lambda(1), \ldots, \Lambda(n)$ and generating vectors $v_1, \ldots, v_n$, respectively. Set $V = V_1 \otimes \cdots \otimes V_n$. Let $z \in \mathbb{C}^n$. We turn V into a Y-module by the homomorphism $\varphi_z^{(n)}: Y \to U(\mathfrak{g})^{\otimes n}$:

$$\varphi_z^{(n)}: X \mapsto (\varphi_{z_1} \otimes \cdots \otimes \varphi_{z_n}) \circ \Delta^{(n-1)}(X).$$

Here $\Delta^{(m)}$ is the m-iterated coproduct ($\Delta^{(0)} = \mathrm{id}$, $\Delta^{(1)} = \Delta$) and $\varphi_z: Y \to U(\mathfrak{g})$ is the homomorphism described in §2. There is a homomorphism of Hopf algebras $\widehat{\varphi}: U(\mathfrak{g}) \to Y$, $E_{ij} \overset{\widehat{\varphi}}{\mapsto} T_{ji}^1$, such that $\varphi_z \circ \widehat{\varphi} = \mathrm{id}$. In this sense, the $\mathfrak{g}$-module and Y-module structures on V are consistent.

Let $e_{ij} \in \mathrm{End}(\mathbb{C}^{N+1})$ be the image of E_{ij} under the natural representation of $\mathfrak{gl}_{N+1}$. Define $R(u,v) \in \mathrm{End}(\mathbb{C}^{N+1} \otimes \mathbb{C}^{N+1})$ and $T_V(u,z) \in \mathrm{End}(V \otimes \mathbb{C}^{N+1})$, $u, v \in \mathbb{C}$, as follows:

(A1.1) $$R(u,v) = 1 + (u-v)^{-1} \sum_{ij} e_{ij} \otimes e_{ji},$$

(A1.2) $$T_V(u,z) = u^n + \sum_{s=1}^{\infty} \sum_{ij} \varphi_z^{(n)}(T_{ij}^s) \otimes e_{ij} u^{n-s}.$$

Since $s \geq n$, we have $\varphi_z^{(n)}(T_{ij}^s) = 0$, so $T_V(u,z)$ is a polynomial in u, z. Fix $\ell \in \mathbb{Z}_{\geq 0}$. Denote by $\iota_a: \mathrm{End}(\mathbb{C}^{N+1}) \to \mathrm{End}((\mathbb{C}^{N+1})^{\otimes \ell})$ the following embedding:

$$\iota_a: x \mapsto 1 \otimes \cdots \otimes x \otimes \cdots \otimes 1,$$

where x stands in the ath place. Set $R^{ab}(u) = 1 \otimes (\iota_a \otimes \iota_b(R(u)))$ and $T_V^a(u,z) = \mathrm{id} \otimes \iota_a(T_V(u,z))$.

Let $t \in \mathbb{C}^\ell$. Define $\mathbb{T}_V(t,z) \in \mathrm{End}(V \otimes (\mathbb{C}^{N+1})^{\otimes \ell})$ as follows:

(A1.3) $$\mathbb{T}_V(t,z) = T_V^1(t_1,z) \cdots T_V^\ell(t_\ell,z) \prod_{a=2}^{\ell} \prod_{b=1}^{a-1} R^{ab}(t_a,t_b),$$

where the last product is taken in lexicographical order: the factor $R^{ab}(t_a,t_b)$ stands on the right side of the factor $R^{cd}(t_c,t_d)$ if $a < c$ or $a = c$ and $b < d$.

Fix $\lambda \in \mathbb{Z}_{\geq 0}^N$ such that $\ell = \sum_{i=1}^N \lambda_i$. Introduce the lexicographical order on the set of pairs $\{(i,j) \mid i = 1, \ldots, N, j = 1, \ldots, \lambda_i\}$:

$$(i,j) < (k,l) \quad \text{if } i < k \text{ or } i = k \text{ and } j < l.$$

Let $(i, j), (k, l), \ldots$ stand for $a, b, \ldots$. Set

$$F_\lambda = 1 \otimes \underbrace{e_{21} \otimes \cdots \otimes e_{21}}_{\lambda_1} \otimes \cdots \otimes \underbrace{e_{N+1,N} \otimes \cdots \otimes e_{N+1,N}}_{\lambda_N} \in \text{End}(V \otimes (\mathbb{C}^{N+1})^{\otimes \ell}).$$

Let $\text{tr}: \text{End}((\mathbb{C}^{N+1})^{\otimes \ell}) \to \mathbb{C}$ be the trace map and let $\text{Tr} = \text{id} \otimes \text{tr}$. Set

$$A(u,v) = \frac{u-v+1}{u-v}, \qquad B(u,v) = \frac{u-v+1}{u-v-1},$$

$$C_{im}(u,v) = \frac{u-v+\Lambda_i(m)}{u-v+\Lambda_{i+1}(m)}.$$

Define

(A1.4)
$$\xi_{\lambda,V}(t,z) = \text{Tr}(F_\lambda \mathbb{T}_V(t,z)) v_1 \otimes \cdots \otimes v_n$$
$$\times \prod_{i=1}^{N} \prod_{j=1}^{\lambda_i} \left(\prod_{m=1}^{n} (t_{ij} - z_m + \Lambda_{i+1}(m)) \prod_{k=1}^{j-1} A(t_{ij}, t_{ik}) \right)^{-1}.$$

For any $i = 1, \ldots, n$, the function $\xi_{\lambda,V}(t,z)$ is symmetric in t_{ij}, $j = 1, \ldots, \lambda_i$. Set

(A1.5)
$$w(t,z) = \xi_{\lambda,V}(t,z) \prod_{i=1}^{N} \prod_{j=2}^{\lambda_i} \prod_{k=1}^{j-1} A^{-1}(t_{ik}, t_{ij}).$$

The vector w is used in integral representations for solutions to qKZ. Relation (2.29) follows from (A1.5).

(A1.6) LEMMA. $w(t,z) \in \mathcal{F} \otimes V_\lambda$.

This lemma is proved at the end of the section.

Let $\Phi(t,z;p)$ be the phase function (2.7). To prove that $\Phi(t,z;p)w(t,z)$ is an integral representation for solutions to qKZ, the following expression for $w(t,z)$ is essential:

(A1.7)
$$w(t,z) = \sum \Bigg(\prod_{l=2}^{n} \prod_{m=1}^{l-1} \prod_{i=1}^{N} \prod_{j \in \Omega_i(l)} C_{im}(t_{i\sigma_i(j)}, z_m)$$

$$\times \prod_{i=1}^{N} \prod_{\substack{1 \leq j < k \leq \lambda_i \\ \sigma_i(j) > \sigma_i(k)}} B(t_{i\sigma_i(j)}, t_{i\sigma_i(k)})$$

$$\times \prod_{l=2}^{n} \prod_{m=1}^{l-1} \prod_{i=2}^{N} \prod_{\substack{j \in \Omega_i(l) \\ k \in \Omega_{i-1}(m)}} \prod_{j \in \Omega_i(l)} A(t_{i\sigma_i(j)}, t_{i-1,\sigma_{i-1}(k)})$$

$$\times \prod_{i=1}^{N} \frac{1}{\lambda_i!} \xi_{\lambda(1),V_1}(t(1),z_1) \otimes \cdots \otimes \xi_{\lambda(n),V_n}(t(n),z_n) \Bigg).$$

Here the sum is taken over all $\lambda(1), \ldots, \lambda(n) \in \mathbb{Z}_{\geq 0}^N$ such that $\lambda = \sum_{m=1}^{n} \lambda(m)$, and over all $\sigma = (\sigma_1, \ldots, \sigma_N) \in \mathbf{S}_{\lambda_1} \times \cdots \times \mathbf{S}_{\lambda_N}$. The notation used in (A1.7) is as follows.

Set $\ell_i(m) = \sum_{l=1}^{m} \lambda_i(l)$. Then
$$\Omega_i(m) = \sigma_i(\{\ell_i(m-1)+1, \ldots, \ell_i(m)\}),$$
$$t(m) = \{t_{ij} \mid i = 1, \ldots, N, \; j \in \Omega_i(m)\}.$$

Whenever the set $t(m)$ is used as an argument in $\xi_{\lambda(m), V_m}(t(m), z_m)$, we order it lexicographically.

Formula (A1.7) follows from the expression for $\xi_{\lambda, V}(t, z)$:

(A1.8)
$$\xi_{\lambda, V}(t, z) = \sum \left(\prod_{l=2}^{n} \prod_{m=1}^{l-1} \prod_{i=1}^{N} \prod_{j \in \Gamma_i(l)} C_{im}(t_{ij}, z_m) \right.$$
$$\times \prod_{l=2}^{n} \prod_{m=1}^{l-1} \prod_{i=1}^{N} \prod_{\substack{j \in \Gamma_i(l) \\ k \in \Gamma_i(m)}} A(t_{ik}, t_{ij})$$
$$\times \prod_{l=2}^{n} \prod_{m=1}^{l-1} \prod_{i=2}^{N} \prod_{\substack{j \in \Gamma_i(l) \\ k \in \Gamma_{i-1}(m)}} A(t_{ij}, t_{i-1,k})$$
$$\left. \times \xi_{\lambda(1), V_1}(t(1), z_1) \otimes \cdots \otimes \xi_{\lambda(n), V_n}(t(n), z_n) \right).$$

Here the sum is taken over all partitions of the set
$$\{(i, j) \mid i = 1, \ldots, N, \; j = 1, \ldots, \lambda_i\}$$
into disjoint subsets $\Gamma(1), \ldots, \Gamma(n)$, and we use the notation
$$\Gamma_i(m) = \Gamma(m) \cap \{(i, j) \mid j = 1, \ldots, \lambda_i\}, \qquad \lambda_i(m) = \#\Gamma_i(m),$$
$$t(m) = \{t_{ij} \mid (i, j) \in \Gamma(m)\}.$$

Formulas (A1.5) and (A1.8) imply (4.19).

Let $t^{ij} \in \mathbb{C}^{\ell-1}$ be obtained from $t \in \mathbb{C}^{\ell}$ by removing the coordinate t_{ij}.

Let $\lambda^i = (\lambda_1, \ldots, \lambda_i - 1, \ldots, \lambda_N)$. The following formula holds for the action of the generators $E_{i,i+1}$ of $\mathfrak{gl}_{N+1}$ on $\xi_{\lambda, V}(t, z)$:

(A1.9)
$$E_{i,i+1} \xi_{\lambda, V}(t, z)$$
$$= \sum_{j=1}^{\lambda_i} \left(\left[\prod_{m=1}^{n} C_{im}(t_{ij}, z_m) \prod_{\substack{k=1 \\ k \neq j}}^{\lambda_i} A(t_{ik}, t_{ij}) \prod_{k=1}^{\lambda_{i-1}} A(t_{ij}, t_{i-1,k}) \right. \right.$$
$$\left. \left. - \prod_{\substack{k=1 \\ k \neq j}}^{\lambda_i} A(t_{ij}, t_{ik}) \prod_{k=1}^{\lambda_{i+1}} A(t_{i+1,k}, t_{ij}) \right] \xi_{\lambda^i, V}(t^{ij}, z) \right).$$

This relation and (A1.5) imply formula (5.2).

Let $V_1^*, \ldots, V_n^*$ be the restricted dual spaces to $V_1, \ldots, V_n$, respectively. Each V_i^* is regarded as a right $\mathfrak{g}$-module. Let $v_i^* \in V_i^*$ be the vector defined at the beginning of §3. Define
$$T_{V^*}(u, z) \in \operatorname{End}(V^* \otimes \mathbb{C}^{N+1}) \quad \text{and} \quad \mathbb{T}_{V^*}(t, z) \in \operatorname{End}(V^* \otimes (\mathbb{C}^{N+1})^{\otimes \ell})$$

by formulas (A1.2) and (A1.3), respectively, where V is replaced by V^*. Set

$$E_\lambda = 1 \otimes \underbrace{e_{12} \otimes \cdots \otimes e_{12}}_{\lambda_1} \otimes \cdots \otimes \underbrace{e_{N,N+1} \otimes \cdots \otimes e_{N,N+1}}_{\lambda_N} \in \operatorname{End}(V^* \otimes (\mathbb{C}^{N+1})^{\otimes \ell}).$$

Define

$$\tilde{\xi}_{\lambda, V}(t,z) = \operatorname{Tr}(E_\lambda \mathbb{T}_{V^*}(t,z)) v_1^* \otimes \cdots \otimes v_n^*$$
$$\times \prod_{i=1}^N \prod_{j=1}^{\lambda_i} \left(\prod_{m=1}^n (t_{ij} - z_m + \Lambda_{i+1}(m)) \prod_{k=1}^{j-1} A(t_{ij}, t_{ik}) \right)^{-1}.$$

For any $i = 1, \ldots, n$, the function $\tilde{\xi}_{\lambda, V}(t,z)$ is symmetric in t_{ij}, $j = 1, \ldots, \lambda_i$. Set

(A1.10) $$\widetilde{w}(t,z) = \tilde{\xi}_{\lambda, V}(t,z) \prod_{i=1}^N \prod_{j=2}^{\lambda_i} \prod_{k=1}^{j-1} A^{-1}(t_{ik}, t_{ij}).$$

We have $\widetilde{w}(t,z) \in \mathcal{F} \otimes V_\lambda^*$. The proof is completely similar to that of Lemma (A1.6). The counterparts of formulas (A1.7) and (A1.8) are

$$\widetilde{w}(t,z) = \sum \left(\prod_{m=2}^n \prod_{l=1}^{m-1} \prod_{i=1}^N \prod_{j \in \Omega_i(l)} C_{im}(t_{i\sigma_i(j)}, z_m) \right.$$
$$\times \prod_{i=1}^N \prod_{\substack{1 \leq j > k \leq \lambda_i \\ \sigma_i(j) > \sigma_i(k)}} B(t_{i\sigma_i(j)}, t_{i\sigma_i(k)})$$
$$\times \prod_{m=2}^n \prod_{l=1}^{m-1} \prod_{i=2}^N \prod_{\substack{j \in \Omega_i(l) \\ k \in \Omega_{i-1}(m)}} A(t_{i\sigma_i(j)}, t_{i-1,\sigma_{i-1}(k)})$$

(A1.11)
$$\left. \times \prod_{i=1}^N \frac{1}{\lambda_i!} \tilde{\xi}_{\lambda(1), V_1}(t(1), z_1) \otimes \cdots \otimes \tilde{\xi}_{\lambda(n), V_n}(t(n), z_n) \right),$$

$$\tilde{\xi}_{\lambda, V}(t,z) = \sum \left(\prod_{m=2}^n \prod_{l=1}^{m-1} \prod_{i=1}^N \prod_{j \in \Gamma_i(l)} C_{im}(t_{ij}, z_m) \right.$$
$$\times \prod_{m=2}^n \prod_{l=1}^{m-1} \prod_{i=1}^N \prod_{\substack{j \in \Gamma_i(l) \\ k \in \Gamma_i(m)}} A(t_{ik}, t_{ij})$$
$$\times \prod_{m=2}^n \prod_{l=1}^{m-1} \prod_{i=2}^N \prod_{\substack{j \in \Gamma_i(l) \\ k \in \Gamma_{i-1}(m)}} A(t_{ij}, t_{i-1,k})$$

(A1.12)
$$\left. \times \tilde{\xi}_{\lambda(1), V_1}(t(1), z_1) \otimes \cdots \otimes \tilde{\xi}_{\lambda(n), V_n}(t(n), z_n) \right).$$

The notation is the same as in (A1.7) and (A1.8), respectively. Formulas (A1.10) and (A1.12) imply (4.20).

Let $\widetilde{\Phi}(t,z;p)$ be the phase function (3.6). Formula (A1.11) implies that $\widetilde{\Phi}(t,z;p)\,\widetilde{w}(t,z)$ is an *integral representation* for solutions to the dual qKZ. The proof is completely similar to the qKZ case.

The $U_q(\mathfrak{gl}_{N+1})$ case is completely similar to the $\mathfrak{gl}_{N+1}$ case. Let $V_1, \ldots, V_n$ be $U_q(\mathfrak{g})$-modules with highest weights $\Lambda(1), \ldots, \Lambda(n)$ and generating vectors $v_1, \ldots, v_n$, respectively. Set $V = V_1 \otimes \cdots \otimes V_n$. Let $z \in \mathbb{C}^n$. We turn V into a $U_q(\widehat{\mathfrak{g}})$-module by the homomorphism

$$\varphi_z^{(n)}: U_q(\widehat{\mathfrak{g}}) \to U_q(\mathfrak{g})^{\otimes n}, \qquad X \mapsto (\varphi_{z_1} \otimes \cdots \otimes \varphi_{z_n}) \circ \Delta^{(n-1)}(X).$$

Here $\Delta^{(m)}$ is the m-iterated coproduct ($\Delta^{(0)} = \mathrm{id}$, $\Delta^{(1)} = \Delta$) and $\varphi_z: U_q(\widehat{\mathfrak{g}}) \to U_q(\mathfrak{g})$ is the homomorphism described in §6. Define $R(u,v) \in \mathrm{End}\,(\mathbb{C}^{N+1} \otimes \mathbb{C}^{N+1})$ as follows:

(A1.13)
$$R(u,v) = (uq - vq^{-1})\sum_{i}^{N} e_{ii} \otimes e_{ii} + (u-v)\sum_{i \neq j} e_{ij} \otimes e_{ij}$$
$$+ (q - q^{-1})\sum_{i<j}(u e_{ij} \otimes e_{ji} + v e_{ji} \otimes e_{ij}).$$

Consider the current type generators of $U_q(\widehat{\mathfrak{g}})$:

$$L_{ij}^{(+0)}, \quad i=1,\ldots,N+1,\ j=1,\ldots,i,$$
$$L_{ji}^{(-0)}, \quad i=1,\ldots,N+1,\ j=1,\ldots,i,$$
$$L_{ij}^{(s)}, \quad i,j=1,\ldots,N+1,\ s \in \mathbb{Z}_{\neq 0};$$

they are related to $\{k_i, e_i, f_i\}$ as follows:

$$L_{ii}^{(+0)} = k_i^{-1}, \quad L_{ii}^{(-0)} = k_i, \qquad i=1,\ldots,N+1,$$
$$L_{i+1,i}^{(+0)} = -k_i^{-1}e_i, \quad L_{i,i+1}^{(-0)} = f_i k_i, \qquad i=1,\ldots,N,$$
$$L_{1,N+1}^{(1)} = -k_0^{-1}e_0, \quad L_{N+1,1}^{(-1)} = f_0 k_0$$

(cf. [**FRT, RS, DF**]). Introduce $T_V(u,z) \in \mathrm{End}\,(V \otimes \mathbb{C}^{N+1})$:

(A1.14) $\quad T_V(u,z) = u^n \sum_{i<j} \varphi_z^{(n)}(L_{ij}^{(-0)}) \otimes e_{ij} + \sum_{s=1}^{\infty}\sum_{ij} \varphi_z^{(n)}(L_{ij}^{(-s)}) \otimes e_{ij} u^{n-s}.$

We also have

(A1.15) $\quad T_V(u,z) = (-1)^n \left(\sum_{i>j} \varphi_z^{(n)}(L_{ij}^{(+0)}) \otimes e_{ij} + \sum_{s=1}^{\infty}\sum_{ij} \varphi_z^{(n)}(L_{ij}^{(s)}) \otimes e_{ij} u^s \right).$

Since $s \geqslant n$, we have $\varphi_z^{(n)}(L_{ij}^{(\pm s)}) = 0$, so $T_V(u,z)$ is a polynomial in u,z.

All the following is almost the same as in the $\mathfrak{gl}_{N+1}$ case. We shall mention only formulas that differ from the $\mathfrak{gl}_{N+1}$ case.

$$A(u,v) = \frac{uq - vq^{-1}}{u-v}, \quad B(u,v) = \frac{uq - vq^{-1}}{uq^{-1} - vq},$$

$$C_{im}(u,v) = \frac{q^{\Lambda_i(m)}u - q^{-\Lambda_i(m)}v}{q^{\Lambda_{i+1}(m)}u - q^{-\Lambda_{i+1}(m)}v},$$

$$\xi_{\lambda,V}(t,z) = \mathrm{Tr}(F_\lambda \mathbb{T}_V(t,z)) v_1 \otimes \cdots \otimes v_n$$

$$\times \prod_{i=1}^{N} \prod_{j=1}^{\lambda_i} \left((q - q^{-1}) t_{ij} \prod_{m=1}^{n} \frac{q^{\Lambda_{i+1}(m)} t_{ij}/z_m - q^{-\Lambda_{i+1}(m)}}{q - q^{-1}} \prod_{k=1}^{j-1} A(t_{ij}, t_{ik}) \right)^{-1},$$

$$\tilde{\xi}_{\lambda,V}(t,z) = \mathrm{Tr}(E_\lambda \mathbb{T}_{V^*}(t,z)) v_1^* \otimes \cdots \otimes v_n^*$$

$$\times \prod_{i=1}^{N} \prod_{j=1}^{\lambda_i} \left((q - q^{-1}) t_{ij}^{-1} \prod_{m=1}^{n} \frac{q^{\Lambda_{i+1}(m)} t_{ij}/z_m - q^{-\Lambda_{i+1}(m)}}{q - q^{-1}} \prod_{k=1}^{j-1} A(t_{ij}, t_{ik}) \right)^{-1}.$$

In addition, $\lambda_i!$ in formulas (A1.7), (A1.11) should be replaced by $[\lambda_i]_q! = \prod_{j=1}^{\lambda_i} [j]_q$.

Let $z_i = q^{\zeta_i}$, $i = 1, \ldots, n$. Define $d_i = \prod_{j=1}^{N+1} k_j^{j\zeta_i} \in \mathrm{End}(V_i)$. Set $d = d_1 \otimes \cdots \otimes d_n \in \mathrm{End}(V)$. Let $e_i' = de_i d^{-1}$. Then

$$e_i' \xi_{\lambda,V}(t,z) = \sum_{j=1}^{\lambda_i} \Biggl(\Biggl[q^{\Lambda_i - \Lambda_{i+1} + \lambda_{i-1} - 2\lambda_i + \lambda_{i+1} + 2}$$

$$\times \prod_{m=1}^{n} C_{im}(t_{ij}, z_m) \prod_{\substack{k=1 \\ k \neq j}}^{\lambda_i} A(t_{ik}, t_{ij}) \prod_{k=1}^{\lambda_{i-1}} A(t_{ij}, t_{i-1,k})$$

$$- \prod_{\substack{k=1 \\ k \neq j}}^{\lambda_i} A(t_{ij}, t_{ik}) \prod_{k=1}^{\lambda_{i+1}} A(t_{i+1,k}, t_{ij}) \Biggr] \xi_{\lambda^i, V}(t^{ij}, z) \Biggr).$$

This is a counterpart of formula (A1.9).

PROOF OF LEMMA (A1.6). It follows from (A1.3) and (A1.4) that $\xi_{\lambda,V}(t,z)$ can have singularities only on the hyperplanes

(A1.16) $\quad t_{ij} - z_m + \Lambda_{i+1}(m) = 0, \quad j = 1, \ldots, \lambda_i, \ m = 1, \ldots, n,$

(A1.17) $\quad t_{ij} - t_{ik} + 1 = 0, \quad j = 1, \ldots, \lambda_i, \ k = 1, \ldots, j-1,$

(A1.18) $\quad t_{ij} - t_{kl} = 0, \quad j = 1, \ldots, \lambda_i, \ k = 1, \ldots, i-1, \ l = 1, \ldots, \lambda_k,$

$i = 1, \ldots, N$. To prove the lemma, we must show that $\xi_{\lambda,V}(t,z)$ is regular on the hyperplanes (A1.17) and on the hyperplanes (A1.18) for $i - k > 1$. For the first case, this was proved in [TV]. Below we consider the second case.

Let $a, b, \ldots$ stand for $(i,j), (k,l), \ldots$. We use the Yang–Baxter equation for $R(u,v)$:

(A1.19) $\quad R^{ab}(t_a, t_b) R^{ac}(t_a, t_c) R^{bc}(t_b, t_c) = R^{bc}(t_b, t_c) R^{ac}(t_a, t_c) R^{ab}(t_a, t_b),$

the unitarity relation

(A1.20) $$R^{ab}(t_a, t_b) R^{ba}(t_b, t_a) = A(t_a, t_b) A(t_b, t_a),$$

and the commutation relations in Y, which imply that

(A1.21) $$R^{ab}(t_a, t_b) T_V^a(t_a, z) T_V^b(t_b, z) = T_V^b(t_b, z) T_V^a(t_a, z) R^{ab}(t_a, t_b).$$

Fix a couple a, b, $a > b$. Using (A1.19) and (A1.21), we write (A1.3) as follows:

$$\mathbb{T}_V = \prod_{\substack{c>d \\ c>a}} R^{cd} T_V^\ell \cdots T_V^{a+1} T_V^1 \cdots T_V^a \prod_{\substack{a \geqslant c > d \\ d < b}}{}'' R^{cd} \prod_{a>c>d>b} R^{cd} \prod_{a>c>b}{}' R^{ac} R^{ab} \prod_{a>c>b} R^{cb}.$$

All unprimed products are taken in lexicographical order (cf. (A1.3)). The prime means that this product is taken in the reverse lexicographical order. The double prime means that in this product the factor R^{cd} stands to the right of the factor R^{ef} if $c > e$ or $c = e$ and $d < f$. For the residue $\operatorname{res}_{t_a = t_b} \mathbb{T}_V(t, z)$ some factors cancel each other because of (A1.20), and we get

$$\operatorname*{res}_{t_a = t_b} \mathbb{T}_V = \prod_{\substack{c>d \\ c>a}} R^{cd} T_V^\ell \cdots T_V^{a+1} T_V^1 \cdots T_V^a$$

$$\times \prod_{\substack{a \geqslant c > d \\ d < b}}{}'' R^{cd} \prod_{a>c>d>b} R^{cd} P^{ab} \prod_{a>c>b} (A(t_a, t_c) A(t_c, t_a)),$$

where $P = \sum_{ij} e_{ij} \otimes e_{ji}$.

Assume now that $a = (i, j)$, $b = (k, l)$, $i - k > 1$. The special matrix structure of $R(u, v)$ implies that $\operatorname{res}_{t_a = t_b} \operatorname{Tr}(F_\lambda \mathbb{T}_V(t, z))$ is a sum of monomials, each of them having the right-most factor of the form $\varphi_z^{(n)}(T_{i'k'}^s)$, $i' \geqslant i > k+1 \geqslant k'$. Every such factor annihilates the vector $v_1 \otimes \cdots \otimes v_n$. Hence, $\operatorname{res}_{t_a = t_b} \xi_{\lambda, V}(t, z) = 0$. □

Appendix 2

Let $V = \mathbb{C}[x]$, $\partial = \partial/\partial x$. Set $Xf(x) = f(xq)$ for any $f(x) \in \mathbb{C}[x]$. Let $\Lambda \in \mathbb{C}^2$. Define the action of the $\mathfrak{gl}_2$ generators $\{E_{ij}\}$ and the $U_q(\mathfrak{gl}_2)$ generators k_1, k_2, e_1, f_1 in V as follows:

$$E_{11} = \Lambda_1 - x\partial, \quad E_{22} = \Lambda_2 + x\partial, \quad E_{21} = x, \quad E_{12} = (\Lambda_1 - \Lambda_2 - x\partial)\partial,$$
$$k_1 = q^{\Lambda_1} X^{-1}, \quad k_2 = q^{\Lambda_2} X, \quad f_1 = x,$$
$$e_1 = \frac{(q^{\Lambda_1 - \Lambda_2} X^{-1} - q^{\Lambda_2 - \Lambda_1} X) x^{-1} (X - X^{-1})}{(q - q^{-1})^2}.$$

For generic Λ, V is a highest weight $\mathfrak{gl}_2$-module and a highest weight $U_q(\mathfrak{gl}_2)$-module with generating vector $v = 1$. Formulas (4.17) and (6.51) can now be checked by direct calculations.

References

[AGV] V. I. Arnold, S. M. Gusein-Zade, and A. N. Varchenko, *Singularities of differentiable maps*, Vol. II, Birkhäuser, Basel, 1988.

[DF] J. Ding and I. Frenkel, *Isomorphism of two realizations of quantum affine algebra $U_q(\widehat{\mathfrak{gl}(n)})$*, Comm. Math. Phys. **156** (1993), 277–300.

[FR] I. Frenkel and N. Reshetikhin, *Quantum affine algebras and holonomic difference equations*, Comm. Math. Phys. **146** (1992), 1–60.

[FRT] L. D. Faddeev, N. Yu. Reshetikhin, and L. A. Takhtajan, *Quantization of Lie groups and Lie algebras*, Algebra i Analiz **1** (1989), 178–206; English transl., Leningrad Math. J. **1** (1990), no. 1, 193–225.

[FT] L. D. Faddeev and L. A. Takhtajan, *Quantum inverse problem method and the Heisenberg XYZ-model*, Uspekhi Mat. Nauk **34** (1979), no. 5, 11–68; English transl. Russian Math. Surveys **34** (1979).

[FT2] _____, *The spectrum and scattering of excitations in the one-dimensional isotropic Heisenberg model*, Zap. Nauchn. Sem. Leningrad. Otdel. Mat. Inst. Steklov. (LOMI) **109** (1981), 134–178; English transl., J. Soviet Math. **24** (1984), 241–267.

[G] M. Gaudin, *Diagonalization d'une classe d'hamiltoniens de spin*, J. Physique **37** (1976), 1087–1098.

[Kid] A. N. Kirillov, *Representation of quantum groups, combinatorics, q-orthogonal polynomials and link invariants*, Thesis, LOMI, Leningrad, 1990. (Russian)

[K] V. E. Korepin, *Calculation of norms of Bethe wave functions*, Comm. Math. Phys. **86** (1982), 391–418.

[KiR] A. N. Kirillov and N. Yu. Reshetikhin, *The Yangians, Bethe ansatz and combinatorics*, Lett. Math. Phys. **12** (1986), 199–208.

[KR] P. P. Kulish and N. Yu. Reshetikhin, *Diagonalization of $GL(N)$ invariant transfer-matrices and quantum N waves (Lee model)*, J. Phys. A **16** (1983), L591–L596.

[M] A. Matsuo, *Quantum algebra structure of certain Jackson integrals*, Comm. Math. Phys. **157** (1993), no. 3, 479–498.

[R] N. Yu. Reshetikhin, *Calculation of Bethe vector norms for models with $SU(3)$ symmetry*, Zap. Nauchn. Sem. Leningrad. Otdel. Mat. Inst. Steklov. (LOMI) **150** (1986), 196–213; English transl. in J. Soviet Math. **46** (1989), no. 1.

[R2] _____, *Jackson type integrals, Bethe vectors, and solutions to a difference analog of the Knizhnik–Zamolodchikov system*, Lett. Math. Phys. **26** (1992), 153–165.

[RS] N. Yu. Reshetikhin and M. A. Semenov-Tyan-Shansky, *Central extensions of quantum current groups*, Lett. Math. Phys. **19** (1990), 133–142.

[RV] N. Reshetikhin and A. Varchenko, *Quasiclassical asymptotics of solutions to the KZ equations*, Preprint (1993).

[S] F. A. Smirnov, *Form factors in completely integrable models of quantum field theory*, Adv. Series in Math. Phys., vol. 14, World Scientific, Singapore, 1992.

[TV] V. Tarasov and A. Varchenko, *Jackson integral representations for solutions to the quantized Knizhnik–Zamolodchikov equation*, Algebra i Analiz **6** (1994), no. 2, 90–137; English transl. in St. Petersburg Math. J. **6** (1995).

[V] A. Varchenko, *Quantized Knizhnik–Zamolodchikov equations, quantum Yang–Baxter equation, and difference equations for q-hypergeometric functions*, Preprint (1993).

[V2] _____, *Critical points of the product of powers of linear functions and families of bases of singular vectors*, Preprint (1993).

Translated by THE AUTHORS

RESEARCH INSTITUTE FOR THEORETICAL PHYSICS P.O. BOX 9 (SILTAVUORENPENGER 20 C), SF-00014, UNIVERSITY OF HELSINKI, FINLAND
E-mail address: `tarasov@phcu.helsinki.fi`

DEPARTMENT OF MATHEMATICS, UNIVERSITY OF NORTH CAROLINA, CHAPEL HILL, NC 27599, USA
E-mail address: `av@math.unc.edu`